MTP International Review of Science

Main Group Elements
Group VII and Noble Gases

MTP International Review of Science

Publisher's Note

The MTP International Review of Science is an important new venture in scientific publishing, which we present in association with MTP Medical and Technical Publishing Co. Ltd. and University Park Press, Baltimore. The basic concept of the Review is to provide regular authoritative reviews of entire disciplines. We are starting with chemistry because the problems of literature survey are probably more acute in this subject than in any other. As a matter of policy, the authorship of the MTP Review of Chemistry is international and distinguished; the subject coverage is extensive, systematic and critical; and most important of all, new issues of the Review will be published every two years.

In the MTP Review of Chemistry (Series One), Inorganic, Physical and Organic Chemistry are comprehensively reviewed in 33 text volumes and 3 index volumes, details of which are shown opposite. In general, the reviews cover the period 1967 to 1971. In 1974, it is planned to issue the MTP Review of Chemistry (Series Two), consisting of a similar set of volumes covering the period 1971 to 1973. Series Three is planned for 1976, and so on.

The MTP Review of Chemistry has been conceived within a carefully organised editorial framework. The over-all plan was drawn up, and the volume editors were appointed, by three consultant editors. In turn, each volume editor planned the coverage of his field and appointed authors to write on subjects which were within the area of their own research experience. No geographical restriction was imposed. Hence, the 300 or so contributions to the MTP Review of Chemistry come from many countries of the world and provide an authoritative account of progress in chemistry.

To facilitate rapid production, individual volumes do not have an index. Instead, each chapter has been prefaced with a detailed list of contents, and an index to the 10 volumes of the MTP Review of Inorganic Chemistry (Series One) will appear, as a separate volume, after publication of the final volume. Similar arrangements will apply to the MTP Review of Physical Chemistry (Series One) and to subsequent series.

Butterworth & Co. (Publishers) Ltd.

Inorganic Chemistry
Series One
Consultant Editor
H. J. Emeléus, F.R.S.
Department of Chemistry
University of Cambridge

Volume titles and Editors

Physical Chemistry
Series One
Consultant Editor
A. D. Buckingham
Department of Chemistry
University of Cambridge

Volume titles and Editors

1 **THEORETICAL CHEMISTRY**
Professor W. Byers Brown, *University of Manchester*

2 **MOLECULAR STRUCTURE AND PROPERTIES**
Professor G. Allen, *University of Manchester*

3 **SPECTROSCOPY**
Dr. D. A. Ramsay, *National Research Council of Canada*

4 **MAGNETIC RESONANCE**
Professor C. A. McDowell, *University of British Columbia*

5 **MASS SPECTROMETRY**
Professor A. Maccoll, *University College, University of London*

6 **ELECTROCHEMISTRY**
Professor J. O'M Bockris, *University of Pennsylvania,*

7 **SURFACE CHEMISTRY AND COLLOIDS**
Professor M. Kerker, *Clarkson College of Technology, New York*

8 **MACROMOLECULAR SCIENCE**
Professor C. E. H. Bawn, F.R.S., *University of Liverpool*

9 **CHEMICAL KINETICS**
Professor J. C. Polanyi, F.R.S., *University of Toronto*

10 **THERMOCHEMISTRY AND THERMODYNAMICS**
Dr. H. A. Skinner, *University of Manchester*

11 **CHEMICAL CRYSTALLOGRAPHY**
Professor J. Monteath Robertson F.R.S., *University of Glasgow*

12 **ANALYTICAL CHEMISTRY —PART 1**
Professor T. S. West, *Imperial College, University of London*

13 **ANALYTICAL CHEMISTRY — PART 2**
Professor T. S. West, *Imperial College, University of London*

INDEX VOLUME

Organic Chemistry
Series One
Consultant Editor
D. H. Hey, F.R.S.
Department of Chemistry
King's College, University of London

Volume titles and Editors

1 **STRUCTURE DETERMINATION IN ORGANIC CHEMISTRY**
Professor W. D. Ollis, *University of Sheffield*

2 **ALIPHATIC COMPOUNDS**
Professor N. B. Chapman, *Duke University, North Carolina*

3 **AROMATIC COMPOUNDS**
Professor H. Zollinger, *Swiss Federal Institute of Technology*

4 **HETEROCYCLIC COMPOUNDS**
Dr. K. Schofield, *University of Exeter*

5 **ALICYCLIC COMPOUNDS**
Professor W. Parker, *University of Stirling*

6 **AMINO ACIDS AND PEPTIDES**
Professor D. H. Hey, F. R. S. and Dr. D. I. John, *King's College, University of London*

7 **CARBOHYDRATES**
Professor G. O. Aspinall, *University of Trent, Ontario*

8 **STEROIDS**
Dr. W. D. Johns, *G. D. Searle & Co., Chicago*

9 **ALKALOIDS**
Professor K. Wiesner, *University of New Brunswick*

10 **FREE RADICAL REACTIONS**
Professor W. A. Waters, F.R.S., *University of Oxford*

INDEX VOLUME

Inorganic Chemistry
Series One

Consultant Editor
H. J. Emeléus, F.R.S.

MTP International Review of Science

Volume 3

Main Group Elements
Group VII and Noble Gases

Edited by **V. Gutmann**
Technical University of Vienna

Butterworths · London
University Park Press · Baltimore

THE BUTTERWORTH GROUP

ENGLAND
Butterworth & Co (Publishers) Ltd
London: 88 Kingsway, WC2B 6AB

AUSTRALIA
Butterworth & Co (Australia) Ltd
Sydney: 586 Pacific Highway 2067
Melbourne: 343 Little Collins Street, 3000
Brisbane: 240 Queen Street, 4000

NEW ZEALAND
Butterworth & Co (New Zealand) Ltd
Wellington: 26–28 Waring Taylor Street, 1

SOUTH AFRICA
Butterworth & Co (South Africa) (Pty) Ltd
Durban: 152–154 Gale Street

ISBN 0 408 70222 2

UNIVERSITY PARK PRESS

U.S.A. and CANADA
Universitv Park Press Inc
Chamber of Commerce Building
Baltimore, Maryland, 21202

Library of Congress Cataloging in Publication Data
Gutmann, Viktor
Main group elements
(Inorganic chemistry, series one, v. 3) (MTP international review of science)
Includes bibliographies
1. Halogens. 2. Gases, Rare. I. Title
QD151.2.15 vol. 3 [QD165] 546′.73 71-160325
ISBN 0-8391-1006-5

MTP MEDICAL AND TECHNICAL PUBLISHING CO. LTD.
Seacourt Tower
West Way
Oxford, OX2 OJW
and
BUTTERWORTH & CO. (PUBLISHERS) LTD.

Filmset by Photoprint Plates Ltd., Rayleigh, Essex
Printed in England by Redwood Press Ltd., Trowbridge, Wilts
and bound by R. J. Acford Ltd., Chichester, Sussex

Consultant Editor's Note

The problem of keeping abreast of research literature on as broad a front as possible is one that confronts all chemists. In the past this difficulty has been met, in the main, by literature surveys and by several uncorrelated reviews of progress in certain subject areas. There are obvious inadequacies in this approach, which have become increasingly apparent in recent years. I was, therefore, grateful for the opportunity of helping to plan this new series, which has been designed to provide a comprehensive, critical survey of each of the main branches of chemistry.

This section of the MTP International Review of Science deals with progress in Inorganic Chemistry. The subject is developing at an astonishing rate and in many directions. Fortunately, however, it lends itself to a systematic treatment. Ten volumes have been prepared, three dealing with the main group elements and two with the general chemistry of the transition metals. Organometallic derivatives of the main group elements and lanthanides and actinides are covered separately, as is the subject of reaction mechanisms. The two remaining volumes on radiochemistry and solid state chemistry have been planned to avoid, as far as possible, overlap with those that have gone before.

It is a pleasure to thank the many experts who have collaborated as authors and volume editors in making this publication possible. While working to a pre-arranged over-all plan, they have been able to assess and interpret the literature in terms of their own experience in specialised fields. I believe that in this way they will not only provide a record of what has been done, but will stimulate further exploration in this fascinating branch of chemistry.

Cambridge — H. J. Emeléus

Preface

Following the discovery of the compounds of the noble gases, a considerable number of them have been prepared. Within the last few years advanced methods of preparation have been less noticeable, and Chapter 1 shows that emphasis in this volume has been laid on characterisation of the compounds by modern techniques and elucidation of their co-ordinating properties.

Halogen Chemistry (Academic Press, 1968) covers many of the recent advances in halogen chemistry up to 1968. The present volume deals with those topics which were not covered in the said review, such as the spectrochemical properties of the halogens, as well as with advances in halogen chemistry within the last three years.

Considerable progress has been made in the elucidation of the chemistry of astatine; one of the most striking advances is the recent discovery of the perbromates, and much interesting work has also been reported on the oxides and oxyacids of the other halogens. Recent work on interhalogen compounds and polyhalide ions is also reviewed in detail as well as the chemistry of hydrogen halides, which may be used as 'non-aqueous' solvents.

The reader will, however, not find information on the halides of most of the elements, since their chemistry is treated in other volumes in connection with the chemistry of the respective elements.

Charge-transfer complexes of the halogens and the interhalogen compounds have not been included, since no important contributions have appeared in this field within the last few years. On the other hand, advances in the field of selected pseudohalides, such as the chemistry of perfluoroalkyl compounds and that of the $(CF_3)_2NO$ radical will be found in this volume.

Vienna V. Gutmann

Contents

1
Noble Gases

F. SLADKY

University of Innsbruck

1.1 INTRODUCTION

The first years immediately following the discovery of the first noble gas compound by Bartlett[1], can be characterised by an almost hectic competition to partake in the investigation of this interesting new field of chemistry.

In the subsequent, calmer atmosphere many physical and molecular data of noble gas compounds have been measured and determined either for the first time or with greater accuracy. Other new developments can be seen in the characterisation of the noble gas fluoride cations like $[XeF]^+$ [2,3], $[Xe_2F_3]^+$ [4] and $[XeF_5]^+$ [5], which to some extent also clarify the Xe/PtF_6 reaction.

Since noble gas chemistry was previously thought to be restricted to fluorides, oxide–fluorides, oxides and their hydrolysis products xenates and perxenates, the discovery of a range of compounds with other electronegative ligands like the FSO_2O– and F_5TeO–groups[6,7] capable of forming bonds to xenon, seems to open a new field worth further exploration.

The question of xenon hexafluoride gas-phase symmetry appears to be on the verge of solution. Certainly few simple molecules have been more thoroughly investigated in recent years. Though the exact shape of the xenon hexafluoride molecule is still not established, it is definitely clear that all attempts have failed to analyse relevant data on the assumption that it is an octahedrally symmetrical molecule like other known hexafluorides. Since no other single molecular geometry has proved capable of accounting for the electron diffraction pattern, intramolecular rearrangement between several molecular geometries has been proposed. A good agreement between experimental and synthetic radial distribution curves is obtained for steric activity of the valence electron lone pair (7-coordination) and the following molecular rearrangement: $C_{2v} \leftrightarrow C_s \leftrightarrow C_{3v}$ [8].

The simple molecular orbital approach to bonding in noble gas compounds, considering Xe 5p and F 2p orbitals only, has proved to be relevant for explaining stereochemical and spectroscopic features. Including the 6s orbital in similar calculations for xenon hexafluoride affords a nonoctahedral symmetry, as does the valence bond approach by emphasising outer d-orbital contributions. Recent calculations[9] indeed seem to prove that such contributions are involved to a degree which is chemically significant, whereas participation of 4f orbitals is probably best seen as a polarisation effect[10].

Compound formation is restricted to the heavy noble gases krypton, xenon and radon. Methods either very effective in synthesising all the known noble gas fluorides like proton bombardment[11], or very sensitive in detecting compound formation like the matrix isolation technique, proved incapable of establishing the existence of an argon fluoride even at 20 K [12].

Calculations[13], however, suggest the existence of $[ArF]^+$, though the main problem will be to find an unusually oxidation-resistant counter anion to withstand the enormous electron affinity of this ion.

Although much of the chemistry of noble gas compounds and especially their use for synthetic purposes remains to be explored, some encouraging results have been obtained. The first synthesis of perbromates was achieved by using xenon difluoride as an oxidant[14].

The preparation and handling of noble gas compounds need not necessarily be restricted to laboratories with special equipment as may be illustrated with the preparation of xenon difluoride by exposure of a mixture of xenon and fluorine in a glass bulb to ordinary sunlight[15].

1.2 COMPOUNDS OF XENON

1.2.1 Fluorides

1.2.1.1 Preparations

A. The xenon–fluorine system

All three known xenon fluorides can be prepared by direct combination of the elements. The synthesis proceeds stepwise according to the following reactions[16]:

$$Xe + F_2 \rightleftharpoons XeF_2;\ K_1$$
$$XeF_2 + F_2 \rightleftharpoons XeF_4;\ K_2$$
$$XeF_4 + F_2 \rightleftharpoons XeF_6;\ K_3$$

Some experimental equilibrium constants are given in Table 1.1. Figures 1.1 and 1.2 show the equilibrium pressures of the xenon fluorides plotted as a function of temperature for a given ratio of the starting elements. It will be appreciated that low temperature and high fluorine pressures favour XeF_6

Table 1.1 Experimental equilibrium constants for the xenon—fluorine system[16]

		Temperature (°C)
K_1	29.8	501
K_2	1431	250
	154.9	300
	27.22	350
	4.857	400
	0.502	501
K_3	0.9435	250
	0.2112	300
	0.05582	350
	0.01822	400

formation while high temperatures and low fluorine pressures enhance the yield of XeF_2. It is not feasible, however, to obtain pure XeF_4 by this method alone.

(a) XeF_8 and xenon fluorides of odd oxidation number* – In this extensive study, no evidence was found for xenon fluorides of odd oxidation number nor for the formation of XeF_8 even at fluorine pressures as high as 500 atm. There is theoretically, however, a marginal chance that XeF_8 is sufficiently stable to be isolated. The extrapolated mean thermo-chemical bond energy may be as high as 21 kcal mol^{-1}, though steric inhibition because of ligand crowding has to be considered as well. Since XeF_8 is expected to be an excellent fluorine ion donor, isolation of $[XeF_7]^+[AsF_6]^-$ looks much more promising. Persistent efforts have been made to synthesise XeF_8 [17].

(b) Kinetic features – Rate studies of the interaction of xenon with fluorine have shown that the reaction is zero order in xenon and first order in

*XeF and KrF radicals have been generated in exceedingly small concentrations in XeF_4 and KrF_2 crystals, subjected to γ-radiation at 77 K and characterised by electron spin resonance[16a, 16b]

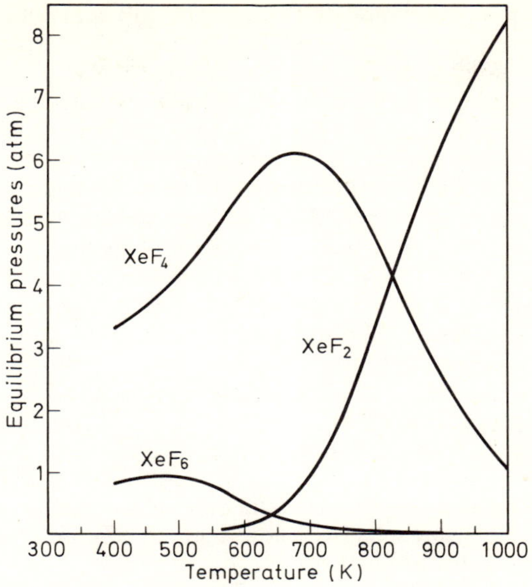

Figure 1.1 Equilibrium pressures of xenon fluorides as a function of temperature. Initial conditions: 125 m mol Xe, 275 m mol F_2 per 1000 ml. ((From Selig[196], by courtesy of Academic Press)

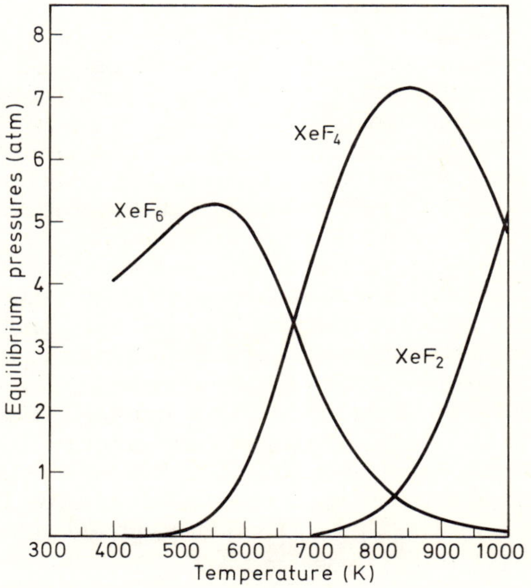

Figure 1.2 Equilibrium pressures of xenon fluorides as a function of temperature. Initial conditions: 125 m mol Xe, 1225 m mol F_2 per 1000 ml. (From Selig[196], by courtesy of Academic Press)

fluorine[18–20]. For F_2/Xe ratios of 16 or more and total pressures of 10–20 mmHg in the temperature range 190–250 °C, the XeF_2 formation was found to be zero order in F_2 and first order in Xe (the reverse of the findings mentioned above for the conditions which favour exclusive XeF_2 formation) and the XeF_4 formation, zero order in F_2 and first order in XeF_2 [21].

The thermal reactions are heterogeneous and a mechanism involving adsorption and dissociation of fluorine on the nickel fluoride surface of the reaction vessel has been proposed. It has been shown that heated filaments of Pd, Ni, Cu and Al catalyse the combination of Xe and F_2, whereas Ti, Zr, Mo, Ta, W, Re, Ir, Fe, Cr, V, Rh and Pt are ineffective[22,23]. Since atomic fluorine, generated in a glow discharge, is capable of converting condensed xenon (at 77 K) to xenon difluoride (45% yield in 75 min) it appears that xenon activation is not necessary for xenon difluoride formation[24].

(c) Preparation of XeF_2 – The static thermal method in reacting xenon and fluorine (reactant ratio 2:1) is the best way to make large quantities of xenon difluoride[25]. Using the reported[16] equilibrium constants for $Xe/F_2 = 2$, at 400 °C, the product will be 99.7% XeF_2 and 0.3% XeF_4 with negligible XeF_6 content. The gaseous mixture is heated in a nickel or Monel vessel at 400 °C for about 2 h, quenched to room temperature and the formed XeF_2 transferred by vacuum sublimation.

A convenient preparation on a smaller scale which avoids the special metal equipment used in the former procedure simply involves exposure of Xe/F_2 mixtures (ratio 1:1, 1 atm total pressure) contained in Pyrex glass bulbs to sunlight, at room temperature[15,26,179]. After 2 or 3 h crystals of pure XeF_2 form on the cooler top of the Pyrex vessel and within a day, the reaction is at least 60% complete.

(d) Preparation of XeF_4 – In agreement with equilibrium data, preparation of pure XeF_4 is not possible by the static thermal method. Moreover, it is difficult to separate it from XeF_2 by physical means. The vapour-pressure relationships are similar and they also form a 1:1 adduct. Since the more volatile impurity, XeF_6 can be rather easily separated, either by fractionated sublimation or reaction with NaF, it seems best to use an excess of fluorine (1:5 or more, 450–500 °C) and at least restrict the formation of XeF_2 to a minimum[27].

A method of purifying XeF_4 chemically has been devised. It effectively eliminates XeF_2 and XeF_6 simultaneously, by making use of the inferior fluoride ion donor ability of XeF_4 compared to XeF_2 and XeF_6 [28]. XeF_4, prepared by the static thermal method, which will contain small but significant quantities of XeF_2 and XeF_6, is dissolved in bromine pentafluoride and treated with an excess of arsenic pentafluoride. XeF_2 and XeF_6 form involatile salts and BrF_5 and excess AsF_5 are removed quantitatively at 0 °C or below, at which temperatures the vapour pressure of XeF_4 is very low. Subsequent vacuum sublimation at room temperature affords pure XeF_4.

$$\left.\begin{array}{l} XeF_2 \\ XeF_4 \\ XeF_6 \end{array}\right\} \text{in } BrF_5 \text{ soln.} + AsF_5 \text{ excess} \xrightarrow{0\,°C} \begin{array}{l} [Xe_2F_3]^+[AsF_6]^- \\ XeF_4 \\ \xrightarrow[-BrF_5]{-AsF_5} [XeF_5]^+[AsF_6]^- \end{array}$$

$$\downarrow 20\,°C, \text{ vacuum}$$

$$XeF_4$$

Table 1.2 Some properties of the xenon fluorides

	XeF_2	XeF_4	XeF_6
Molecular weight	169.30	207.29	245.29
Appearance: solid	colourless	colourless	colourless
liquid	colourless	colourless	yellow green
vapour	colourless	colourless	—
Triple point (°C)	129.03 [36]	117.10 [36]	49.48 [31]
Boiling point (°C)	—	—	75.57 [31]
Sublimation point (°C)	114.30 [36]	115.74 [36]	—
Vapour pressure (Torr)	$\log P(\text{solid}) = -3057.67/T - 1.23521 \log/T + 13.969736$ [36]	$\log P(\text{solid}) = -3226.21/T - 0.43434 \log/T + 12.301738$ [36]	$\log P(\text{solid}) = -X/T - Y$ [31] Temperature range (K) — X — Y 273.19–295.82 — 3400.12 — 12.86125 254.00–291.80 — 3313.5 — 12.5923 291.8–322.38 — 3098.0 — 11.8397 $\log P(\text{liquid}) = -6170.88/I - 23.67815 \log I \pm 80.77778$ [31]
Density ($g\ cm^{-3}$)	4.32 [37]	4.04 [37]	solid (242.97) 3.668 (0.044) [31] liquid (328.34) 3.173 (0.03) [31]
ΔH_{sub} ($kcal\ mol^{-1}$)	13.2 ± 0.2 [36]	14.8 ± 0.2 [36]	15.6 [16]
Diamagnetic susceptibility (m.u.)	$40–50 \times 10^{-6}$ [38]	52×10^{-6} [39]	44.5×10^{-6} [40]
Specific conductance ($\Omega^{-1}\ cm^{-2}$)	—	—	$1.45 \pm 0.05 \times 10^{-6}$ [41]
Dielectric constant	—	—	4.10 ± 0.05 (55 °C) [41]

(e) Preparation of XeF_6—A 95% conversion to XeF_6 is obtained with F_2/Xe ratios of 20 at 50 atm or above and temperature of about 300 °C for 10–20 h[29].

Complete separation of impurities can be effected by taking advantage of the reversible reaction of XeF_6 with NaF[30]. The likely impurities XeF_2, XeF_4 and $XeOF_4$ do not form stable complexes with NaF and can be pumped off at temperatures up to 50 °C. Subsequent heating of the $2NaF{\cdot}XeF_6$ complex up to 125 °C under vacuum yields pure XeF_6.

Since XeF_6 readily reacts with water vapour or oxides in the metal vacuum line, it is necessary to season the line with fluorine and small batches of XeF_6 to avoid $XeOF_4$ formation. As with XeF_4, the usual care should be taken to guard against the formation of the highly explosive XeO_3.

1.2.1.2 Physical properties (Table 1.2)

XeF_2 and XeF_4 are colourless as solid, liquid or gas. XeF_6 is colourless only as a solid, the liquid and vapour are yellow-green. The high entropy of vaporisation (32.74 cal deg^{-1} mol^{-1})[31] is indicative of polymerisation in the liquid state. Two independent vapour density measurements[16,32] show the hexafluoride to be monomeric in the gas phase.

The hexafluoride is much more volatile than either XeF_2 or XeF_4, although much less volatile than other hexafluorides. The respective vapour pressures at 25 °C (Torr) are: XeF_6 27, XeF_2 4.6, XeF_4 2.5, which allows convenient handling by vacuum sublimation of all three xenon fluorides.

A. Solvents for xenon fluorides

Several organic and inorganic solvents have been explored, especially for XeF_2. Some quantitative results are compiled in Table 1.3. An approximate

Table 1.3 Solubility of XeF_2 in some solvents

HF[35]	Temperature (°C)	−2	12.25	29.95		
	moles/1000 g	6.38	7.82	9.88		
CH_3CN[39]	Temperature (°C)	0	21			
	moles/l	0.99	1.83			
$ONF{\cdot}3HF$[42]	Temperature (°C)	16.8	33.2	49.2	61.0	80.0
	moles/1000 g	16.3	19.56	22.17	25.38	30.55

order of relative solubilities of XeF_2 in the following solvents has been given[33]: BrF_5, very good; BrF_3, very good; HF, very good; IF_5, good (but with adduct formation); CH_3CN, good, $(CF_3CO)_2O$, good; SO_2, fair; WF_6, poor; NH_3, sparingly. Conductivity measurements and spectroscopic analysis established the molecular nature of XeF_2 in some of these solutions[34].

With the exception of HF[35], solvents for XeF_4 and XeF_6 have been studied only to a smaller extent. Bromine pentafluoride, however, proved to be a convenient solvent for all xenon fluorides, especially for fluoride ion transfer reactions[28].

The high conductivity of solutions of XeF_6 in HF has been suggested to be due to the ionisation equilibrium[41]:

$$XeF_6 + HF \rightarrow [XeF_5]^+ + [HF_2]^-$$

1.2.1.3 Thermodynamic properties (Table 1.4)

The error in the measurement of the enthalpies of formation still appears to be greater than the expected differences in bond strength between the xenon fluorides. Without question however, the mean thermochemical bond energy of 32 ± 1 or 2 kcal mol^{-1} is higher than the limiting value of 18.3 kcal mol^{-1} required for the exothermic formation of xenon fluorides.

Unlike the Xe^{IV} oxide and hydroxide systems, xenon tetrafluoride is stable towards disproportionation. This is consistent with the reported enthalpies and entropies of formation. Thus for the disproportionations:

$$2\,XeF_4(g) \rightarrow XeF_2(g) + XeF_6(g) \qquad \Delta H^\circ = +5 \text{ kcal mol}^{-1}$$
$$\Delta S^\circ = 0 \text{ cal deg}^{-1}\text{ mol}^{-1}$$
$$3\,XeF_4(g) \rightarrow Xe(g) + 2XeF_6(g) \qquad \Delta H^\circ = +9 \text{ kcal mol}^{-1}$$
$$\Delta S^\circ = -9 \text{ cal deg}^{-1}\text{ mol}^{-1}$$

Table 1.4 Some thermodynamic data for xenon fluorides

	(kcal mol^{-1}; cal $deg^{-1} mol^{-1}$)		
	XeF_2	XeF_4	XeF_6
ΔH_f°	−41.5 [43]	−72.4 [44]	−98.5 [44]
ΔH_f°(g)	−28.3	−57.6	−82.9
ΔS_f°(g)	−26.5 [16]	−61 [16]	−96 [16]
$\Delta H_{sub.}$	13.2 [36]	14.8 [36]	15.6 [16]
S°(g)	25 °C:62.057 [16]	25 °C:73 [36]	62 °C:96.27 [31]
S°(s)	57 °C:29.4 [36]	—	25 °C:50.33 [31]
S°(l)	—	—	62 °C:61.10 [31]
E (Xe—F)	32.5	32.7	32.1

1.2.1.4 Spectroscopic properties, crystal structures

A. Infrared and Raman spectra

Infrared and Raman spectra of XeF_2 ($D_{\infty h}$) and XeF_4 (D_{4h}) have previously been given in detail elsewhere[196]. A high-resolution infrared study of the rotational fine structure of the ν_3 stretching mode in XeF_2 has provided a bond length of 1.9773 ± 0.0015 Å in the vapour phase[45]. The value of 2.00 ± 0.01 Å for the bond length in the crystalline phase derived by single-crystal neutron diffraction is similar[46]. The Xe–F bond length in XeF_4 is 1.94 ± 0.01 Å and 1.953 ± 0.02 Å as determined by electron diffraction[47] and neutron diffraction[48] respectively. The mean Xe–F bond length in XeF_6, determined by electron diffraction, is 1.890 ± 0.005 Å, but the radial distribution function for Xe–F bonds corresponds to that of a composite for non-equivalent bonds[49]. Some data on bond lengths and force constants are compiled in Table 1.5.

Infrared and Raman spectra of XeF_6 (Figures 1.3, 1.4 and 1.5) are of considerable interest in view of the much-disputed molecular symmetry of this compound. The infrared spectrum of XeF_6 vapour shows three distinct

Table 1.5 Bond lengths and force constants in xenon fluorides

	XeF_2	XeF_4	XeF_6
Xe—F (vapour, Å)	1.9773 ± 0.0015 [45]	1.94 ± 0.01 [47]	1.890 ± 0.005 [49]
Xe—F (crystal, Å)	2.00 ± 0.01 [46]	1.953 ± 0.002 [48]	(see 1.2.1.4.B.(c)
k_r(mdyn/Å)	2.82 [51]	2.860	
k_{rr}	0.247	0.30_0	
k'_{rr}	—	0.03_4 [52]	
k_α	0.197	0.21_8	

absorption maxima in the bond stretching region at 616, 563 and 520 cm^{-1} [50], whereas an O_h molecule has only one active stretching fundamental. Moreover, there is a very close coincidence in frequency between infrared and Raman for these stretching bands[51]. In the far-infrared region, where other hexafluorides show sharp absorption by their f_{1u} bending modes, no definitive absorption band could be located. These results seem to support either a model that is non-rigid with respect to angles between bonds or a model of

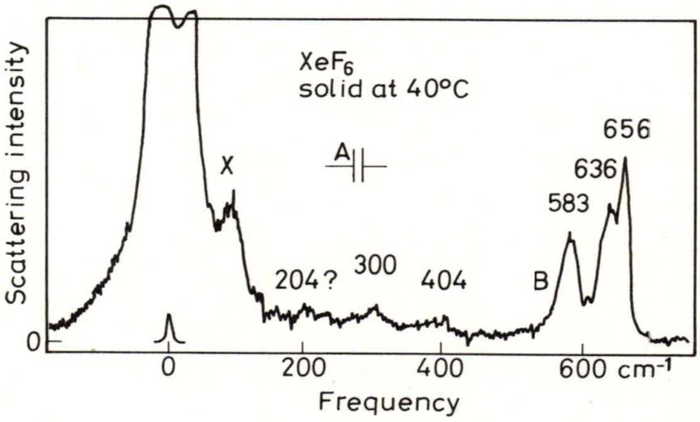

Figure 1.3 Raman spectrum of solid XeF_6 at 40°. (From Gasner and Claassen[51], by courtesy of The American Chemical Society)

more rigidity but with unusually large anharmonicities in the bending potential function[50].

The millimetre wave region was scanned in an attempt to give support to the inverting or pseudo-rotator model proposed[53]. In this model the f_{1u} bending potential is a double minimum type, having a relatively high barrier, but owing to the triply degenerate character of the f_{1u} mode the inversion levels lie quite low. It was predicted that the first transition of the inversion levels would be about an energy of 6 cm^{-1}. No microwave absorption, however, could be detected in the range 3.7–8.6 cm^{-1} [50].

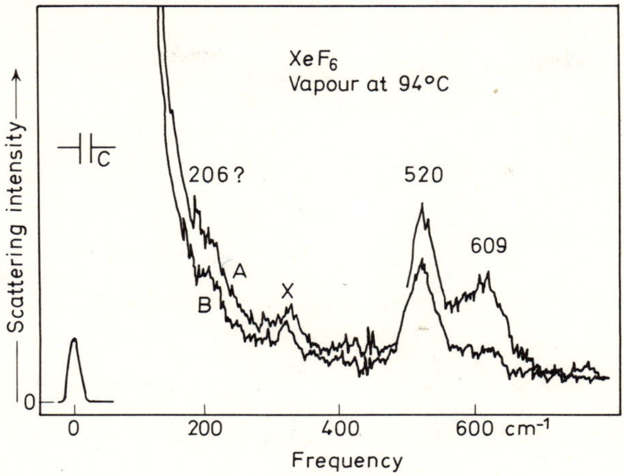

Figure 1.4 Raman spectrum of XeF_6 vapour at 94°C: A, incident polarisation perpendicular, B, incident polarisation parallel, C, equivalent slit width. (From Gasner and Claassen[51], by courtesy of the American Chemical Society)

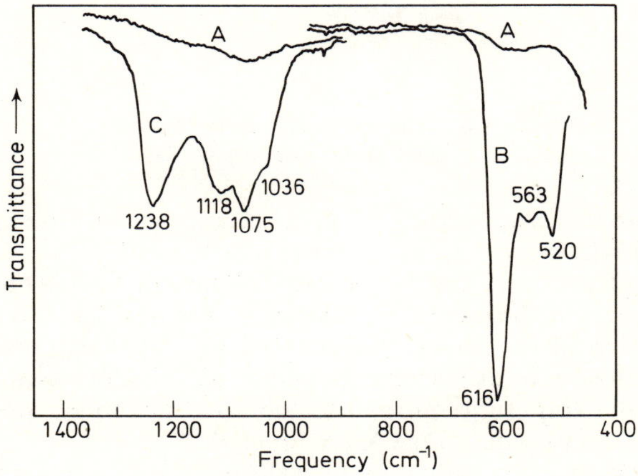

Figure 1.5 Infrared absorption spectrum of xenon hexafluoride (From Kim *et al.*[50], by courtesy of The American Chemical Society)

B. Crystal structures:

(a) XeF_2 – In crystalline XeF_2, the molecules are aligned parallel in a body centred tetragonal array, the unit cell of which is shown in Figure 1.6. There are strong interactions between molecules, since each xenon has not only its two bound fluorine atoms at 2.00 Å, but also eight fluorine atoms at

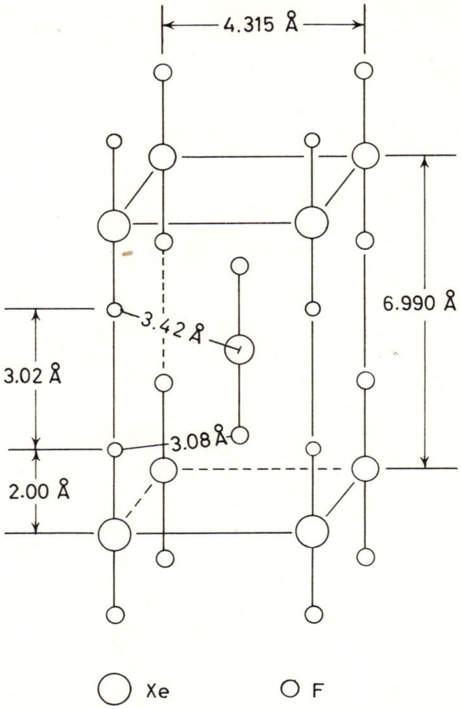

Figure 1.6 The unit cell of xenon difluoride
(From Levy and Agron[46] by courtesy of the American Chemical Society)

3.41 Å[46]. This structural arrangement is compatible with the high enthalpy of sublimation ($\Delta H_{Subl} = 12.3$ kcal mol^{-1}) of XeF_2 [54].

(b) XeF_4 – The crystal and molecular structure of XeF_4 as determined by x-ray and neutron diffraction, is shown in Figures 1.7 and 1.8 [48,56]. The thermal motion of the atoms is illustrated by ellipsoids constructed from the root-mean-square displacements along the principal axes of motion. It is seen that the amplitudes normal to the bond directions are greater than the stretching amplitudes of the Xe–F bonds[51].

(c) XeF_6 – In the solid state xenon hexafluoride is known to exist in at least four polymorphic forms[56] (Table 1.6). The existence of three modifications was demonstrated from heat capacity measurements[57,31]. These were designated phases I (crystallographically characterised by Agron *et al.*[58]) II and III in order of decreasing temperature with transitions at 292 and 254 K.)

A cubic phase stable from the melting point to at least 93 K (the limit of instrumental range) was characterised, and designated phase IV[59]. There

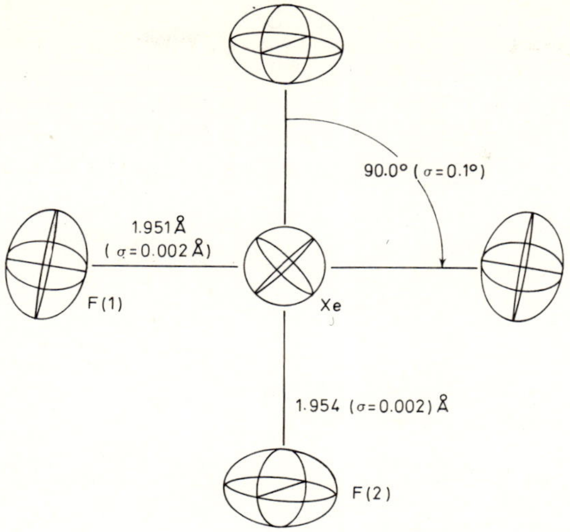

Figure 1.7 Geometry and thermal motion of the XeF_4 molecule
(From Burns *et al.*[48] by courtesy of The American Asscn. for the Advancement of Science)

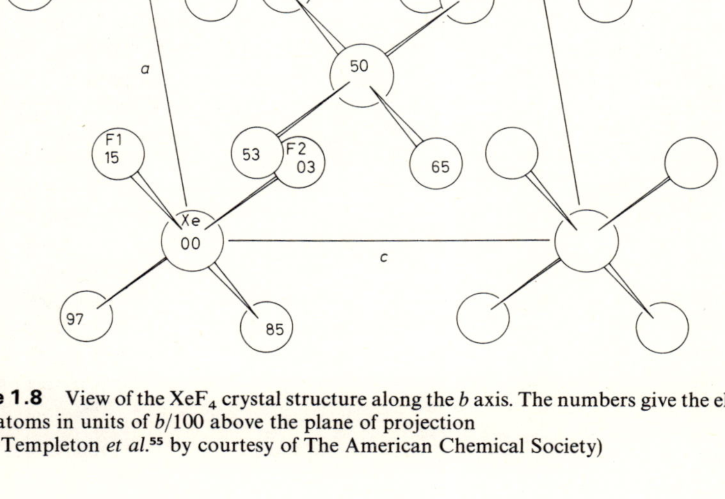

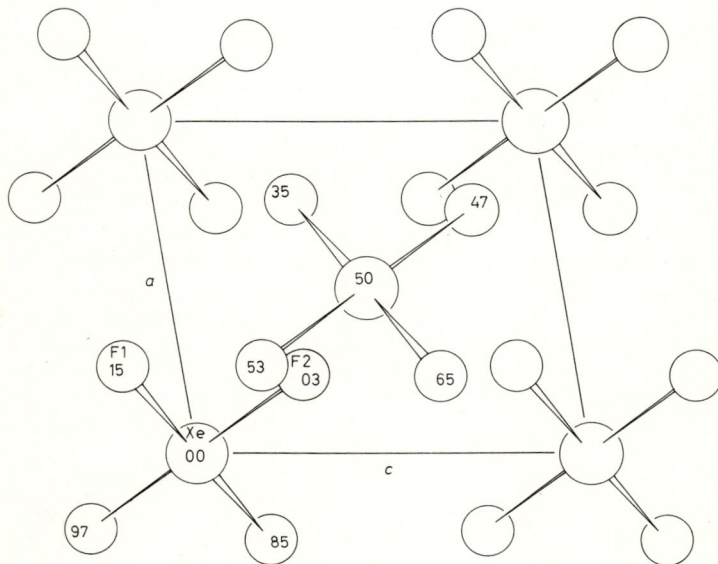

Figure 1.8 View of the XeF_4 crystal structure along the *b* axis. The numbers give the elevation of the atoms in units of *b*/100 above the plane of projection
(From Templeton *et al.*[55] by courtesy of The American Chemical Society)

are no simple molecules in the complex structure which involves 1008 atoms distributed over 1600 positions per unit cell. The ions of $[XeF_5]^+$ and F^- are associated in tetrameric and hexameric rings, of point group symmetries $\bar{4}$ and 32, respectively. The structure contains right- and left-handed conformations of both tetramers and hexamers. The handedness of the tetramers is disordered but the orientation is ordered, the handedness of the hexamers

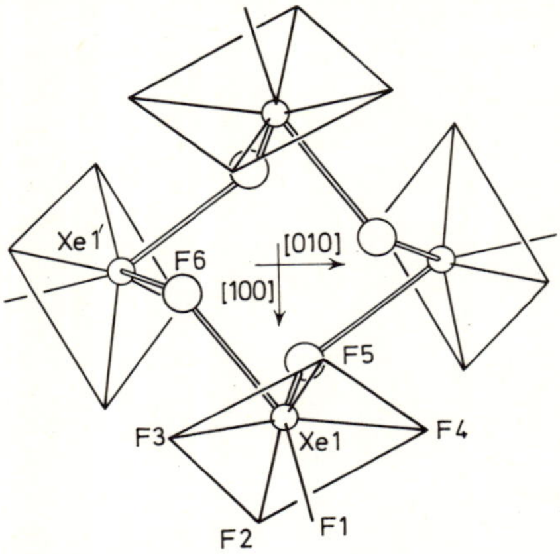

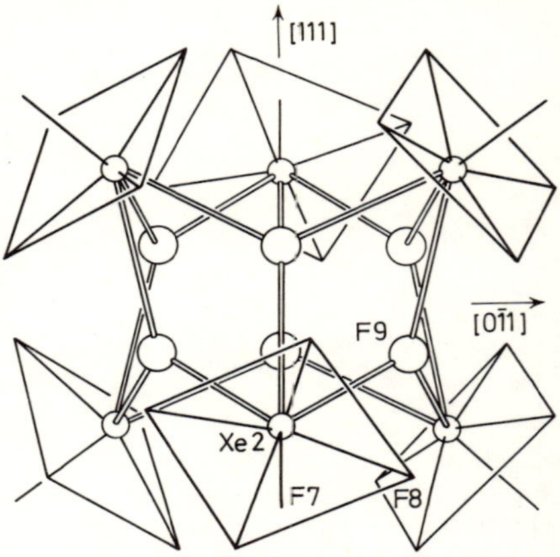

Figure 1.9 (left) Tetramer of $XeF_6^+F^-$ centred at $00\frac{1}{4}$ with the $\bar{4}$-axis parallel to [001]. Xenon atoms are indicated by small circles and bridging fluoride ions by large circles: XeF_5^+ ions are drawn in skeletal form to preserve clarity

Figure 1.10 (right) Hexamer of $XeF_5^+F^-$ centred at $\frac{1}{4}\frac{1}{4}\frac{1}{4}$ and oriented with the 3-axis parallel to (111). The three 2-axes are parallel to [0$\bar{1}$1], [10$\bar{1}$], and [$\bar{1}$10]
(From Burbank and Jones[59], by courtesy of The American Asscn. for the Advancement of Science)

is ordered but the orientation is disordered. There are four structural components in each type of polymer. In the tetramer (Figure 1.9) there are the xenon atom Xe(1), the apical fluorine F(1), the basal fluorines F(2),(3),(4),(5) and the bridging fluorine F(6). The bridging fluorine has a short contact to Xe(1) and a longer one to Xe(1′). An eight-membered ring results, consisting of four $[XeF_5]^+$ and four F^- ions. In the hexamer (Figure 1.10) the components are xenon atom Xe(2), apical fluorine F(7), basal fluorine F(8) and bridging fluorine F(9). The bridging fluorine makes equal contacts to three $[XeF_5]^+$ ions, with three equal bridging angles. The important bond lengths and angles are (estimated standard deviations are given in parenthesis):

$[XeF_5]^+$	Hexamer	Tetramer
$Xe—F_{ap}$ (Å)	1.76(3)	1.84(4)
$Xe—F_{bas}$ (Å)	1.92(2)	1.86(3)
$F_{ap}—F_{bas}$ (Å)	2.33(3)	2.29(6)
$F_{bas}—F_{bas}$ (Å)	2.63(3)	2.54(13)
$F_{ap}—Xe—F_{bas}$	80.0(0.6)°	77.2(1.8)°av.
$F_{bas}—Xe—F_{bas}$	88.3(0.2)°	87.2(4.5)°av.
$F_5Xe^{\delta+}—F^{\delta-}$ 'bridge' (Å)	2.56(2)	2.23(3) 2.60(3)
$F_5Xe^{\delta+}—F^{\delta-}—XeF_5^{\delta+}$ 'bridge'	118.8(0.3)	120.7(1.2)°

It should be noted, that the bridging fluoride ions are not close to the fourfold axis of the $[XeF_5]^+$ group. This is consistent with the location of the non-bonding Xe^{VI} valence-electron pair on the fourfold axis, *trans* to the apical fluorine atom of the $[XeF_5]^+$ ion. Thus the xenon appears pseudo-octahedrally coordinated. Another interesting feature is, that the $[XeF_5]^+$

Table 1.6 Crystalline modifications of Xenon hexafluoride

Phase I:	monoclinic	2
	$P2_1/m$ or $P2_1$	tetrameric rings of XeF_5^+ and F^-
	$Z = 2$, $d_{x\text{-ray}} = 3.56\ g\ cm^{-3}$	
Transition	$a = 9.33$ Å	per unit cell
temperature	$b = 10.96$ Å	
285 K	$c = 8.95$ Å	
	$\beta = 91.9°$	
Phase II:	orthorhombic	4
	$Pnma$ or $Pn2_1a$	tetrameric rings of XeF_5^+ and F^-
	$Z = 16$, $d_{x\text{-ray}} = 3.71\ g\ cm^{-3}$	
Transition	$a = 17.01 \pm 0.04$ Å	per unit cell
temperature	$b = 12.04 \pm 0.03$ Å	
250 K	$c = 8.57 \pm 0.02$ Å	
Phase III:	monoclinic (A centred)	16
	$P2_1/b$	tetrameric rings of XeF_5^+ and F^-
	$Z = 64$, $d_{x\text{-ray}} = 3.82_5\ g\ cm^{-3}$	
	$a = 16.80 \pm 0.04$ Å	per unit cell
	$b = 23.93 \pm 0.05$ Å	
	$c = 16.95 \pm 0.04$ Å	
	$\alpha = 90°40' \pm 10'$	
Phase IV:	cubic	24
no	$Fm3c$	tetrameric and
	$Z = 144$, $d_{x\text{-ray}} = 3.73 \pm 0.02\ g\ cm^{-3}$	8
transformations	$a(-80\ °C) = 25.06 \pm 0.05$ Å	hexameric rings of XeF_5^+ and F^-
observed to		per unit cell
phase I, II, or III		

unit in this cubic phase is almost indistinguishable from that observed in $[XeF_5]^+[PtF_6]^-$ [6].

Transformations between phase I and II or II and III occur with relative ease, since these three modifications all contain only tetramers. By contrast, transformation of Phase IV which contains both tetramers and hexamers into any of phase I, II or III would require an energetically unfavourable reconstructive process. Consequently a transformation of the cubic phase IV has not been observed.

1.2.1.5 *Bonding and molecular structure*

A. XeF_2 *and* XeF_4

A comprehensive review of bonding models for noble-gas compounds has been given previously[61]. There is a considerable weight of expert opinion that the bonding in noble gas compounds does not involve outer d or f orbitals of the noble gas atom, at least not to an extent which could significantly affect stability and other features of chemical interest. Thus xenon is considered to use only its 5p and possibly 5s orbitals in bonding which is essentially of σ-type. This can be accommodated either in the simple molecular orbital scheme[62, 63] or in the valence bond scheme[61].

Taking XeF_2 as an example the former model generates three three-centre MO's from the Xe 5p and a 2p orbital of each fluorine ligand. Since the noble-gas atom contributes two electrons and each fluorine one electron to the pσ MO system, the antibonding $3\sigma_u$ remains empty. In the first approximation the $2\sigma_g$ MO is mostly found associated with the fluorine atoms and is rather non-bonding. Obviously this bonding picture must result in a net negative charge on each of the fluorines, thus leaving the

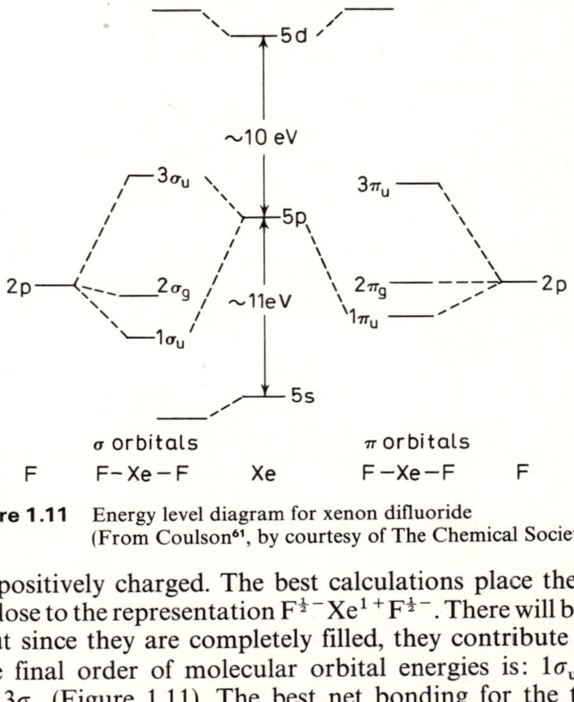

Figure 1.11 Energy level diagram for xenon difluoride (From Coulson[61], by courtesy of The Chemical Society)

xenon atom positively charged. The best calculations place the net charge distribution close to the representation $F^{\frac{1}{2}-}Xe^{1+}F^{\frac{1}{2}-}$. There will be analogous π orbitals, but since they are completely filled, they contribute little to the bonding. The final order of molecular orbital energies is: $1\sigma_u < 1\pi_u < 2\sigma_g < 2\pi_g < 3\pi_u < 3\sigma_u$ (Figure 1.11). The best net bonding for the three atoms occurs, when the arrangement is centrosymmetric which is in harmony with the observed $D_{\infty h}$ molecular symmetry of XeF_2.

Bonding in XeF_4 may be dealt with similarly by considering two three-centre orbitals at right angles, giving a square-planar (D_{4h}) geometry, in full accord with the observed molecular symmetry in the solid and vapour phase.

The valence bond scheme primarily involves resonance between the structures $F{-}Xe^+F^-$ and $F^-Xe^+{-}F$ with minor contributions from doubly-charged formulations.

A more recent investigation, however, indicates that the halogens may perturb the atomic potential at the xenon sufficiently to expand the octet and allow the use of 5d orbitals in bonding models[64, 65]. These model calculations, following the extended valence-bond method of Hurley, Lennard-

Jones and Pople are based on four types of wave functions $\psi_{I, II, II, IV}$. ψ_I corresponds to the structure X—M—X, ψ_{II} accounts for resonance between the structures X—M^+ X^- and X^-M^+—X, ψ_{III} is the completely ionic structure $X^-M^{2+}X^-$ and ψ_{IV} describes the resonance between the two ionic-covalent structures. The configuration interaction was restricted to mixing of the normalised wave functions according to

$$\psi = C_I\psi_I + C_{II}\psi_{II} + C_{III}\psi_{III} + C_{IV}\psi_{IV}.$$

Results are given for several existent and non-existent noble-gas dihalides. Energies of valence-bond structures and results of configuration interaction calculations for XeF_2 are given in Table 1.7. These calculations suggest that the covalent bond structure F—Xe—F, involving 5d interactions, is energetically more favourable than structures depending on ionic contributions and contributes approximately 69% to the total wave function. Moreover, the $F^-Xe^{2+}F^-$ structure is claimed to be lower in energy than F—Xe^+F^-, the valence bond structure favoured by Coulson.

Table 1.7 XeF_2: energies of valence-bond structures and results of configuration interaction calculation[65]

	ψ_I	ψ_{II}	ψ_{III}	ψ_{IV}	ψ
Energies (a.u.*)	−1.99	−2.03	−1.66	−1.92	−2.33
Coefficients	0.481	0.329	0.246	0.209	
% Contributions to ψ	38.9	29.7	15.1	16.2	

*1 a.u. −27.205 eV

Clearly the extent and chemical significance of outer orbital participation in bonding in noble-gas halides has not yet been satisfactorily established. In the following, results of a number of techniques employed in recent years to elucidate bonding conditions in noble-gas compounds will be summarised.

(a) Mössbauer effect—Experimental evidence for charge migration in the xenon–halogen bond has been obtained from Mössbauer-effect studies. In addition $XeCl_2$ and $XeCl_4$ have been detected in the β-decay of $[ICl_2]^-$ and $[ICl_4]^-$ containing radioactive ^{129}I. The electric quadruple interaction in the 39.6 keV excited state of the ^{129}Xe nucleus was then employed to study these compounds by the Mössbauer effect at 4.2 K [66]. There is also similar evidence for $XeBr_2$ but not for $XeBr_4$ [67]. The data were analysed on the assumption that the only xenon orbitals participating in the bonding are the Xe 5p orbitals. Fluorine ligand charges of −0.75 and −0.72 are assigned for XeF_4 and XeF_2, respectively. These values seem rather high and represent probably an upper limit because of the over-simplification of the applied bonding model. The Mössbauer data for xenon halides are compared in Table 1.8.

From the core-electron chemical shifts derived from x-ray electron spectroscopy (ESCA) of the gaseous xenon fluorides, Karlsson *et al.*[68] have concluded that the negative fluorine ligand charge is in the range 0.3–0.5. A coulombic model was used in this study, assuming the central atom and the ligands to be charged spheres.

(b) Nuclear magnetic resonance—Findings from nuclear magnetic resonance have also generally been interpreted in terms of considerable bond ionicity. A fluorine ligand charge of about −0.50 was evaluated from chemical shifts[69], either by assuming the bonding to involve primarily F 2p and Xe 5p orbitals or on the basis of a localised bond-description using spd-hybrid xenon orbitals[70,71].

Recent broad-line n.m.r.-studies on XeF_2 and XeF_4 show that the previous quantitative evaluations may be considerably in error, since much of the earlier experimental work was apparently carried out using rather impure

Table 1.8 Mössbauer data for the xenon halides[66,67]

Halide	*Splitting* (mm/s)	e^2qQ_{exc} (MHz)	5p *Electron transfer*	*Electron transfer per bond*
XeF_4	41.04±0.07	2620	3.00	0.75
XeF_2	39.0±0.1	2490	1.43	0.72
$XeCl_4$	25.6±0.1	1640	1.88	0.47
$XeCl_2$	28.2±0.1	1800	1.03	0.52
$XeBr_2$	22.2±0.4	1415	0.81	0.41

samples[72]. Shielding results for XeF_2 and XeF_4, referred to the bare ^{19}F nucleus, are given in Table 1.9. The total absolute shielding, $\sigma = \frac{1}{3}(\sigma_x+\sigma_y+\sigma_z)$ was determined from the field-dependent contributions to the total rigid-lattice second moment: $M_2 = M_2$ (field independent)$+\frac{4}{45}$ $[(\sigma_z-\sigma_x)^2+(\sigma_y-\sigma_z)(\sigma_y-\sigma_x)]H_0^2$, where σ_x is the out-of-plane shielding, σ_y the in-plane

Table 1.9 Shielding results for XeF_4 and XeF_2

	XeF_4		XeF_2	
	Experiment[72]	*Theory*[71]	*Experiment*[73]	*Theory*[71]
σ	$218\pm5\times10^{-6}$	168×10^{-6}	$358\pm8\times10^{-6}$	233×10^{-6}
σ_x	0±8	−33	393±11	64
σ_y	261±25* or 394±25	−33	393±11	64
σ_z	394±25* or 261±25	570	288±15	570

*If σ_y is 261 then σ_z is 394 and vice versa

shielding perpendicular to the Xe—F bond and σ_z the shielding parallel to the Xe—F bond.

A comparison of the experimental and theoretical shielding values (Table 1.9) supports the contention that previous semiempirical localised-orbital calculations are not satisfying quantitatively and that the previous assumption of axial symmetry about the Xe—F bond in XeF_4 was an oversimplification.

(c) Electronic spectra – A direct test for calculated molecular orbital energy levels is the comparison with observed electronic transitions.

The principal feature in the ultraviolet absorption spectrum of XeF_2 is a singlet–singlet allowed transition from the non-bonding a_{1g} orbital to the antibonding a_{2u} orbital. Using p orbitals only for the molecular orbital calculations 8.1 eV were derived for this transition which compares favourably with the 7.9 eV observed[54].

A high-resolution He^{I} and He^{II} photoelectron spectroscopic study of XeF_2, involving valence electron promotion or ionisation has been reported[74]. The vertical ionisation potentials of the first eight ionisations in this molecule compare favourably with the results of Gaussian-type orbital calculations. The first two ionic states of XeF_2 are the $^2\Pi_{\frac{3}{2}}$ (12.42 eV) and $^2\Pi_{\frac{1}{2}}$ (12.89 eV) spin–orbit components, formed by ionisation from the highest filled orbital ($5\Pi_u$). The vibronic structure of the transitions indicates that the two upper states are linear with Xe—F distances very much like those of the ground

Table 1.10 Observed and calculated ionisation potentials (eV) of XeF_2 [74]

Adiabatic obs.	*Vertical obs.*	0.92 *KT calc.*
12.35 ± 0.01	12.42 ± 0.01 }	12.51 ($5\pi_u$)
12.89 ± 0.01	12.89 ± 0.01 }	
≈ 13.5	13.65 ± 0.05	11.79 ($10\sigma_g$)
14.00 ± 0.05	14.35 ± 0.05	14.71 ($3\pi_g$)
15.25 ± 0.05	15.60 ± 0.05 }	15.92 ($4\pi_u$)
	16.00 ± 0.05 }	
16.80 ± 0.05	17.35 ± 0.05	16.93 ($6\sigma_u$)
	≈ 22.5	25.24 ($9\sigma_g$)
		37.10 ($5\sigma_u$)
		37.20 ($8\sigma_g$)

state. This demonstrates an essentially non-bonding character of the $5\pi_\mu$ orbital. Table 1.10 lists the empirically adjusted theoretical values (all Koopmans' theorem ionisation potentials are reduced uniformly by multiplication by 0.92) together with adiabatic and vertical ionisation potentials observed experimentally. Similar to the observation in other heavy atom molecules, the molecular Rydberg-term values of XeF_2 are strikingly like

Table 1.11 Term energies in Xe and XeF_2 [74]

Xe		XeF_2	
Upper state	*Term* (cm^{-1})	*Upper state*	*Term* (cm^{-1})
$5p(^2P_{\frac{3}{2}})6s$	30 400	$5\pi_u(^2\pi_{\frac{3}{2}})6s$	30 865
$5p(^2P_{\frac{1}{2}})6s$	31 433	$5\pi_u(^2\pi_{\frac{1}{2}})6s$	30 080
$5p(^2P_{\frac{3}{2}})6p$	19 322		
$5p(^2P_{\frac{1}{2}})6p$	19 317		
$5p(^2P_{\frac{3}{2}})5d$	16 628	$5\pi_u(^2\pi_{\frac{3}{2}})5d$	17 860
$5p(^2P_{\frac{1}{2}})5d$	16 567	$5\pi_u(^2\pi_{\frac{1}{2}})5d$	16 600
$5p(^2P_{\frac{3}{2}})7s$	12 551		
$5p(^2P_{\frac{1}{2}})7s$	12 590		

those of xenon itself (Table 1.11) and correlate well with those derived from the XeF_2 ultraviolet spectral data[54].

The more complicated electronic absorption spectrum of XeF_4 is summarised in Table 1.12.Using analytical SCF wave functions for xenon instead of Slater orbitals which are less suited to describe orbitals of high principal quantum numbers, molecular orbital energies of XeF_4 have been calculated in the Wolfsberg–Helmholz semi-empirical approximation[75]. By properly normalising the ligand symmetry orbitals, ligand–ligand overlaps were included in the calculations, since their magnitude is comparable to certain Xe—F overlaps. Instead of the usual geometric mean the reciprocal mean was used as the proportionality constant in the evaluation of resonance integrals. 38 electrons were considered, 18 coming from the xenon ($4d^{10}5s^2 5p^6 6s^0 5d^0$) and five from each fluorine ($2sp^1 2p_x^2 2p_y^2$). The resulting energy

Table 1.12 Ultraviolet absorption spectrum of XeF_4

Experimental (eV)[76]	*Calculated* (eV)[75]	*Assignment*
4.81	—	Singlet–triplet
5.43	5.4	$2a_{2u} \rightarrow 3e_u$
6.8	6.9	$3a_{1g} \rightarrow 3e_u$
8.3 (shoulder)	8.0	$2a_{2u} \rightarrow 4a_{1g}$
9.4	≈8.8	$\begin{cases} 2a_{2u} \rightarrow 5a_{1g} \\ 2a_{2u} \rightarrow 3e_g \end{cases}$
11.3	12.3	$2a_{2g} \rightarrow 3e_u$

level diagram is shown in Figure 1.12, giving the following ground state configuration for XeF_4:

$$(1b_{1g})^2(1a_{1g})^2(1e_u)^4(1b_{2g})^2(1e_g)^4(2a_{1g})^2(2b_{1g})^2(2b_{2g})^2(2e_g)^4(1a_{2u})^2(2e_u)^4 (b_{2u})^2(a_{2g})^2(3a_{1g})^2(2a_{2u})^2,$$

with a $^1A_{1g}$ ground state. It is interesting to note that two of the five a_{1g}MO's, the $2a_{1g}$ and $4a_{1g}$ remain essentially atomic in character. This is expected since the overlaps of both 6s and 4d xenon orbitals with the $\Sigma\,(a_{1g})$ ligand orbital are quite small (Table 1.13).

Table 1.13 XeF_4: Diatomic overlaps obtained from analytical SCF wave functions[75]

$\langle 6s \mid \sigma \rangle$	0.04036
$\langle 5p_\sigma \mid \sigma \rangle$	0.32468
$\langle 6p_\sigma \mid \sigma \rangle$	−0.00271
$\langle 5d_\sigma \mid \sigma \rangle$	0.27342
$\langle 5p_\pi \mid \pi \rangle$	0.09361
$\langle 6p_\pi \mid \pi \rangle$	0.00621
$\langle 5d_\pi \mid \pi \rangle$	0.25301
$\langle 5s \mid \sigma \rangle$	0.19392
$\langle 4d_\sigma \mid \sigma \rangle$	0.01928
$\langle 4d_\pi \mid \pi \rangle$	0.00676

B. XeF_6

Models which correctly predicted the structures of xenon difluoride and xenon tetrafluoride have led to contradictory predictions for xenon hexafluoride. A simple extension of the theoretical picture of three-centre, four-electron bonds, using Xe 5p orbitals only, was widely interpreted as forecasting O_h symmetry for XeF_6. On the other hand the Gillespie–Nyholm–Sidgwick–Powell–Tsuchida valence-shell electron-pair repulsion (VSEPR) theory[77–79] led to a prediction that the molecule should be distorted from O_h symmetry. The xenon valence shell in XeF_6 contains eight Xe electrons and six F electrons for a total of seven pairs. Six pairs are considered to be bonding electron pairs, the remaining one is taken as a localised, stereo-

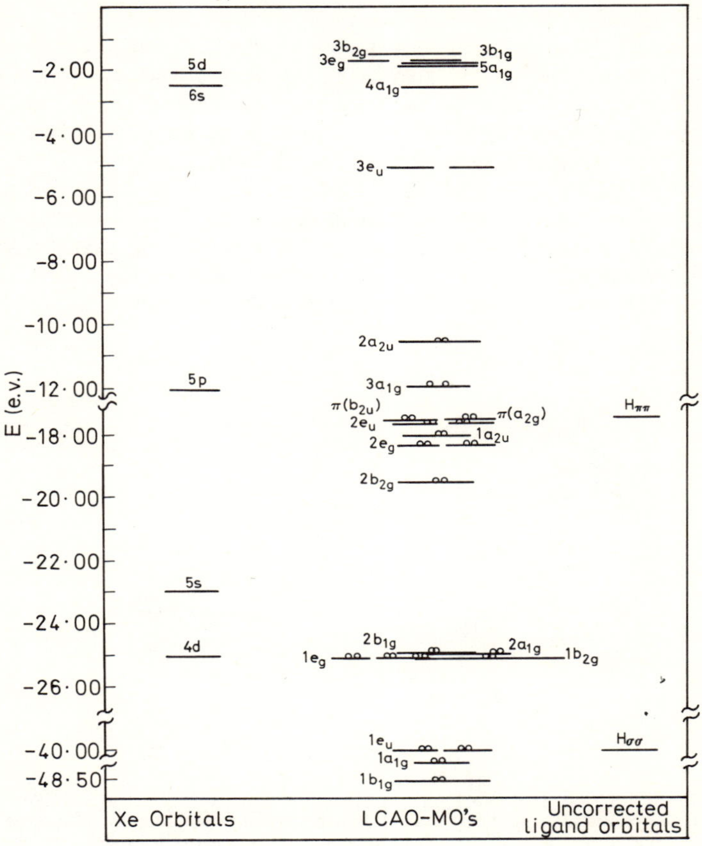

Figure 1.12 One-electron molecular orbital energies of XeF_4 (From Yeranos[75], by courtesy of Academic Press)

chemically active lone pair. On the basis of VSEPR theory rules XeF_6 would then be considered to exhibit C_{2v} or C_{3v} symmetry.

It is worth mentioning, however, that an elementary Hückel LCAO–MO calculation for XeF_6 including Xe 5p *and* Xe 5s electrons (and a $2p_\sigma$ orbital of each fluorine ligand) affords a non-octahedral geometry as well, the most stable one being the C_{3v} model[79, 80]. Inclusion of Xe 5d orbitals and of non-bonded interactions gives virtually the same variation of energy with geometry[81]. These findings are, of course, not in account with the suggestion that d orbital involvement is what destabilises O_h symmetry for XeF_6 [82].

Since XeF_6 is polymeric in condensed phases, most of the argument about its vapour phase structure rests on electron diffraction data[49, 83, 84].

The diffraction data are not compatible with an O_h molecule vibrating in independent, uncorrelated normal modes. The mean bond length is 1.890 ± 0.005 Å, but the radial distribution function for Xe—F bonds corresponds to that of a composite of non-equivalent bonds.

The diffraction data, however, are also not compatible with a single molecular configuration as derived by the above cited LCAO–MO calculations or VSEPR considerations. Both models overemphasise the tendency of XeF_6 to deform. It should also be noted, in this respect, that an electrostatic-deflection molecular-beam experiment established the dipole moment of XeF_6 to be $\leqslant 0.03$ D [85]. This excludes a rigid, polar molecule or more specifically is difficult to reconcile with non-centric structures.

Bartell and Gavin[79a], concluded that the instantaneous molecular configurations encountered by the incident electrons are predominantly in the broad vicinity of C_{3v} structures conveniently described as distorted octahedra in which the xenon lone pair avoids the bonding pairs. In those distorted structures the Xe—F bond lengths are distributed over a range of

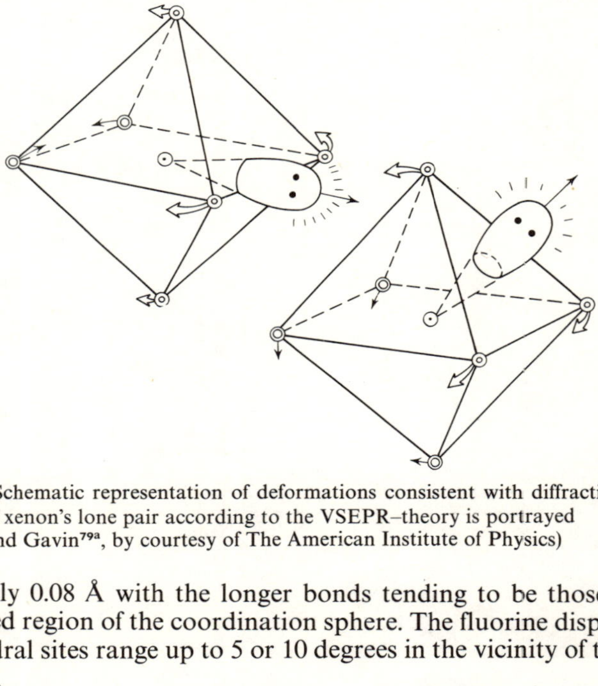

Figure 1.13 Schematic representation of deformations consistent with diffraction patterns. The influence of xenon's lone pair according to the VSEPR–theory is portrayed (From Bartell and Gavin[79a], by courtesy of The American Institute of Physics)

approximately 0.08 Å with the longer bonds tending to be those adjacent to the avoided region of the coordination sphere. The fluorine displacements from octahedral sites range up to 5 or 10 degrees in the vicinity of the region. (Figure 1.13).

The non-octahedral geometry and the exceptional flexibility of the XeF_6 molecule in the gas phase (the t_{1u} bending amplitudes are enormous and correlated in a certain way with substantial t_{2g} deformations), have also been interpreted in terms of a pseudo-Jahn–Teller effect[79a]. The Hamiltonian operator can be expanded as a Taylor series[86]:

$$H = H^{\circ} + H_i' S_i + \tfrac{1}{2} H_{ii}'' S_i^2 + \ldots$$

in the symmetry coordinate S_i for molecular deformations. An application of perturbation theory yields, for the ground electronic state, the result:

$$E = E^{\circ} + \langle\psi_0 | H'_i | \psi_0\rangle S_i + \{\tfrac{1}{2}\langle\psi_0 | H^0_{ii} | \psi_0\rangle - \Sigma[|\psi_0 | H'_i | \psi_n|^2/(E_n - E_0)]\}S_i^2 + \ldots$$

The first-order term is the Jahn–Teller term which vanishes for non-degenerate ground states. The second order terms describe the force constant for S_i. A small value of $(E_n - E_0)$ coupled with a non-vanishing matrix element $\langle\psi_0 | H'_i | \psi_n\rangle$ can lead to a low or even negative value of the force constant.

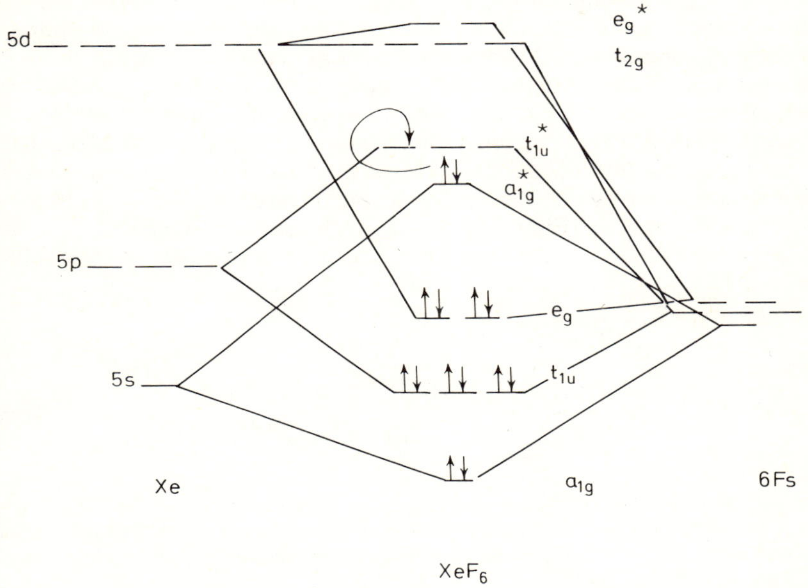

Figure 1.14 Schematic correlation diagram illustrating MO energy levels for an O_h molecule

stant. When the mixing between ground and excited states on deformation is large enough to make a distortion energetically favourable, in other words, by making the force constant negative, the molecule is said to suffer a pseudo-Jahn–Teller effect.

The relevance of this to XeF_6 is probably best understood with the aid of Figure 1.14. This model supposes that the Xe 5d orbitals partake in the bonding, the d orbital degeneracy having been removed by the ligand field. A pair of electrons of the O_h symmetry XeF_6 molecule is placed in the antibonding a_{1g}^* orbital, supposedly being very close in energy to the triply degenerate t_{1u}^* orbitals and indeed so close that considerable mixing of the ground and excited states occurs. This implies, according to the above mentioned formalism, an especially low force constant for t_{1u} deformations. In accordance with diffraction data, t_{1u} deformations indeed appear to be unusually large. To visualise this, it should be appreciated that a t_{1u} deformation is equivalent, geometrically, to a 'lone pair' protruding into the coordination sphere and pushing aside the ligands.

The same correlation diagram is appropriate for alternative conjectures of the diffraction data[87]. XeF_6 of O_h symmetry is assumed to be the ground-state species. Further assuming the a_{1g}^*—t_{1u}^* energy gap to be small enough, an extensive population of the orbitally-degenerated triplet state would result at room temperature and hence these molecules would be subject to a (first order) Jahn–Teller deformation to a non-degenerate, distorted state. Thus, according to this view, the XeF_6 population should contain octahedral and D_{4h} (by $e_g$deformation) or D_{3d} (by t_{2g}deformation) symmetry species. A molecular-deflection molecular-beam experiment, however, has provided no evidence for a paramagnetic XeF_6 species[88], although this does not necessarily disprove the existence of a triplet species if the spin–orbit coupling is sufficiently strong. Whereas a D_{4h} distortion is not compatible with diffraction data, it is possible to construct a D_{3d} model with sizeable a_{2u} oscillations which gives a reasonably good fit even with no O_h molecules present[79a].

Another model which accounts for the electron diffraction data has been given by Burbank and Bartlett[8], and involves intramolecular rearrangement between several molecular geometries. Using uniform weights for configurations from C_{2v} to C_{3v} (Figure 1.15) the synthetic radial distribution curve fits the observed one within the limits of experimental error. This facile

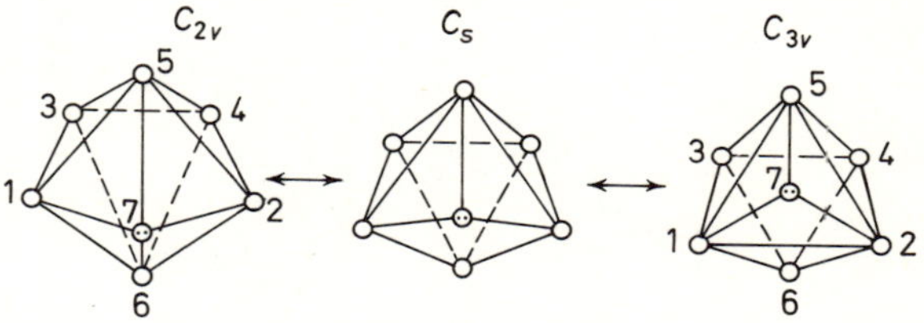

Figure 1.15 XeF_6: Transformation of configurations. The seventh ligand represented by an open circle containing two dots is the electron lone fair. (There are 7! ways in which the ligands can be numbered. If the numbering is permuted after two or more transformations, then a rearrangement has occurred
(From Burbank and Bartlett[8], by courtesy of The Chemical Society)

ligand rearrangement process is also consistent with the slight directional quality in the bonding and suggests that re-orientation of the valence lone pair in XeF_6 is very easy.

As in the case of the other xenon fluorides, the ^{19}F n.m.r. data have been interpreted on the basis of bond polarities close to $Xe^{3+}(F^{0.5-})_6$, either by assuming Xe 5d orbital involvement in bonding or not[70, 89].

Soft x-ray photoionisation studies, involving the ejection of core electrons from the xenon atom in XeF_6, have shown that the charge withdrawn per fluorine ligand is in the range -0.3 to -0.5[68].

1.2.1.6 Non-aqueous chemistry

A. Fluoride ion donor and acceptor properties ($[XeF]^+$, $[Xe_2F_3]^+$, $[XeF_5]^+$ etc.)

(a) XeF_2—Compounds of the composition $XeF_2{\cdot}2MF_5$, like $XeF_2{\cdot}2SbF_5$ and $XeF_2{\cdot}2TaF_5$ have been known since 1963[90], but only recently they have been fully characterised and shown to be salts containing the $[XeF]^+$ ion[2, 4, 91].

Table 1.14 Some physical and spectroscopic properties of $[XeF]^+[MF_6]^-$, $[XeF]^+[M_2F_{11}]^-$ and $[Xe_2F_3]^+[MF_6]^-$ salts

$[XeF]^+[MF_6]^-$							
M =	As [91]	Ru [91]	Os* [91]	Ir [91]	Pt [91]	Ta [92]	Nb [92]
Colour	pale yellow-green	pale yellow-green	brown	yellow-green	orange-red	pale yellow	pale yellow
m.p. (°C)	—	110–111	—	152–153	82–83	52–53	30–35
Xe—F stretch [91] (Raman, cm^{-1})	—	604,599	—	608,602	609,602	—	—
Unit cell[91]	—	←——— isomorphous ———→ $a = 11.1$ Å, $b = 7.96$ Å, $c = 7.24$ Å, $Z = 4$; Space group: *Pnma*				—	—

*Unstable

$[XeF]^+[M_2F_{11}]^-$						
M =	Sb [2]	Ru [91]	Ir [91]	Pt [91]	Ta [92]	Nb [92]
Colour	yellow	bright green	orange-yellow	dark red	pale yellow	pale yellow
m.p. (°C)	63 [2, 93]	49–50	69–70	—	82–83	42–47
Xe—F stretch[3] (Raman, cm^{-1})	621	604,598	612,601	—	—	—
Unit cell	$a = 8.07$, $b = 9.55$, $c = 7.33$ Å, $\beta = 105.8°$, $Z = 2$, $P2_1$	←——— isomorphous ———→			←——— isomorphous ———→	

$[Xe_2F_3]^+[MF_6]^-$				
M =	As [91, 4]	Ru [91]	Os [91]	Ir [91]
Colour	pale yellow-green	pale yellow-green	pale yellow-green	pale yellow-green
m.p. (°C)	99–100	98–99	—	92–93
Xe—F stretch[3] (Raman, cm^{-1})	600,588	593,579	593,582	592,578
Unit cell	$a = 15.443$, $b = 8.678$, $c = 20.888$ Å, $\beta = 90.13°$, $Z = 12$, $I\,2/a$	←——— *isomorphous* ———→		

In addition, a thorough study of related systems revealed that strong fluoride ion acceptors, such as AsF_5 and the noble metal pentafluorides readily abstract a fluoride ion from XeF_2. Depending on reactants ratios three classes of salts have been established[91].

$$XeF_2 + MF_5 \rightarrow [XeF]^+[MF_6]^-$$
$$XeF_2 + 2\,MF_5 \rightarrow [XeF]^+[M_2F_{11}]^-$$
$$2\,XeF_2 + MF_5 \rightarrow [Xe_2F_3]^+[MF_6]^-$$

Some of these salts can be prepared by fusing XeF_2/MF_5 mixtures[2,92]. The best preparative method, however, is to dissolve xenon difluoride and

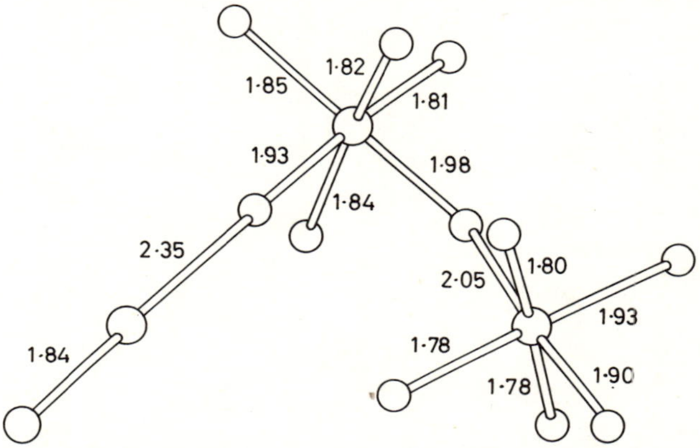

Figure 1.16 The molecular structure of $[XeF]^+[Sb_2F_{11}]^-$ (From McRae *et al.*[2], by courtesy of The Chemical Society)

the fluoride ion acceptor in the desired molar ratio in bromine pentafluoride. By distilling off BrF_5 at room temperature or below, crystalline, homogenous products may be obtained[91]. Some physical and spectroscopic properties of these salts are given in Table 1.14.

A phase study of the XeF_2–SbF_5 system established the existence of the compounds $XeF_2{\cdot}SbF_5$, $XeF_2{\cdot}\frac{3}{2}\,SbF_5$, $XeF_2{\cdot}2\,SbF_5$ and $XeF_2{\cdot}6\,SbF_5$ [93].

The molecular structure of $[XeF]^+[Sb_2F_{11}]^-$ (Figure 1.16) establishes the existence of a $[XeF]^+$ species with a Xe—F bond length of 1.84 Å[2]. This and a force constant of about 3.58 mdyn/Å for other $[XeF]^+$ salts[3] indicate a considerably shorter and stronger bond than in XeF_2 (Xe—F, XeF_2: 2.00 Å; K_r, XeF_2: 2.82 mdyn/Å).

From the thermochemical cycle in Figure 1.17 it can be seen that the total bond energy of XeF_2 is related to the bond energy of $[XeF]^+$ by the equation:

$$\text{T. B. E. } (XeF_2,g) = \text{B.E. } (XeF^+) + \text{A.P.}(XeF^+, XeF_2) - \text{I.P.}(Xe)$$

Since the appearance potential of XeF^+, XeF_2 and the ionisation potential of xenon are 12.8 eV and 12.1 eV, respectively[94]:

$$65\ \text{kcal mol}^{-1} = \text{B.E.}(XeF^+) + 16\ \text{kcal mol}^{-1}\ (0.7\ \text{eV});$$

therefore the Xe—F bond energy in $[XeF]^+$ is 49 kcal mol^{-1} in accord with the greater force constant and shortness of the Xe—F bond in the cation,

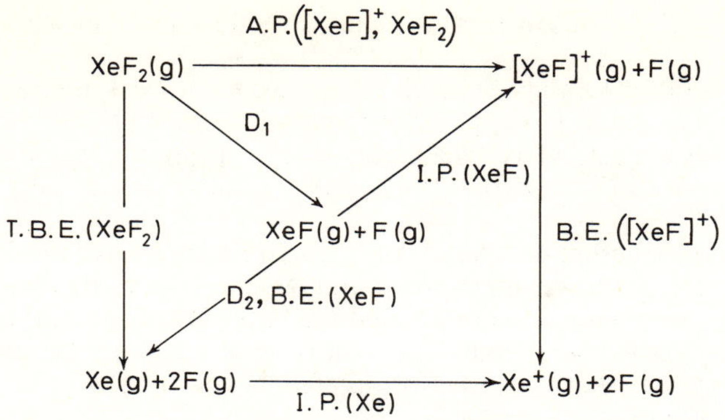

Figure 1.17 Thermochemical cycles for XeF_2 and related species

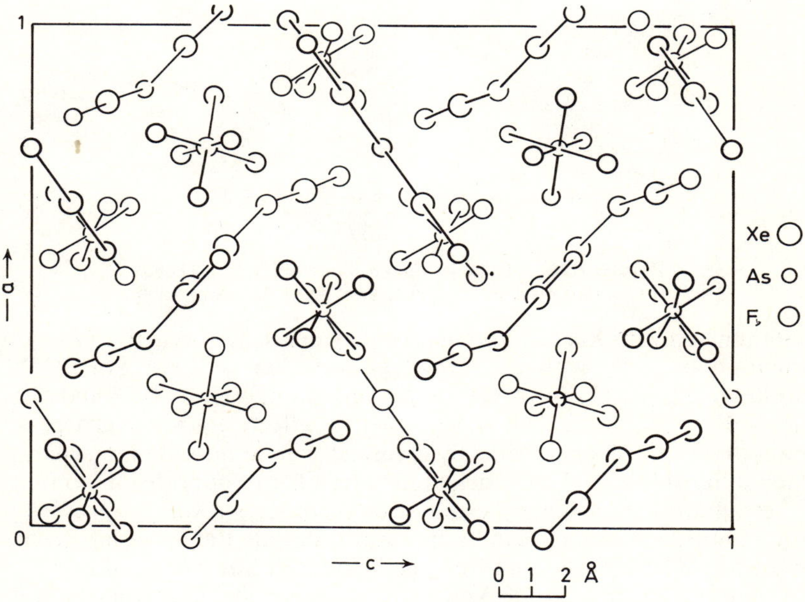

Figure 1.18 Unit cell of $[Xe_2F_3]^+[AsF_6]^-$
(From Sladky *et al.* by courtesy of The Chemical Society)

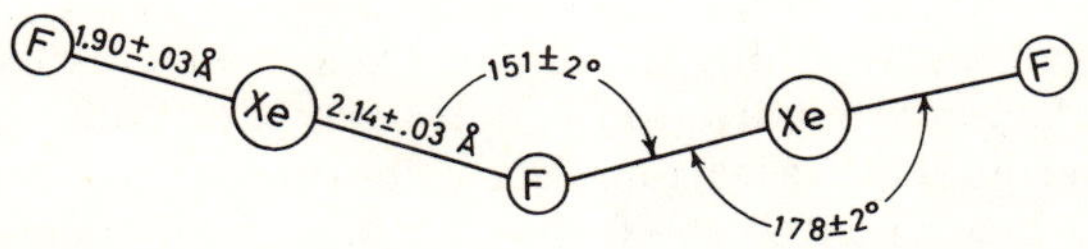

Figure 1.19 Bond lengths and bond angles in $[Xe_2F_3]^+$
(From Sladky *et al.*, by courtesy of The Chemical Society)

compared to XeF_2 (mean thermochemical bond energy in XeF_2: 32.5 kcal mol^{-1}).

Raman data for $[XeF]^+$ salts give evidence for a weak interaction of the $[XeF]^+$ ion with the $[MF_6]^-$ entity and with other $[XeF]^+$ ions[91]. That the $[XeF]^+$ ion must indeed be a highly polarising species is underlined with the characterisation[4,91] of the $2XeF_2{\cdot}MF_5$ adducts as the salts $[Xe_2F_3]^+[MF_6]^-$ [3,4].

The crystal structure of $2\,XeF_2{\cdot}AsF_5$ (Figure 1.18) proves the formulation $[Xe_2F_3]^+[AsF_6]^-$ to be appropriate and Raman data of this salt and other $2\,XeF_2{\cdot}MF_5$ compounds establish that the latter also contain the $[Xe_2F_3]^+$ ion. This cation is planar and V-shaped (Figure 1.19) and the bond length disparity suggests that it can be formulated as $[F{-}Xe]^+F^-[Xe{-}F]^+$ to a first approximation.

(b) XeF_4 – Reports that XeF_4 interacts with SbF_5 or TaF_5 to form XeF_2 adducts[90] are erroneous and probably a consequence of use of XeF_4 grossly

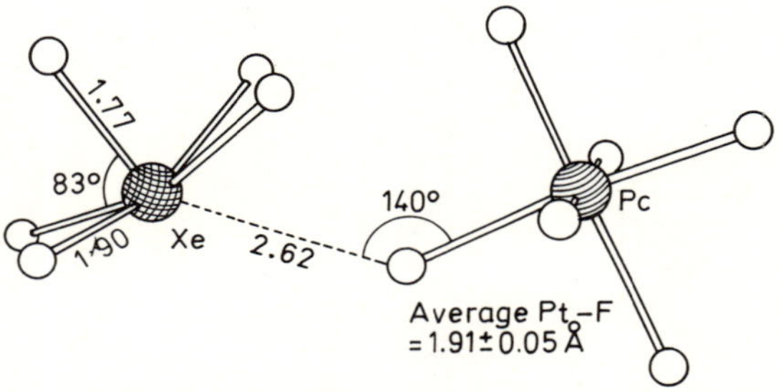

Figure 1.20 The molecular structure of $[XeF_5]^+[PtF_6]^-$ (From Bartlett *et al.*[5], by courtesy of The Chemical Society)

contaminated with XeF_2. Although scant evidence has been given for adduct formation of XeF_4 with the strong fluoride ion acceptor SbF_5 [95,96], the fluoride-ion donor ability of XeF_4 is certainly far inferior to XeF_2 and XeF_6 [28].

(*c*) XeF_6 – On the basis that the higher the effective positive charge of the xenon atom, the less energetically favoured will be fluoride ion separation, xenon difluoride would be anticipated to be a better fluoride ion donor than the tetrafluoride and xenon hexafluoride the poorest. XeF_6, however, parts with a fluoride ion more readily than XeF_2, despite the lower lattice energy for the larger cation case and XeF_4 is the poorest fluoride ion donor[28].

A number of adducts of XeF_6 with fluoride ion acceptors have been reported. They include the 1:1 adducts with BF_3 [97], AsF_5 [97,98,99], SbF_5 [100], GeF_4 [101], VF_5 [102], TaF_5 [103], PtF_5 [104], IrF_5 and RuF_5 [28].

X-ray single-crystal structure analyses of the 1:1 adducts with PtF_5 and AsF_5 revealed the ionic structures $[XeF_5]^+[PtF_6]^-$ and $[XeF_5]^+[AsF_6]^-$ [5,105]. (Figures 1.20 and 1.21). The $[XeF_5]^+$ ion ($\approx C_{4v}$) is considered to contain pseudo-octahedrally coordinated xenon which appears energetically especially favoured relative to the non-octahedral XeF_6 molecule.

Several other adducts in various XeF_6/fluoride-ion acceptor ratios have been prepared as well, either by fusing the neat components or dissolving them in solvents (BrF_5, HF) or from $Xe/F_2/MF_5$ mixtures where fluorine pressure and reaction temperature favour XeF_6 formation: $2XeF_6{\cdot}AsF_5$,

$2XeF_6{\cdot}PF_5$ [99], $2XeF_6{\cdot}SbF_5$, $XeF_6{\cdot}2SbF_5$ [100], $2XeF_6{\cdot}TaF_5$ [103], $2XeF_6{\cdot}PtF_5$ [104], $2XeF_6{\cdot}IrF_5$ [28], $4XeF_6{\cdot}GeF_4$, $2XeF_6{\cdot}GeF_4$ [101], $4XeF_6{\cdot}SnF_4$ and $2XeF_6{\cdot}SnF_4$ [106]. All these adducts are reactive and readily hydrolysed solids, with melting points around 100 °C. Though not adequately characterised yet, the observation that XeF_6 is essentially a polymer, containing $XeF_5{}^+F^-$ units, suggests similar formulations for these adducts. A 2:1 adduct is therefore expected to be the salt $[F_5XeFXeF_5]^+[MF_6]^-$.

XeF_2 and XeF_4 do not exhibit any fluoride ion acceptor properties, whereas a number of compounds have been reported involving XeF_6 in combination with recognised fluoride-ion donors like alkali fluorides[107]. Infrared and Raman studies of a related compound $2NOF{\cdot}XeF_6$ indicate the presence of the $[NO]^+$ cation[108], and it may be expected that they are salts of the general formulae $A^+[XeF_7]^-$ and $(A^+)_2[XeF_8]^{2-}$.

B. The Xe–PtF_6 *reaction*

The investigations of the fluoride ion donor properties of the xenon fluorides have greatly clarified[91] the nature of the products of the interaction

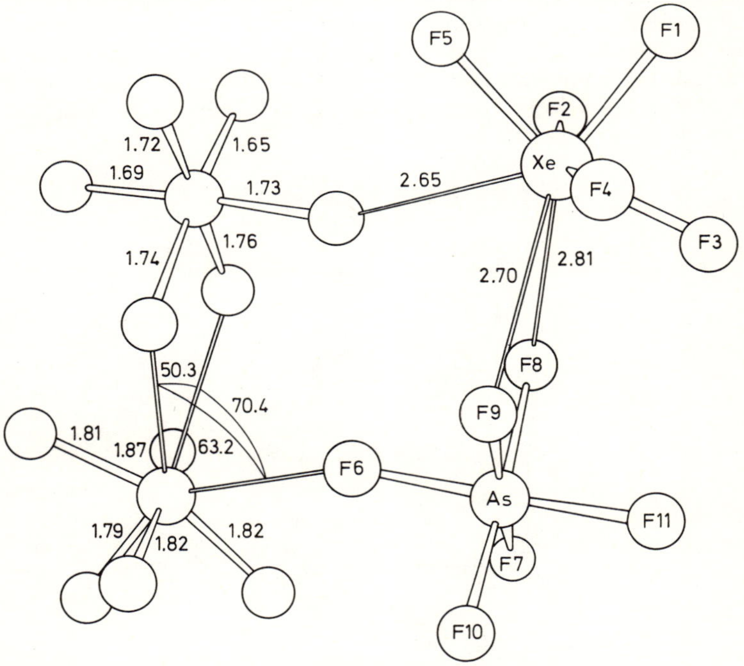

Figure 1.21 The molecular structure of $[XeF_5]^+[AsF_6]^-$
(From Hollander *et al.*[105], by courtesy of The Chemical Society)

of xenon with platinum hexafluoride. One of these products, $XePtF_6$, is of particular interest since it was the first xenon compound to be reported, in which the xenon valence electron configuration was unequivocally different from the supposed 'ideal' octet[1].

$Xe[PtF_6]_x$ $(1<x\leqslant 2)$, obtained at *c.*20 °C contains $[XeF]^+[PtF_6]^-$ as

shown by x-ray powder photographs, presumably produced as follows:

$$Xe + PtF_6 \rightarrow XePtF_6$$
$$XePtF_6 + PtF_6 \rightarrow [XeF]^+[PtF_6]^- + PtF_5$$

The presence of amorphous platinum pentafluoride in the product is indicated by the formation of $[XeF]^+[Pt_2F_{11}]^-$, when the material of empirical composition $Xe(PtF_6)_2$, is warmed to 60 °C:

$$[XeF]^+[PtF_6]^- + PtF_5 \xrightarrow{60°C} [XeF]^+[F_5PtFPtF_5]^-$$

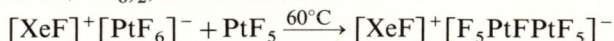

For the formation of the compound of 1:1 stoichiometry it is essential to maintain a large excess of xenon. Unfortunately $XePtF_6$ is amorphous to x-rays; it contains Pt^{IV}, this suggesting the formulation $Xe^+[PtF_6]^-$. So far there is no reliable magnetic (the Xe^+ species should be paramagnetic) or structural data to support the designation of the xenon as Xe^+ rather than Xe_2^{2+} or $[Xe_2F]^+$ which would be appropriate for the formulation $[Xe_2F]^+$ $[Pt_2F_{11}]^-$.

C. Molecular adducts ($XeF_2{\cdot}XeF_4$, $XeF_2{\cdot}IF_5$, $XeF_2{\cdot}XeOF_4$ *etc.*)

A single-crystal x-ray structural analysis of a xenon fluoride, which was initially thought to be a high-temperature modification of XeF_4, proved it an ordered, 1:1 molecular adduct of XeF_2 and XeF_4 [109]. Bond lengths and angles of XeF_2 and XeF_4 in the adduct are not significantly different from those in the pure components.

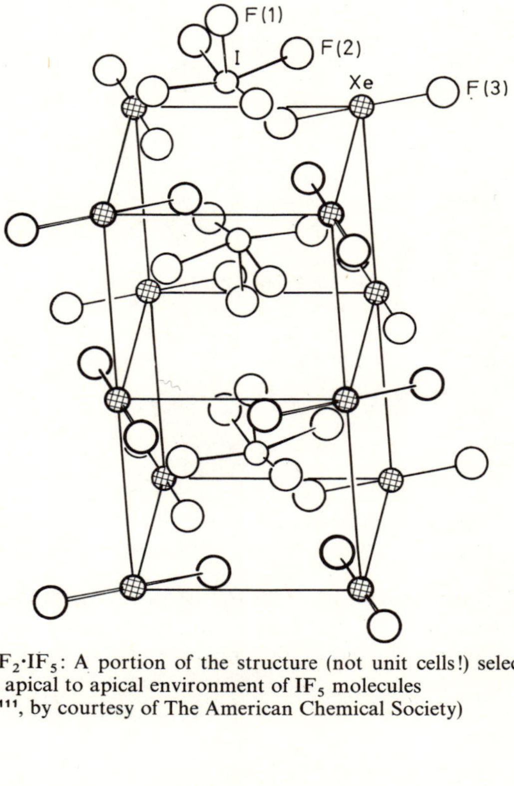

Figure 1.22 $XeF_2{\cdot}IF_5$: A portion of the structure (not unit cells!) selected to indicate the basal to basal and apical to apical environment of IF_5 molecules
(From Jones *et al.*[111], by courtesy of The American Chemical Society)

Recently it became obvious that the formation of molecular adducts is a rather common feature of xenon compounds, especially of xenon difluoride. Beside the 1:1 adducts $XeF_2{\cdot}XeF_4$ [109,110], $XeF_2{\cdot}IF_5$, m.p.98 °C [3,111], $XeF_2{\cdot}XeOF_4$, m.p.29 °C and $XeF_2{\cdot}[XeF_5]^+[AsF_6]^-$ [112] the 1:2 adducts $XeF_2{\cdot}2IF_5$ [34] and $XeF_2{\cdot}2[XeF_5]^+[AsF_6]^-$ [112] have been reported as well.

These adducts may be readily prepared by fusing the components in the appropriate molar ratio or in the case of $XeF_2{\cdot}IF_5$ by dissolving XeF_2 in

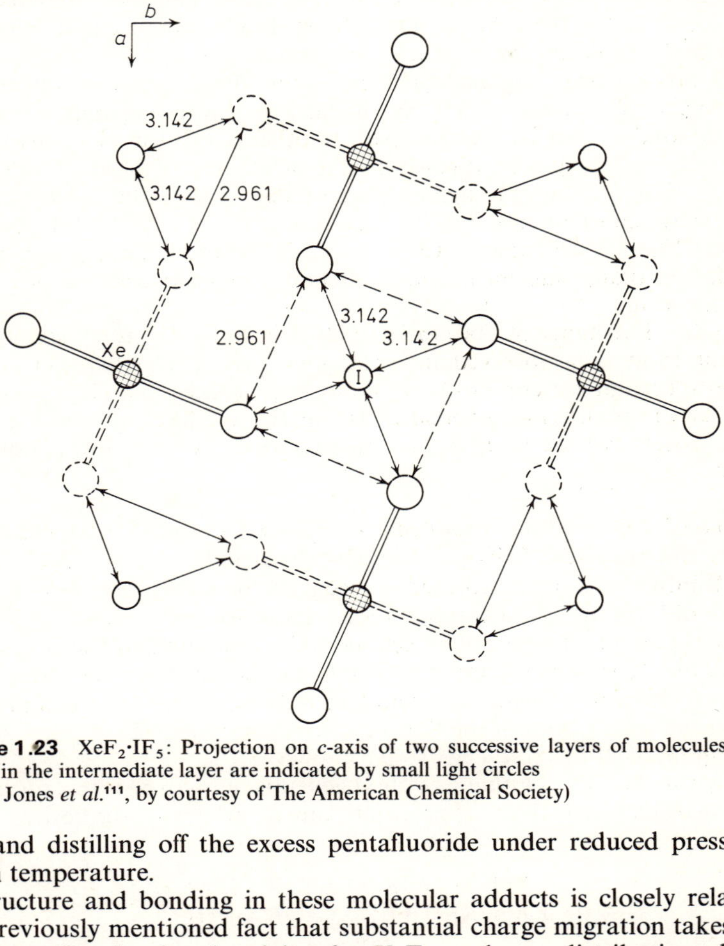

Figure 1.23 $XeF_2{\cdot}IF_5$: Projection on *c*-axis of two successive layers of molecules. Iodine atoms in the intermediate layer are indicated by small light circles
(From Jones *et al.*[111], by courtesy of The American Chemical Society)

IF_5 and distilling off the excess pentafluoride under reduced pressure at room temperature.

Structure and bonding in these molecular adducts is closely related to the previously mentioned fact that substantial charge migration takes place in xenon–fluorine bonds, giving for XeF_2 a charge distribution close to $F^{-\frac{1}{2}}$—Xe^{+1}—$F^{-\frac{1}{2}}$. Excellent agreement has been obtained between the experimental heat of sublimation of XeF_2 and the calculated one, based on coulombic interactions between XeF_2 molecules, according to these net charges and pertinent crystal structure data[113].

The specific nature of these intermolecular interactions is illustrated with the crystal structure of the 1:1 $XeF_2{\cdot}IF_5$ adduct[111] (Figures 1.22 and 1.23). Each molecule is surrounded by an approximately cubic arrangement of molecules of the other kind. The structural order is consistent with close packing of the molecules and with electrostatic attraction of the negatively charged fluorine ligands of one molecular species for the positively charged central atom of the other; the attraction of the fluorine ligands of XeF_2 for the iodine atom of IF_5 being particularly important. The disposition of the fluorine ligands in a layer of XeF_2 molecules is determined by the orientation of the nearest IF_5 molecules. If IF_5 molecules in adjacent layers are base to base, the sandwiched XeF_2 molecules orient to make short I—F contacts. On the other hand, XeF_2 molecules are oriented away from adjacent IF_5 molecules where they are apex to apex.

This arrangement suggests that the iodine atom bears an appreciable positive charge which is effectively shielded by fluorine ligands but not by the non-bonding valence electron pair. In other words this is a consequence of the theory, recognised already in the structures of $[XeF_5]^+[PtF_6]^-$ [5] and also XeF_6 [59], that although an electron pair in the valence shell exerts a larger steric effect than a fluorine ligand (e.g. IF_5:F, apical–I–F, basal:81 degrees; $[XeF_5]^+$:F, apical–Xe–F, basal:79 degrees) the electron pair is less effective than a fluorine ligand at shielding the positive charge on a xenon or iodine atom.

The Xe—F distance of 2.007 ($\sigma = 0.009$) Å in XeF_2. IF_5 is not significantly different from that observed in crystalline XeF_2 (2.00 ± 0.01 Å) and the Raman spectrum is a composite of the spectra of XeF_2 and IF_5.

In view of this, it is not surprising that $XeOF_4$ and $[XeF_5]^+$, both geometrically closely related to IF_5, also interact with XeF_2 to form molecular adducts.

D. Xenon(II) fluoride fluorosulphate, $FXeOSO_2F$, *xenon(II) bispentafluoro-orthotellurate*, $Xe(OTeF_5)_2$ *and related compounds*

Only fluorine and unsubstituted oxygen as ligand atoms were thought capable of forming xenon compounds available for use at ordinary temperature and pressure. In the case of xenon(II) the situation had been even more limited, the fluoride being the only stable compound, even the oxide being unknown. Recently, however, several ligands have been successfully substituted for one or both of the fluorines in xenon difluoride. Table 1.15 lists all compounds of this new class so far known.

(a) Synthesis – The synthetic method generally employed is the interaction of xenon difluoride with the appropriate anhydrous acid in the desired molar ratio:

$$XeF_2 + HOR \rightleftharpoons FXeOR + HF$$
$$FXeOR + HOR \rightleftharpoons Xe(OR)_2 + HF$$
or
$$XeF_2 + 2\,HOR \rightleftharpoons Xe(OR)_2 + 2\,HF$$

The synthesis is usually carried out at the lowest possible temperature (−75 °C to room temperature), as is the removal of the formed HF. In the case of the xenon(II) trifluoroacetates, it proved very useful to run the reaction in a solvent like HF, CH_3CN or $(CF_3CO)_2O$ [7c].

The main driving force for these reactions seems to be the exceptionally high enthalpy of formation of HF. However, $FXeOTeF_5$ is best prepared in the reaction of equimolecular amounts of XeF_2 and $Xe(OTeF_5)_2$:

$$XeF_2 + Xe(OTeF_5)_2 \rightarrow F\,XeOTeF_5$$

A different synthetic approach is the use of acid displacement reactions, especially in displacing pentafluoro-orthotelluric acid, $HOTeF_5$, which has a convenient vapour pressure at room temperature ($p_{25\,°C} = 130$ Torr)[114]:

$$Xe(OTeF_5)_2 + \text{excess } HF \rightarrow XeF_2 + 2\,HOTeF_5$$
$$Xe(OTeF_5)_2 + \text{excess } CF_3COOH \rightarrow Xe[OC(O)CF_3]_2 + 2HOTeF_5$$
$$Xe(OTeF_5)_2 + 2\,HOSO_2F \rightarrow Xe(OSO_2F)_2 + 2HOTeF_5$$

(b) Thermal stability—With the remarkable exception of the xenon(II) pentafluoro-orthotellurates, all other compounds of this type have a half-life of only hours or at most days at room temperature. Xenon(II) fluoride fluorosulphate is relatively stable with a half-life of about one week but it can be kept for many weeks at 0 °C without decomposition. The xenon(II)

Table 1.15 Compounds of the type FXeOR and Xe(OR)$_2$

FXeOR *or* Xe(OR)$_2$	M.p. (°C)	*Approximate decomposition temperature* (°C)	*Reference*
$FXeOSO_2F$	36.6		
$FXeOClO_3$	16.5	room temperature	6, 28
$Xe(OSO_2F)_2$	43–45		
$Xe(OClO_3)_2$	–		
$FXeOC(O)CF_3$	–	explosive above	7, 182, 183
$Xe[OC(O)CF_3]_2$	–	−20	
$FXeOTeF_5$	−15	150	
	(b.p.$_{0.001}$: 53 °C)		7
$Xe(OTeF_5)_2$	35–37	150	
$[XeOTeF_5]^+[AsF_6]^-$	160	200	

perchlorates and -trifluoroacetates are dangerously explosive and may detonate if thermally or mechanically shocked. The decomposition reactions (Table 1.16) generally proceed according to the equations:

$$2FXeOR \rightarrow Xe + XeF_2 + ROOR\ (\rightarrow ROR + \tfrac{1}{2}\,O_2)$$
$$Xe(OR)_2 \rightarrow Xe + ROOR$$

Presumably, this involves radical formation, e.g.

$$Xe(OSO_2F)_2 \rightarrow Xe + 2\,SO_3F$$
$$FXeOSO_2F \rightarrow XeF + SO_3F$$

The equilibrium $2SO_3F \rightleftharpoons S_2O_6F_2$ is well known[115] and all evidence indicates that the XeF radical disproportionates:

$$2XeF \rightarrow Xe + XeF_2$$

In contrast, $FXeOTeF_5$ and $Xe(OTeF_5)_2$ are thermally stable up to 150 °C and decompose smoothly above this temperature. $FXeOTeF_5$ can be distilled in a glass apparatus *in vacuo* and $Xe(OTeF_5)_2$ readily sublimes. The most stable compound of this new class of compounds is the salt $[XeOTeF_5]^+$ $[AsF_6]^-$ which melts at 160 °C and does not show any sign of decomposition up to at least 200 °C in prefluorinated Monel vessels.

(c) Structural features—A single-crystal x-ray analysis of $FXeOSO_2F$ [6]

Table 1.16 FXeOR, $Xe(OR)_2$: Decomposition reactions

$2FXeOSO_2F \rightarrow Xe + XeF_2 + S_2O_6F_2$
$2FXeOClO_3 \rightarrow Xe + XeF_2 + Cl_2O_6 + \frac{1}{2}O_2$ (+some ClO_3F, Cl_2O_7, ClO_2)
$Xe(OSO_2F)_2 \rightarrow Xe + S_2O_6F_2$
$Xe(OClO_3)_2 \rightarrow Xe + Cl_2O_6 + \frac{1}{2}O_2$
$2FXeOC(O)CF_3 \rightarrow Xe + XeF_2 + 2CO_2 + C_2F_6$
$Xe[OC(O)CF_3]_2 \rightarrow Xe + 2CO_2 + C_2F_6$
$Xe(OTeF_5)_2 \rightarrow Xe + F_5TeOTeF_5 + \frac{1}{2}O_2$ (+some TeF_6, $F_5TeOOTeF_5$, $F_5TeOTeF_4OTeF_5$, $F_5TeOTeF_4OTeF_4OTeF_5$)

shows that the linear three-centre atomic feature of XeF_2 is maintained when one of the fluorine ligands is substituted (Figure 1.24). The Xe—O bond is longer (and presumably weaker) than the Xe—F bond. The shortening of the Xe—F bond in $FXeOSO_2F$ (1.94 Å) relative to XeF_2 (2.00 Å) may be indicative of a tendency to impose greater weight on the resonance form $F—Xe^+SO_3F^-$ than $F^- \, ^+Xe—OSO_2F$. This is in line with the finding that XeF_2 is a better fluoride ion donor than $FXeOTeF_5$. The

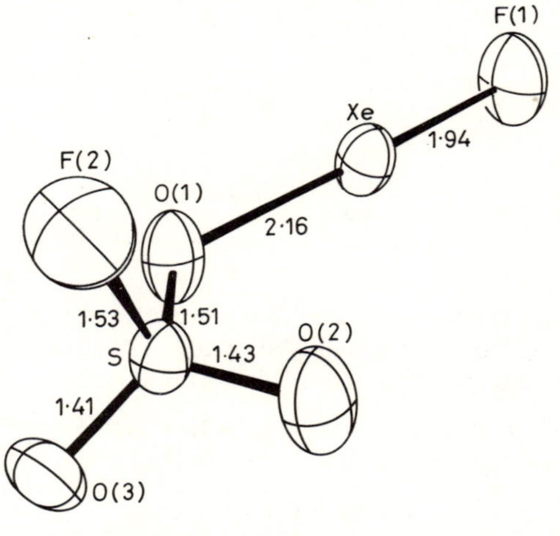

Figure 1.24 Molecular structure of $FXeOSO_2F$. Precision of bond lengths is *c.* 0.01 Å (uncorrected for thermal motion). The angles F(1)—Xe—O(1) and Xe—O(1)—S are $177.5 \pm 4°$ and $123.4 \pm 0.6°$
(From Bartlett *et al.*[6], by courtesy of The Chemical Society)

similarity of spectroscopic features between all of the known FXeOR and $Xe(OR)_2$ compounds suggests that bonding and structure must be akin. Some Raman data for xenon(II) pentafluoro-orthotellurates are given in Table 1.18.

X-ray single-crystal data for $Xe(OTeF_5)_2$ (space group *Cmca*, four molecules per unit cell) require $2/m$ symmetry for the equivalent positions (000, 0 $\frac{1}{2}$ $\frac{1}{2}$) which means that the TeOXeOTe entity is planar and the F_5Te groups are in *trans*-position. ^{19}F n.m.r. data on $Xe(OTeF_5)_2$ and $FXeOTeF_5$ show the usual AB_4 pattern of the F_5TeO groups and for the

Table 1.17 $FXeOSO_2F$ and $Xe(OTeF_5)_2$: Crystallographic data

$FXeOSO_2F$ [6] *orthorhombic*	$Xe(OTeF_5)_2$ [7] *orthorhombic*
$Pbca\text{–}D_{2h}^{15}$	$Cmca\text{–}D_{2h}^{18}$
$Z = 8$	$Z = 4$
$a = 9.88$ Å	$a = 9.83$ Å
$b = 10.00$ Å	$b = 8.73$ Å
$c = 10.13$ Å	$c = 12.97$ Å

latter a $\delta_F(FXeOTeF_5)$ of 573.4 p.p.m. about midway in between $\delta_F(XeF_2)$ of 630.3 and $\delta_F(XeF_4)$ of 452 p.p.m. (all relative to δ_F, $F_2 = 0$). The ion of highest relative intensity in the mass-spectrum of $Xe(OTeF_5)_2$ is at $m/e = 377$, corresponding to $XeOTeF_5^+$ [7].

There is no obvious correlation between thermal stability of FXeOR and $Xe(OR)_2$ compounds and various chemical or physical properties of the corresponding acid. The acid strength certainly is not the governing factor. In this respect, $HOTeF_5$, giving the most stable xenon(II) compounds, lies about midway in between fluorosulphuric acid and acetic acid[116]. Xenon(II) fluorosulphates and as preliminary results show, xenon(II) acetates, however are thermally unstable or dangerously explosive at room temperature. Without question the electronegativity of the substituting group is of importance, but reported values of relative group electronegativities are

Table 1.18 Xenon(II) pentafluoro-orthotellurates: Raman data (cm^{-1})[7]

	ν(Xe—F)	ν(Xe—O)	δ(Xe—O—Te)
$Xe(OTeF_5)_2$	—	434(s)	131(vs)
$FXeOTeF_5$	520(s)	457(vs)	153(s)
$[XeOTeF_5]^+[AsF_6]^-$	—	477(s)	174(vs)

not complete enough to be able to use them as a guide in determinating relative stabilities of such compounds. It might well be that the remarkably high stability of the xenon(II) pentafluoro-orthotellurates derives from the ability of tellurium d orbitals to interact with oxygen p orbitals thus removing negative charge from the oxygen. This in turn would facilitate negative charge transfer from xenon to oxygen, an essential requirement for bond stabilisation in noble-gas compounds as discussed in Section 1.2.1.5.

Attempts to exchange chlorine for fluorine in XeF_2 have failed. XeF_2 is soluble in HCl at −75 °C without reaction, but at higher temperatures chlorine is evolved rapidly[117]:

$$XeF_2 + 2HCl \rightarrow Xe + Cl_2 + 2HF.$$

Whereas attempts have failed to substitute F_5TeO-groups for fluorine in XeF_4 and XeF_6 because of redox reactions and formation of tellurium hexafluoride, there is some evidence for $F_4Xe(OSO_2F)_2$ and $F_2Xe(OSO_2F)_2$ [118].

(d) Reactions – Carbon tetrachloride and acetonitrile are excellent solvents for $Xe(OTeF_5)_2$ and xenon(II) trifluoroacetates. Explosive or vigorous reactions occur with acetone, benzene and ethanol. $Xe(OTeF_5)_2$ seems to be only sparingly soluble in H_2O and hydrolytic reactions are slow. In alkaline solutions, however, decomposition is complete within seconds, accompanied by vigorous xenon and oxygen evolution:

$$Xe(OTeF_5)_2 + H_2O \rightarrow Xe + \tfrac{1}{2}O_2 + 2HOTeF_5$$
$$(+10H_2O \rightarrow 2H_6TeO_6 + 10HF)$$
$$FXeOTeF_5 + H_2O \rightarrow Xe + \tfrac{1}{2}O_2 + HF + HOTeF_5$$

Attempts to prepare xenates(II) via interaction of the xenon(II) pentafluoro-orthotellurates with CsF have not been successful so far:

$$FXeOTeF_5 + CsF \xrightarrow{90\%} TeF_6 + [FXeO^-Cs^+] \rightarrow Xe + \tfrac{1}{2}O_2 + CsF$$
$$FXeOTeF_5 + CsF \xrightarrow{10\%} XeF_2 + F_5TeO^-\,Cs^+$$

Interactions of xenon(II)-fluoride pentafluoro-orthotellurate, $FXeOTeF_5$ with the fluoride ion acceptors BF_3, GeF_4, PF_5, VF_5 and AsF_5 have been studied. With arsenic pentafluoride an adduct with a molar ratio of 1:1 is formed. The Raman spectrum indicates the salt-like structure $[XeOTeF_5]^+[AsF_6]^-$ [7]. The pale-yellow solid (m.p. 160 °C) can be sublimed and is thermally stable up to at least 200 °C. The fluoride ion donor strength of $FXeOTeF_5$ and XeF_2 is comparable. XeF_2 which also does not form adducts with BF_3, GeF_4 and VF_5 which are stable at room temperature is, however, capable of displacing $FXeOTeF_5$ out of its $[AsF_6]^-$ salt:

$$[XeOTeF_5]^+[AsF_6]^- + 2XeF_2 \rightarrow [Xe_2F_3]^+[AsF_6]^- + FXeOTeF_5.$$

E. Fluorination reactions

Xenon difluoride is easy to prepare and relative to other fluorides safe to handle. Because of its low average bond energy and the 'inertness' of its reduction product xenon it appears as an attractive reagent for oxidative fluorinations. XeF_2, however, exhibits considerable kinetic stability. As an example, dry acetonitrile solutions show no evidence of interaction between dissolved I_2 and XeF_2. On the other hand, the introduction of a trace of acid leads to rapid oxidation of iodine and to formation of iodine pentafluoride. Catalytic activity is shown by HF, BF_3 and AsF_5. The same is true for SO_2 and SO_3, but since they are effective reducers themselves, they are fluorinated to SO_2F_2 and $S_2O_6F_2$, respectively[33].

These findings may be simply rationalised by supposing that the acid (A) facilitates XeF_2 ionisation:

$$XeF_2 + A — [XeF]^+ + [AF]^-$$

Presumably the electron affinity of $[XeF]^+$ is greater and the kinetic stability smaller than that of XeF_2. Consequently electron transfer to the cation would generate the weakly-bound XeF radical as an effective fluorine atom source[33].

The electron affinity of the cation, $E(XeF^+, g)$ is estimated from the thermochemical cycle in Figure 1.17. The electron affinity is therefore equal to $-I.P.$ (Xe) $= -281$ kcal mol^{-1} plus the difference in bond energy of $[XeF]^+$ and the XeF radical. The bond energy of $[XeF]^+$ was calculated to be 49 kcal mol^{-1} (Section 1.2.1.6.A.(a)) and indications are that the bond energy of XeF is significantly smaller than the average bond energy of XeF_2 (32.5 kcal mol^{-1})[119]. A value of about 20 kcal mol^{-1} seems reasonable and is in accord with the observed (Section 1.2.1.6.D.(a)) disproportionation $2XeF \rightarrow Xe + XeF_2$. This would give $E(XeF^+,g) = -250 \pm 10$ kcal mol^{-1}, also consistent with the instability of the salt $[XeF]^+[OsF_6]^-$ towards decomposition[7]:

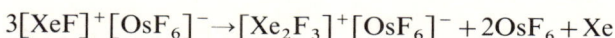

$$3[XeF]^+[OsF_6]^- \rightarrow [Xe_2F_3]^+[OsF_6]^- + 2OsF_6 + Xe$$

In the $[Xe_2F_3]^+$ ion, each $[XeF]^+$ receives electron density from the integrated F^- ion, thus becoming an entity with a lower electron affinity than $[XeF]^+$.

The same mechanism can be applied to fluorine substitution of aromatic hydrocarbons by XeF_2. Solutions of XeF_2 in benzene are stable until trace amounts of hydrogen fluoride are introduced, at which point the solutions become green coloured, xenon is evolved and fluoro-aromatics are formed. The main product is fluorobenzene, beside small amounts of *p*- and *o*-difluorobenzene, biphenyl and fluorinated biphenyls[120, 121].

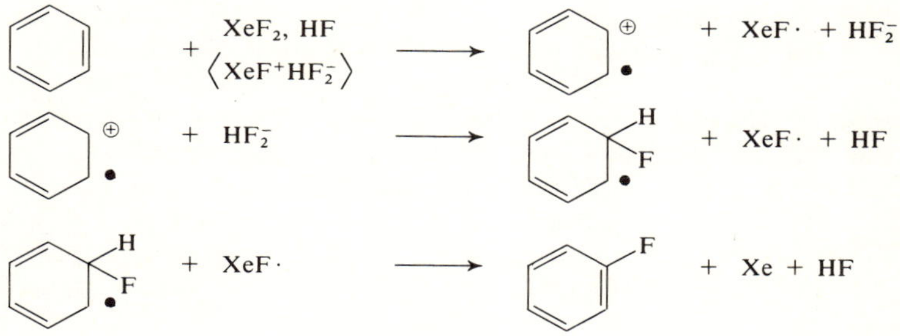

E.S.R. studies of benzene and substituted benzenes, reacting with XeF_2 in CH_2Cl_2 in the presence of HF, have indeed revealed that radical cations are formed as intermediates[121].

Beside being an efficient fluorinating agent in solution, fluorine substitutions of benzene, nitrobenzene, di-, tetra- and penta-fluorobenzenes have been carried out with XeF_2 in the vapour phase (temperature range 100–200 °C) as well. No catalysts are needed in such reactions, but at the reaction temperatures significant amounts of F_2 and xenon are in equilibrium with XeF_2 to initiate reaction which seems likely to be of a free radical mechanism[122].

Related non-aqueous chemistry of XeF_4 and XeF_6 has been studied only to a small extent. The kinetic stability of XeF_4 seems to be comparable to XeF_2, whereas XeF_6 clearly exhibits greater reactivity. Xenon tetrafluoride, fluorinates perfluoropropene at room temperature, whereas xenon difluoride is reactive only to the corresponding hydro-olefin and xenon hexafluoride cleaves perfluoropropene[123].

1.2.1.7 Aqueous chemistry

A. XeF_2

ΔG° for the hydrolysis of XeF_2, according to the equation

$$XeF_2 + H_2O \rightarrow Xe + \tfrac{1}{2}O_2 + 2\,HF$$

has been estimated[125] to be -53.4 kcal mol^{-1}. This would give a value of about 10^{40} for the corresponding equilibrium constant. In neutral or acid solutions, the half-life of XeF_2, however, is ≈ 7 h implying considerable kinetic stability[126]. The only established species in such solutions is molecular xenon difluoride, as indicated by electric conductance[127] and u.v. spectroscopy[126]. Exchange studies with aqueous $H^{18}F$ further showed that equilibria like

$$XeF_2 \rightleftharpoons [XeF]^+ + F^-$$
$$XeF_2 + 2F^- \rightleftharpoons [XeF_4]^{2-}$$
$$XeF_2 + H_2O \rightleftharpoons XeF(OH) + HF$$

can be excluded[127].

Decomposition of XeF_2 in basic solution is virtually instantanous, but interaction of XeF_2 with water is not only catalysed by base but also by species which have an affinity for fluoride ions. Of a variety of metal ions investigated, the order of the accelerating effect is the same as the order of the stability constants of their monofluoro complexes: $Th^{4+} < Al^{3+} < Be^{2+} < La^{3+}$ [128]. These findings suggest that the first step in XeF_2 hydrolysis may involve $[XeF]^+$ formation, paralleling the oxidative behaviour in non-aqueous solvents (Section 1.2.1.6.E).

In 0.01 M perchloric acid, oxidation of water by XeF_2 proceeds with a first order rate constant of 4.2×10^{-4} s^{-1} (25 °C), an activation energy $\Delta H^\ddagger$ of 19.6 kcal mol^{-1} and $\Delta S^\ddagger$ of -8.1 e.u.[129]. In neat aqueous solutions, kinetics of XeF_2 hydrolysis are also of first order with $k_1 = 2.83 \pm 0.02 \times 10^{-5}$s^{-1} (0 °C) and $2.52 \pm 0.01 \times 10^{-4}$ (25 °C)[130]. Independent studies report $k_1 = 1.2 \times 10^{12}\exp(-18400/RT)$ min^{-1} for the temperature range 10 to 40 °C and an activation energy of 18.4 ± 2.1 kcal mol^{-1} [125].

The oxidation potential for XeF_2/Xe in acid solution has been estimated to be 2.2 V. Accordingly, aqueous XeF_2 solutions oxidise chloride to chlorine, iodide to periodate, Ce^{III} to Ce^{IV}, Cr^{III} to Cr^{IV} and Co^{II} to Co^{III} [126, 131].

No evidence for xenon monoxide has been found in the course of XeF_2 hydrolysis. Interestingly enough, however, XeO_3 introduced into an aqueous solution of XeF_2 is extensively consumed in the course of the interaction of XeF_2 with water, although an aqueous solution of XeO_3 itself can be kept almost indefinitely. To rationalise these findings, reduction of XeO_3 by intermediate XeO has been presumed[129]:

$$XeF_2 + H_2O \rightarrow XeO + 2\,HF$$
$$XeO + XeO_3 \rightarrow 2\,XeO_2$$
$$XeO_2 \rightarrow Xe + O_2$$

Involvement of such a species, or even more likely oxygen atoms, is supported by the oxidation of bromate to perbromate by aqueous xenon difluoride[14].

B. XeF_4 *and* XeF_6

(a) Xenates(VI) – The reaction of XeF_4 or XeF_6 with water or dilute acid yields aqueous Xe^{VI} [132]. The solutions are stable and their Raman spectra prove that the primary xenon species is molecular XeO_3 [133]. Aqueous XeO_3 behaves as a weak, monobasic acid and salts like $CsHXeO_4$ have been prepared[134]. Raman spectra of aqueous solutions 1.1 M in sodium xenate(VI),

Table 1.19 $[HXeO_4]^-$: **Raman spectrum and band assignments based on C** $_3$ **symmetry**[139]

Frequency (cm^{-1})	*Intensity or Description*	*Polarisation*	*Assignment*
325	m	dp	ν_6, e fundamental
350	m	p	ν_3, a_1 fundamental
430	w	dp	ν_5, e fundamental
595	vb		?
665	m	p	ν_2, a_1 fundamental
750	vs	p	ν_1, a_1 fundamental
795	w	dp	ν_4, e fundamental
1042	vw	p	$\nu_2+\nu_3$, a_1 combination
1635	b		O—H bend
3425	vb		O—H stretch

$NaHXeO_4$, indicate that $HXeO_4^-$ is the predominant anion present. The Raman spectrum and band assignment, based on C_{3v} symmetry are summarised in Table 1.19 [135].

(b) Fluoro-, chloro, bromo-xenates(VI) – Alkali fluoroxenates(VI), M^+ XeO_3F^-(M = K, Rb, Cs), are the most stable solid, oxygenated compounds

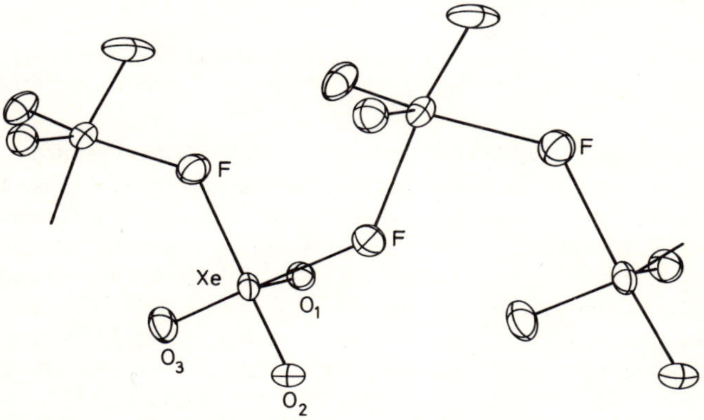

Figure 1.25 A view of the $[XeO_3F]^-$ anion showing the bridged Xe—F—Xe bonds and the polymeric nature of the ion
(From Hodgson and Ibers[138], by courtesy of The American Chemical Society)

of xenon(VI). Decomposition to xenon and oxygen starts at 260 °C, but explosions have occurred when samples were heated above 300 °C.

These salts are best prepared by taking approximately equal volumes of 0.5 M aqueous XeO_3 and 1 M aqueous MF (M = K, Rb, Cs). The solution is acidified with a few drops of 1 M HF and slowly evaporated. The solubility of alkali fluoroxenates(VI) decreases with increasing atom weight of the alkali metal[136]. $Cs^+[XeO_3F]^-$ has also been obtained by atmospheric hydrolysis of the adducts $CsF{\cdot}XeOF_4$ and $CsF{\cdot}XeF_6$ [160].

By substituting CsCl or CsBr for CsF, the corresponding chloro- and bromo-xenates(VI) have been obtained. The stability and ease of preparation of halo-xenates increases with decreasing atomic weight of the halogen: $Cs^+[XeO_3F]^- > CsXeO_3Cl > CsXeO_3Br$. The caesium chloroxenate(VI) is stable to ≈150 °C, but explodes at 205 °C [137].

The crystal and molecular structure of $KXeO_3F$ has been determined from three-dimensional x-ray data[138]. The crystal structure consists of infinite

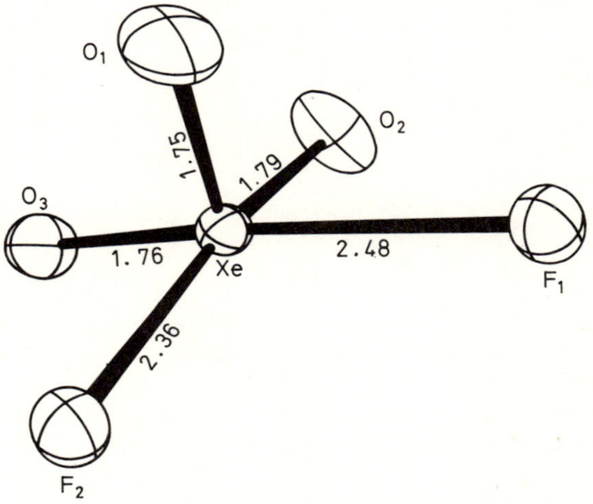

Figure 1.26 The inner coordination geometry around Xe in $K^+[XeO_3F]^-$ (From Hodgson and Ibers[138], by courtesy of The American Chemical Society)

chains of XeO_3 units linked by bridging fluorine atoms (Figure 1.25.). The geometry of the XeO_3 moiety is very similar to that of XeO_3 itself (Figure 1.26). The two independent Xe—F distances of 2.36 and 2.48 Å are considerably longer than the distances for non-bridging Xe—F bonds, but are significantly shorter than the value of 3.5 Å predicted for non-bonded interactions between xenon and fluorine.

(c) Perxenates – Xenates(VIII) – The most efficient synthesis of perxenates is achieved by ozonising alkaline solutions of XeO_3 [132]. Crystal and molecular structures have been reported for $Na_4XeO_6{\cdot}6H_2O$, $K_4XeO_6{\cdot}9H_2O$ [140] and $Na_4XeO_6{\cdot}8H_2O$ [141]. In all of these salts the XeO_6^{4-} ion exists as a nearly symmetrical octahedron with Xe—O bond lengths of 1.84 to 1.86 Å and O—Xe—O angles between 87 and 93 degrees. The Raman and infrared spectra of aqueous solutions 1.8 M in caesium perxenate suggest a high concentration of symmetrical $[XeO_6]^{4-}$ ions, but certain details imply the presence of other ionic forms[139]. The intense, highly polarised Raman band at 685 cm^{-1} in the solution spectrum, corresponds to the band at 683 cm^{-1} in solid $Na_4XeO_6{\cdot}0.4H_2O$ and has been assigned to the totally symmetrical ν_1 (O_h) band. This stretching frequency is compatible with the Xe—O bond length (1.85 Å) and indicates an intrinsically slightly stronger bond than the Xe—F bond in XeF_6 (1.89 Å).

Perxenate solutions are powerful oxidising agents and have been recommended as analytical reagents[142].

1.2.2 Chlorides, bromides

1.2.2.1 $XeCl_2$

Mass spectrometric evidence was given for the first preparation of xenon dichloride[143]. A 1:1:1 mixture of Xe, F_2 and $SiCl_4$ or CCl_4, subjected to a high-frequency discharge (25 MHz, 150–350 mA) at −80 °C yielded colourless crystals, stable up to +80 °C. The mass spectrum of the negative ions of this material showed the presence of $XeCl^-$.

$XeCl_2$ has also been prepared by passing mixtures of xenon and chlorine (Xe/Cl_2 = 200–100) through a microwave discharge and condensing them upon a caesium iodide optical window, maintained at 20 K. A broad, structured absorption in the infrared, centred at 313 cm^{-1}, was shown to be due to the assymetric stretching mode of a linear, symmetric $XeCl_2$ ($D_{\infty h}$) molecule[144]. The corresponding assymetric stretching force constant is 1.317 mdyn/Å, approximately half that of XeF_2, reflecting the expected weakness of the Xe—Cl bond, relative to the Xe—F bond.

1.2.2.2 $XeCl_4$, $XeBr_2$

$XeCl_2$, $XeCl_4$ and $XeBr_2$ have been detected by Mössbauer spectroscopy as beta decay products of their radioactive ^{129}I analogues[66, 67]:

$$^{129}ICl_2^- \xrightarrow{\beta} {}^{129}XeCl_2$$
$$^{129}ICl_4^- \xrightarrow{\beta} {}^{129}XeCl_4$$
$$^{129}IBr_2^- \xrightarrow{\beta} {}^{129}XeBr_2$$

No evidence has been obtained for $XeBr_4$ by the same technique. The timescale of these studies is of such an order that a xenon compound will be detected if its lifetime at 4.2 K is of the order of 10^{-9}s. A comparison of the Mössbauer data (Table 1.8) shows that bond polarity decreases in the sequence $XeF_2 > XeCl_2 > XeBr_2$. This is in harmony with the decrease in electronegativity of the respective ligands. Quantitatively, however, the figures in Table 1.8 probably represent upper limits for bond polarities since Xe 5d orbital participation in bonding was excluded.

1.2.3 Oxides

1.2.3.1 Xenon trioxide

A. Preparation

XeO_3 is formed in the hydrolysis of XeF_4 and XeF_6. It is most efficiently prepared from XeF_6 [134, 145]. Since XeO_3 is a powerful explosive, great care must be exercised in its preparation, and also in the handling of XeF_4 and XeF_6, since the oxide is formed when they interact with moisture.

The enthalpy of formation of solid XeO_3 is 96 ± 2 kcal mol^{-1} [146], ΔH°_f of XeO_3 (aq.) is 99.94 ± 0.24 kcal mol^{-1} [147]. Aqueous solutions of the colourless solid XeO_3 may be kept indefinitely, however, if oxidisable impurities are excluded. Raman spectra of such solutions prove[133] that the primary xenon species is molecular XeO_3.

B. Molecular structure and bonding

The XeO_3 molecule is pyramidal (C_{3v}) and almost identical in shape and size to the IO_3^- ion[148]. The shortness (1.76 Å) and higher force constant of 5.66 mdyn/Å [133] of the Xe—O bond indicate that the intrinsic Xe—O bond energy is greater than the intrinsic Xe—F bond energies. This is probably a consequence of the oxygen valence-state promotion energy, from the ^3P ground state to the ^1D state, of 45.1 kcal mol^{-1}. On this basis the intrinsic Xe—O bond energy would be the mean thermochemical bond energy of ≈ 18 kcal mol^{-1} plus the oxygen valence-state promotion energy, giving ≈ 63 kcal mol^{-1} which is more in accord with the vibrational spectroscopic data and the remarkable kinetic stability of XeO_3.

C. Chemistry

The electrode potential of the Xe—XeO_3 couple in acidic solution was deduced[147] to be $+2.10 \pm 0.01$ V. This shows XeO_3 to be one of the strongest oxidants in aqueous media.

The kinetics of the reaction between Pu^{III} and aqueous XeO_3, according to the equation

$$6Pu^{III} + XeO_3 + 6H^+ \rightarrow 6Pu^{IV} + Xe + 3H_2O$$

have been studied in perchlorate media[149]. The rate law for the reaction is $-d[Pu^{III}]/dt = k[Pu^{III}][XeO_3]$. The following thermodynamic quantities of activation at 25 °C were calculated: $\Delta H^{\ddagger} = 15.3 \pm 2.1$, $\Delta F^{\ddagger} = 20.2 \pm 0.1$ kcal mol^{-1} and $\Delta S^{\ddagger} = -16.0 \pm 6.9$ e.u. A two-electron change mechanism is highly probable. Evidence of a photochemical induced oxidation of Np^V by aqueous XeO_3 has also been obtained[150]: $6NP^V + Xe^{VI} \rightarrow 6Np^{VI} + Xe$ The reaction is first order in XeO_3 and zero order in Np^V: $-d[Np^V]/dt = k_1 [XeO_3]$ with $k_1 \times 10^{-6}$ $(s^{-1}) = 6.28 \pm 0.58$. The formation of excited XeO_3: $XeO_3 + h\nu(\text{u.v.}) \rightarrow XeO_3^*$, appears to be the rate-controlling step.

Xenon trioxide has also been recommended as an analytical reagent for the determination of primary and secondary alcohols in aqueous solution, the products being CO_2 and H_2O [151]. Similarly, carboxylic acids may also be oxidised quantitatively to CO_2 and H_2O [152]. There is some evidence for ester-like species as $Bu^tO—XeO_2—OM \cdot t\text{-}BuOH$ (M = K, Rb) in the oxidation of alcohols. Attempts to isolate such compounds failed, however[153].

1.2.3.2 Xenon tetroxide

This highly endothermic compound can be prepared by reaction of sulphuric acid with sodium or barium perxenate[183]. Decomposition to xenon and oxygen occurs slowly at room temperature, but is occasionally explosive. The instability is consistent with a ΔH_f° of +153.5 kcal mol^{-1} [154]. The molecular structure of XeO_4 has been investigated by vibrational spectroscopy[155], and electron diffraction[156].The data are in complete accord with a tetrahedral molecule of T_d symmetry. The Xe—O bond length in XeO_4 of 1.736 (0.002$_3$) Å is close to the Xe—O bond lengths in XeO_3 and $XeOF_4$, but is

Table 1.20 XeO_4: Observed fundamental frequencies and Urey–Bradley force constants[155]

	vapour (cm^{-1})	*solid* (cm^{-1})
ν_1 (a_1)	773	767.1
ν_2 (e)	—	277
ν_3 (f_2)	877	867
ν_4 (f_2)	305.7	303

Force constants (mdyn/Å)	
K	6.28
H	0.27
F	−0.18
F'	−0.08

much shorter than the 1.84 to 1.86 Å values found in perxenates. Observed fundamental frequencies and Urey–Bradley force constants are listed in Table 1.20.

1.2.4 Oxide–fluorides

Four xenon oxide–fluorides are known: $XeOF_2$, $XeOF_4$, XeO_2F_2 and XeO_3F_2, but only the two containing xenon(VI) have been characterised adequately and only $XeOF_4$ appears to be thermodynamically stable[154].

1.2.4.1 $XeOF_2$

The compound claimed to be $XeOF_2$ is the (bright yellow?) solid formed by hydrolysis of XeF_4 at −80 °C. This product neither gives an e.s.r. spectrum, nor an infrared D shift, and contains only one oxygen atom. Assuming C_{2v} symmetry (i.e. FXe(O)F would be planar and the O—Xe—F angle 90 degrees) the observed infrared absorption bands have been assigned: 747 (ν_1, A_1, Xe—O stretch), 520 (ν_2, A_1, Xe—F sym.stretch) and 490 cm^{-1} (ν_4, B_2, Xe—F asym.stretch) [156].

1.2.4.2 $XeOF_4$

Xenon oxide tetrafluoride is prepared by controlled partial hydrolysis of XeF_6 [158, 159, 187]. Great care has to be taken to avoid formation of the highly explosive XeO_3 by excessive hydrolysis. $XeOF_4$ is colourless, low melting

(−46.2 °C) and volatile[160].

Microwave spectroscopy shows the molecule to be square-pyramidal, with the oxygen in apical position(C_{4v}) with bond lengths and angles of[162]:

$$\text{Xe—O}: 1.703 \pm 0.015 \text{ Å}$$
$$\text{Xe—F}: 1.900 \pm 0.005 \text{ Å}$$
$$\text{O—Xe—F}: 91.8 \pm 0.5 \text{ degrees}$$

The Xe—F bond length is very close to that found for XeF_6 (1.890 Å), the Xe—O bond length, however, is shorter than those found for XeO_3 (1.76 Å) and XeO_4 (1.74 Å). The force constants, derived from vibrational data are consistent with these observations[158, 165, 166].

The x-ray photo-electron spectrum of $XeOF_4$(g)[161] yields a xenon core-electron chemical shift intermediate between XeF_4 and XeF_6 (Table 1.21). The shifts have been interpreted in terms of a coulombic model. This assumes

Table 1.21 X-ray photo-electron chemical shifts for XeM$_v$ core-electrons in some gaseous xenon compounds[161]

Compound	*Shift*, ΔE (eV)	*Shift per fluorine atom* (eV)
Xe	0	—
XeF_2	2.95 ± 0.13	1.48
XeF_4	5.47 ± 0.18	1.37
$XeOF_4$	7.02 ± 0.13	—
XeF_6	7.88 ± 0.18	1.31

a spherical positively charged xenon atom of charge $+q$ and radius r_v and spherical, negatively-charged ligands, distant R_L from the xenon atom. Thus, for $XeOF_4$

$$\Delta E_{XeOF_4} = -(q - q_O)(1/r_v - 1/R_F) - q_O(1/r_v - 1/R_O)$$

and by equating the first term with ΔE_{XeF_4} or $\frac{2}{3}\Delta E_{XeF_6}$

$$\Delta E_{XeOF_4} = \Delta E_{XeF_4} \text{ (or } \tfrac{2}{3}\Delta E_{XeF_6}) - q_O(1/r_v - 1/R_O).$$

The dependence of q_O upon r_v has been evaluated and compared with a similar interdependence for the xenon fluorides. The findings are as follows:

Xenon valence shell radius (Å), r_v	1.5	1.4	1.3	1.2	1.1	1.0	0.9
Oxygen ligand charge, q_O	1.5	0.91	0.64	0.47	0.36	0.28	0.22
Fluorine ligand charge, q_F*	0.63	0.48	0.37	0.30	0.20	0.16	0.13

*essentially constant for all xenon fluorides

They show, no matter what the choice of the xenon valence-shell radius, that the oxygen withdraws more electron density than a fluorine ligand. A xenon valence-shells radius between 1.2 to 1.4 Å appears to be a reasonable choice and for r_v 1.4 Å the charge distribution of $Xe^{3+}(F^{\frac{1}{2}-})_4O^{1-}$ is in harmony with the simple three-centre molecular orbital model for $XeOF_4$.

$XeOF_4$ yields adducts with fluoride ion acceptors and also fluoride ion donors[102, 108, 160]. Though not characterised adequately, these compounds may be salts containing $[XeOF_3]^+$ or $[XeOF_5]^-$ ions, respectively.

1.2.4.3 XeO_2F_2

Interaction of XeO_3 with XeF_6 or XeF_4 provides a method for the preparation of xenon dioxide difluoride[163, 164]

$$XeO_3 + XeOF_4 \rightarrow 2XeO_2F_2$$
$$2XeO_3 + XeF_6 \rightarrow 3XeO_2F_2$$

(With excess XeF_6, $XeOF_4$ is formed: $XeO_2F_2 + XeF_6 \rightarrow 2XeOF_4$). XeO_2F_2 forms colourless crystals (m.p.: 30.8 °C) which are slightly less volatile than

Table 1.22 XeO_2F_2: Vibrational spectra[167]

Raman (cm^{-1}) *solid*	*liquid*	*Infrared* (cm^{-1}) (*low temperature matrix*)	*Assignment*
205 ms	198 w		ν_4 (a_1)
224 w	223 vw		ν_5 (a_2)
315 vs	313 ms	317 ms	ν_9 (b_2)
		324 s	ν_7 (b_1)
350 ms	333 ms		ν_3 (a_1)
537 vs	490 s		ν_2 (a_1)
		537 vw	$\nu_5+\nu_9$ 541 (B_1)
		550 w	$\nu_5+\nu_7$ 547 (B_2)
		574 w	impurity?
	578 w	585 vs	ν_8 (b_2)
769 w	788 vw		$\nu_8+\nu_4$ 783 (B_2)
814 w			?
850 vs	845 vs	848 ms	ν_1 (a_1)
882 s	902 w	905 s	ν_6 (b_1)
		1023 w	$\nu_1+\nu_4$ 1046 (A_1)
		1444 w	$\nu_1+\nu_8$ 1433 (B_2)
		1496 vw	impurity?

XeF_2. ΔH_f° of XeO_2F_2(g) has been estimated[153] to be about +56 kcal mol^{-1} [154]. It can decompose to XeF_2 and O_2, the rate depending on previous conditioning of the containing apparatus, and hydrolyses in moist air to xenon trioxide.

The vibrational spectra[167], represented in Table 1.22, indicate that XeO_2F_2 is a trigonal-bipyramid (C_{2v} symmetry) with the two oxygen atoms and the lone pair occupying the equatorial positions.

1.2.4.4 XeO_3F_2

Xenon trioxide difluoride has been generated by interaction of XeF_6 with solid sodium perxenate at room temperature[168] along with a larger quantity of other xenon compounds, principally, $XeOF_4$. XeO_3F_2 was detected mass spectroscopically, $[XeO_3F_2]^+$ being observed. Its volatility at −78 °C, comparable to the most volatile xenon compound, XeO_4, is sufficient to yield characteristic mass spectra.

1.3 COMPOUNDS OF KRYPTON

1.3.1 KrF_2

1.3.1.1 Preparation and thermodynamic properties

The chemistry of krypton is limited to that of krypton difluoride. It is best prepared by passing a mixture of the elements at low pressure (≈20 mmHg) and low temperature (−183 °C) through an electrical discharge[169].

KrF_2 is colourless both in the solid and vapour phase. Its vapour pressure is ≈30 mmHg at 0 °C. The compound can be stored at −78 °C but decomposes spontaneously well below room temperature. Measurement of the heat of dissociation, at 93 °C, of gaseous krypton difluoride has given[181] a standard enthalpy of formation ΔH_f° of 14.4±0.8 kcal mol^{-1} [188]. The calorimetric measurements are supported by mass-spectrometric appearance potential data. Although there is ambiguity of interpretation, it is probable that the appearance potential A.P. (Kr^+, KrF_2) of 13.21±0.25 eV corresponds to $KrF_2(g) + e \rightarrow Kr^+ + F_2(g) + 2\ e$. Since I.P. (Kr) = 14.00 eV, it follows that $\Delta H_f^\circ(KrF_2, g) = 0.79 \pm 0.25$ eV or 18±5 kcal mol^{-1} [189]. Thus the mean thermochemical bond energy in KrF_2 is ≈12 kcal mol^{-1}. This is the lowest bond energy of any known fluoride, lower indeed than the dissociation energy of fluorine. KrF_2 is therefore expected to be an effective fluorine atom source even at low temperatures.

1.3.1.2 Molecular structure and bonding

Like XeF_2, krypton difluoride is a symmetrical linear molecule ($D_{\infty h}$)[12,170]. The Kr—F bond length has been determined for the gaseous species by

Table 1.23 KrF_2, XeF_2: Comparison of frequencies and force constants[170]

	Frequencies (cm^{-1})			*Force constants* (mdyn/Å)		
	ν_1	ν_2	ν_3	f_r	f_{rr}	f_α
KrF_2	449	232.6	588	2.46	−0.20	0.21
XeF_2	515	213.2	558	2.84	0.13	0.20

electron diffraction (1.889±0.010 Å)[84] and rotational infrared spectroscopy (1.875±0.002 Å)[172].

The force constants for KrF_2 (Table 1.23) are defined by the following equation for twice the potential energy:

$$2V = f_r(\Delta r_1^2 + \Delta r_2^2) + 2f_{rr}\Delta r_1 \Delta r_2 + r_0^2 f_\alpha \Delta\alpha^2$$

where Δr_1 and Δr_2 are changes in the bond distances and $\Delta\alpha$ is the change in bond angle. A striking feature of the force constants for KrF_2, in comparison to XeF_2, is the negative value of its bond–bond interaction constant f_{rr}. The negative value indicates that in KrF_2 it is easier to lengthen or shorten both bonds simultaneously than it is to lengthen one bond and to shorten the other[170]. This has been interpreted in terms of a considerable weight of a no-bond F Kr F structure in resonance admixture with $F—Kr^+F^-$, F^- $Kr—F^+$ [173].

Recent *ab initio* all-electron SCF–MO calculations, however, show that krypton 4d orbitals must be included to give an adequate description of the bonding, whereas addition of krypton 5s and 5p orbitals does not change the picture greatly[174]. Similar calculations also suggest that structures such as $F—Kr^+F^-$, $F^-\,^+Kr—F$ and $F^-Kr^{2+}F^-$, which maintain the octet rule and exclude the use of d orbitals, are less stable than the structure F—Kr—F which implies localised electron-pair bonds based on pd hybrids at the krypton atom[9].

These calculations leave some doubt about the charge distribution in KrF_2, derived from ^{19}F-n.m.r. ($q_F = -0.45$) [169] and Mössbauer studies ($q_F \approx -1$) [175] on the assumption that only p_σ orbitals are involved in bonding. It is quite likely that inclusion of outer orbital character of the krypton atom would have yielded a lower charge distribution.

1.3.1.3 Chemistry

The chemistry of KrF_2 has been studied only to a small extent, owing to the difficulty of preparing quantities of this compound. The only established derivative is an adduct with antimony pentafluoride: $KrF_2{\cdot}2SbF_5$ (m.p. ≈ 50 °C) [176]. The formulation as the salt $[KrF]^+[Sb_2F_{11}]^-$ is in agreement with the observation that its thermal stability is greater than that of KrF_2. This would parallel the strengthening of the Xe—F bond on formation of $[XeF]^+$ compared to XeF_2 (Section 1.2.1.6.A.(a)).

1.4 COMPOUNDS OF RADON

The most abundant isotope of radon, ^{222}Rn, has a half-life of 3.82 days and is produced by decay of radium. The isotope decays by α-emission, and by counting the 1.8 MeV γ ray of the ^{214}Bi daughter the position of radon can be determined. Other radon isotopes, ^{219}Rn and ^{220}Rn, have half-lives of only seconds and occur in the decay chains of actinium and thorium, respectively. Experiments with radon have been conducted with curie amounts (one curie of ^{222}Rn produces 3.7×10^{10} disintegrations per second, weighs 6.5 microgrammes and has a volume of 0.66 mm^3). Adequate shielding has to be provided for the very intense γ-radiation.

The first evidence of the existence of a radon fluoride was obtained in 1962[177]. Radon remained fixed in the nickel reaction vessel after reaction with fluorine at 400 °C. Efforts to analyse the compound mass spectrometrically have been unsuccessful, but the reaction of the radon fluoride with water, analogous to that of KrF_2 and XeF_2, suggests that it is RnF_2 [180]:

$$RnF_2 + H_2O \rightarrow Rn + \tfrac{1}{2}O_2 + 2HF$$

Only radon gas has been observed as a product after hydrolysis with water, acid and alkaline solutions. A claim for the oxidation of radon by strong aqueous oxidisers, as hydrogen peroxide, potassium permanganate and potassium persulphate[178] has been refuted[179].

Beside its reactivity with elementary fluorine, solutions of radon fluoride have been prepared through spontaneous reactions of ^{222}Rn with ClF, ClF_3, ClF_5, BrF_5, IF_5 and K_2NiF_6 in HF at 25 °C and lower temperatures[181, 190]. Weak fluorinating agents, such as HF, IF_5, AsF_5 and SbF_5 do not react with elementary radon. Knowledge of the free energies of formation of these fluorides allows the deduction of upper and lower limits for the free energy of formation of RnF_2 of -29 and -51 kcal mol^{-1}, respectively[181].

Electromigration studies in such solutions show that radon is present as a cation in BrF_3 and HF—BrF_3 solutions and that it is present chiefly as a cation in solutions containing alkali metal fluorides, although anionic radon species have also been detected in the latter. Therefore Rn^{II} might be expected to be present chiefly as Rn^{2+}, RnF^+ and RnF_2 and to some extent as RnF_3^-.

There is a remarkable similarity in the behaviour of radon and astatine[185] in fluorination experiments (AtF is isoelectronic with RnF^+) and both elements are probably more metallic than anticipated.

References

1. Bartlett, N. (1962). *Proc. Chem. Soc.*, 218
2. McRae, V. M., Peacock, R. D. and Russell, D. R. (1969). *Chem. Comm.*, 62
3. Sladky, F. and Bartlett, N. (1969). *J. Chem. Soc. (A)*, 2188
4. Sladky, F., Bulliner, P. A. and Bartlett, N. (1968). *Chem. Comm.*, 1048
5. Bartlett, N., Einstein, F., Stewart, D. F. and Trotter, J. (1967). *J. Chem. Soc. (A)*, 1190
6. Bartlett, N., Wechsberg, M., Sladky, F., Bulliner, P. A., Jones, G. R. and Burbank, R. D. (1969). *Chem. Comm.*, 703
7. Sladky, F. (1969). *Angew Chem.*, **81**, 330; *Angew. Chem. Int. Ed.*, **8**, 373

7a. Sladky, F. (1969d). *Angew. Chem.*, **81**, 536; *Angew. Chem. Int. Ed.*, **8**, 523

7b. Sladky, F. (1969b). *5th International Symposium on Fluorine Chemistry.* (Moscow)

7c. Sladky, F. (1970). *Mh. Chem.*, **101**, 1559

7d. Sladky, F. (1970a). *Mh. Chem.*, **101**, 1571

7e. Sladky, F. (1970b). *Mh. Chem.*, **101**, 1578

7f. Sladky, F. (1970c). *Angew. Chem.*, **82**, 357; *Angew. Chem. Int. Ed.*, **9**, 375

8. Burbank, R. D. and Bartlett, N. (1968). *Chem. Comm.*, 645
9. Catton, R. C. and Mitchell, K. A. R. (1970). *Canad. J. Chem.*, **48**, 2695
10. Coulson, C. A. (1969). *Nature*, **221**, 1107
11. MacKenzie, D. R. and Fajer, J. (1966). *Inorg. Chem.*, **5**, 699
12. Turner, J. J. and Pimentel, G. C. (1963). *Science*, **140**, 974
13. Liebman, J. F. and Allen, L. C. (1970). *J. Amer. Chem. Soc.*, **92**, 3539
14. Appelman, E. H. (1968). *J. Amer. Chem. Soc.*, **90**, 1900
15. Streng, L. V. and Streng, A. G. (1965). *Inorg. Chem.*, **4**, 1370
16. Weinstock, B., Weaver, E. E. and Knop, C. P. (1966). *Inorg. Chem.*, **5**, 2189

16a. Morton, J. R. and Falconer, W. E. (1963). *J. Chem. Phys.*, **39**, 427

16b. Falconer, W. E., Morton, J. R. and Streng, A. G. (1964). *J. Chem. Phys.*, **41**, 902

17. Frlec, B., Holloway, J. H., Slivnik, J., Smalc, A., Volavsek, B. and Zemljic, A. (1970). *J. Inorg. Nucl. Chem.*, **32**, 2521
18. Baker, B. G. and Fox, P. G. (1964). *Nature*, **204**, 466
19. Weaver, E. E. Weinstock, B. and Knop, C. P. (1963). *J. Amer. Chem. Soc.*, **85**, 111
20. Davis, B. H., Wishlade, J. L. and Emmett, P. H. (1968). *J. Catal.*, **10**, 266
21. Weaver, C. F. (1966). *Ph. D. Thesis*, University of California, Berkeley

22. Baker, B. G. and Fox, P. G. (1970). *J. Catal.*, **16,** 102
23. Baker, B. G. and Lawson, A. (1970). *J. Catal.*, **16,** 108
24. Sinel'nikov, S. M., Nikitin, I. V. and Rosolovskii, V. Y. (1968). *Izv. Akad. Nauk. SSSR, Ser. Khim.*, 2806
25. Falconer, W. E. and Sunder, W. A. (1967). *J. Inorg. Nucl. Chem.*, **29,** 1380
26. Williamson, S. M. (1968). *Inorganic Syntheses,* Vol. XI, 147. (New York: McGraw-Hill)
27. Malm, J. G. and Chernick, C. L. (1966). *Inorganic Syntheses,* Vol. VIII, 254 (New York: McGraw-Hill)
28. Bartlett, N. and Sladky, F. (1968). *J. Amer. Chem. Soc.*, **90,** 5316
29. Chernick, C. L. and Malm, J. G. (1966). *Inorganic Syntheses,* Vol. VIII, 258 (New York: McGraw-Hill)
30. Sheft, I., Spittler, T. M. and Martin, F. H. (1964). *Science,* **145,** 701
31. Schreiner, F., Osborne, D. W., Malm, J. G. and McDonald, G. N. (1969). *J. Chem. Phys.*, **51,** 4838
32. Serpinet, J. and Rochefort, O. (1968). *Bull. Soc. Chim.*, **10,** 4297
33. Bartlett, N. and Sladky, F. (1968). *Chem. Comm.*, 1046
34. Meinert, H. and Kauschka, G. (1969). *Z. Chem.*, **9,** 35
35. Hyman, H. H. and Quarterman, L. A. (1963). *Noble Gas Compounds,* 275 (Chicago: University of Chicago Press)
36. Schreiner, F., McDonald, G. N. and Chernick, C. L. (1968). *J. Phys. Chem.*, **72,** 1162
37. Siegel, S. and Gebert, E. (1963). *J. Amer. Chem. Soc.*, **85,** 240
38. Hoppe, R., Mattauch, H., Rödder, K. M. and Dähne, W. (1963). *Z. Anorg. Chem.*, **324,** 214
39. Meinert, H. and Rüdiger, S. (1967). *Z. Chem.*, **7,** 239
40. Volavsek, B. (1966). *Mh. Chem.*, **97,** 1531
41. Selig, H. and Mootz, A. (1967). *Inorg. Nucl. Chem. Letters,* **3,** 147
42. Nikolaev, A. V., Nazarov, A. S., Opalovskii, A. A. and Trippel, A. F. (1969). *Dokl. Akad. Nauk. SSSR.*, **186,** 1331
43. Pepekin, V. I., Lebedev, Y. A. and Apin, A. Y. (1969). *Zh. Fiz. Khim.*, **43,** 1564
44. Stein, L. and Plurien, P. L. (1963). *Noble Gas Compounds,* 144 (Chicago: University of Chicago Press)
45. Reichman, S. and Schreiner, F. (1969). *J. Chem. Phys.*, **51,** 2355
46. Levy, H. A. and Agron, P. A. (1963). *J. Amer. Chem. Soc.*, **85,** 241
47. Bohn, R. K., Katada, K., Martinez, J. V. and Bauer, S. H. (1963). *Noble Gas Chemistry,* 238 (Chicago: University of Chicago Press)
48. Burns, J. H., Agron, P. A and Levy, H. A. (1963). *Science,* **139,** 1208
49. Gavin, R. M. and Bartell, L. S. (1968). *J. Chem. Phys.*, **48,** 2460
50. Kim, H., Claassen, H. H. and Pearson, E. (1968). *Inorg. Chem.*, **7,** 616
51. Gasner, E. L. and Claassen, H. H. (1967). *Inorg. Chem.*, **6,** 1937
51a. Yeranos, W. A. (1967). *Mol. Phys.*, **12,** 529
52. Yeranos, W. A. (1965). *Mol. Phys.*, **9,** 449
53. Bartell, L. S. (1967). *J. Chem. Phys.*, **46,** 4530
54. Jortner, J., Wilson, E. G. and Rice, S. A. (1963a). *Noble Gas Compounds,* 358 (Chicago: University of Chicago Press)
55. Templeton, D. H., Zalkin, A., Forrester, J. D. and Williamson, S. M. (1963). *J. Amer. Chem. Soc.*, **85,** 242
56. Jones, G. R., Burbank, R. D. and Falconer, W. E. (1970). *J. Chem. Phys.*, **53,** 1605
57. Malm, J. G., Schreiner, F. and Osborne, D. W. (1965). *Inorg. Nucl. Chem. Letters,* **1,** 97
58. Agron, P. A., Johnson, C. K. and Levy, H. (1965). *Inorg. Nucl. Chem. Letters,* **1,** 145
59. Burbank, R. D. and Jones, G. R. (1970). *Science,* **168,** 248
60. Jones, G. R., Burbank, R. D. and Falconer, W. E. (1970). *J. Chem. Phys.*, **52,** 6450
61. Coulson, C. A. (1964). *J. Chem. Soc.*, 1442
62. Pimentel, G. C. (1951). *J. Chem. Phys.*, **19,** 446
63. Rundle, R. E. (1963). *Survey Progr. Chem.*, **1,** 81
64. Mitchell, K. A. R. (1969). *J. Chem. Soc. (A),* 1637
65. Catton, R. C. and Mitchell, K. A. R. (1970). *Chem. Comm.*, 457
66. Perlow, G. J. and Perlow, M. R. (1968). *J. Chem. Phys.*, **48,** 955
67. Perlow, G. J. and Yoshida, H. (1968). *J. Chem. Phys.*, **49,** 1474
68. Karlsson, S. E., Siegbahn, K. and Bartlett, N. (1969). Lawrence Radiation Laboratory, University of California, Berkeley, Reprint 18502

69. Hindman, J. C. and Svirmickas, A. (1963). *Noble Gas Compounds*, 251 (Chicago: University of Chicago Press)
70. Jameson, C. J. and Gutowsky, H. S. (1964). *J. Chem. Phys.*, **40,** 2285
71. Karplus, M. Kern, C. W. and Lazdins, D. (1964). *J. Chem. Phys.*, **40,** 3738
72. Hindermann, D. K. and Falconer, W. E. (1970). *J. Chem. Phys.*, **52,** 6198
73. Hindermann, D. K. and Falconer, W. E. (1969). *J. Chem. Phys.*, **50,** 1203
74. Brundle, C. R., Robin, M. B. and Jones, G. R. (1970). *J. Chem. Phys.*, **52,** 3383
75. Yeranos, W. A. (1966). *Mol. Phys.*, **11,** 85
76. Jortner, J., Wilson, E. G. and Rice S. A. (1963). *J. Amer. Chem. Soc.*, **85,** 815
77. Gillespie, R. J. and Nyholm, R. S. (1957). *Quart. Rev.*, **11,** 339
78. Sidgwick, N. V. and Powell, H. M. (1940). *Proc. Roy. Soc., A* **176,** 153
79. Tsuchida, R. (1939). *Rev. Phys. Chem. Japan,* **13,** 31 and 61
79a. Bartell, L. S. and Gavin, R. M. (1968). *J. Chem. Phys.*, **48,** 2466
80. Gavin, R. M. (1969). *J. Chem. Ed.*, **46,** 413
81. Lohr, L. L. (1967). *Bull. Amer. Phys. Soc.*, **12,** 295
82. Willett, R. D. (1966). *Theoret. Chim. Acta,* **6,** 186
83. Hedberg, K., Peterson, S. H. and Ryan, R. R. (1966). *J. Chem. Phys.*, **44,** 1726
84. Harshbarger, W., Bohn, R. K. and Bauer, S. H. (1967). *J. Amer. Chem. Soc.*, **89,** 6466
85. Falconer, W. E., Büchler, A., Stauffer, J. L. and Klemperer, W. (1968). *J. Chem. Phys.*, **48,** 312
86. Longuet-Higgins, H. C. and Nicholson, B. J. (1965). *Mol. Phys.*, **9,** 461
87. Goodman, G. (1967). *Bull. Amer. Phys. Soc.*, **12,** 296
88. Code, R. F., Falconer, W. E., Klempner, W. and Ozier, I. (1970). *J. Chem. Phys.*, **47,** 4955
89. Lazdins, D., Kern, C. W. and Karplus, M. (1963). *J. Chem. Phys.*, **39,** 1611
90. Edwards, A. J., Holloway, J. H. and Peacock, R. D. (1963). *Proc. Chem. Soc.*, 275
91. Sladky, F., Bulliner, P. A. and Bartlett, N. (1969). *J. Chem. Soc. (A)*, 2179
92. Holloway, J. H. and Knowles, J. G. (1969). *J. Chem. Soc. (A)*, 756
93. Maslov, O. D., Legasov, V. A., Prusakov, V. N. and Chaivanov, B. B. (1967). *Zh. Fiz. Khim.*, **141,** 1832
94. Morrison, J. O., Nicholson, A. J. C. and O'Donnell, T. A. (1968). *J. Chem. Phys.*, **49,** 959
95. Cohen, B. and Peacock, R. D. (1966). *J. Inorg. Nucl. Chem.*, **28,** 3056
96. Martin, D. (1969). *C. R. Acad. Sci. Paris, C*, 1145
97. Selig, H. (1964). *Science*, **144,** 537
98. Bartlett, N., Beaton, S. and Iha, N. K. (1964). *148th Meeting of the American Chemical Society,* K3. (Chicago)
99. Pullen, K. E. and Cady, G. H. (1967). *Inorg. Chem.*, **6,** 2267
100. Gard, G. L. and Cady, G. H. (1964). *Inorg. Chem.*, **3,** 1745
101. Pullen, K. E. and Cady, G. H. (1967). *Inorg. Chem.*, **6,** 1300
102. Moody, G. J. and Selig, H. (1966). *J. Inorg. Nucl. Chem.*, **28,** 2429
103. Aubert, J. and Cady, G. H. (1970). *Inorg. Chem.*, **9,** 2600
104. Bartlett, N., Einstein, F., Stewart, D. F. and Trotter, J. (1966). *Chem. Comm.*, 550
105. Hollander, F., Templeton, D., Wechsberg, M. and Bartlett, N. (1969). (Private communication)
106. Pullen, K. E. and Cady, G. H. (1966). *Inorg. Chem.*, **5,** 2057
107. Peacock, R. D., Selig, H. and Sheft, I. (1966). *J. Inorg. Nucl. Chem.*, **28,** 2561
108. Moody, G. J. and Selig, H. (1966). *Inorg. Nucl. Chem. Letters,* **2,** 319
109. Burns, J. H., Ellison, R. D. and Levy, H. A. (1965). *Acta Crystallogr.*, **18,** 11
110. Allamagny, P., Langignard, M. and Dognin, P. (1968). *C. R. Acad. Sc. Paris, C,* 226
111. Jones, G. R., Burbank, R. D. and Bartlett, N. (1970). *Inorg. Chem.*, **9,** 2264
112. Wechsberg, M. and Bartlett, N. (1969). (Private communication)
113. Jortner, J., Wilson, E. G. and Rice, S. A. (1963). *J. Amer. Chem. Soc.*, **85,** 814
114. Engelbrecht, A. and Sladky, F. (1965). *Mh. Chem.*, **96,** 159
115. Nutkowitz, P. M. and Vincow, G. (1969). *J. Amer. Chem. Soc.*, **91,** 5956
116. Porcham, W. and Engelbrecht, A. (1970). (Private communication)
117. Shaw, M. J., Holloway, J. H. and Hyman, H. H. (1970). *Inorg. Nucl. Chem. Letters,* **6,** 32
118. Eisenberg, M. and Des Marteau, D. D. (1970). *J. Amer. Chem. Soc.*, **92,** 4759
119. Johnston, H. S. and Woolfolk, R. (1964). *J. Chem. Phys.*, **41,** 269
120. Shaw, M. J., Hyman, H. H. and Filler, R. (1969). *J. Amer. Chem. Soc.*, **91,** 1563
121. Shaw, M. J., Weil, J. A., Hyman, H. H. and Filler, R. (1970). *J. Amer. Chem. Soc.*, **92,** 5096

122. MacKenzie, D. R. and Fajer, J. (1970). *J. Amer. Chem. Soc.*, **92,** 4994
123. Shieh, T., Yang, N. C. and Chernick, C. L. (1964). *J. Amer. Chem. Soc.*, **86,** 5021
124. Not allocated
125. Legasov, V. A., Prusakov, V. N. and Chaivanov, B. B. (1968). *Russ. J. Phys. Chem.*, **42,** 610
126. Appelman, E. H. and Malm, J. G. (1964). *J. Amer. Chem. Soc.*, **86,** 2297
127. Appelman, E. H. (1967). *Inorg. Chem.*, **6,** 1268
128. Beck, M. T. and Dozsa, L. (1967). *J. Amer. Chem. Soc.*, **89,** 5413
129. Appelman, E. H. (1967). *Inorg. Chem.*, **6,** 1305
130. Feher, I. and Lörinc, M. (1968). *Magy. Kem. Folyoir.*, **74,** 232
131. Schneer Erdeyne, A. and Kozmulza, K. (1969). *Magy. Kem. Folyoir.*, **75,** 378
132. Appelman, E. H. and Malm, J. G. (1964). *J. Amer. Chem. Soc.*, **86,** 2141
133. Claassen, H. H. and Knapp, G. (1964). *J. Amer. Chem. Soc.*, **86,** 2341
134. Jaselskis, B., Spittler, T. M. and Huston, J. L. (1966). *J. Amer. Chem. Soc.*, **88,** 2149
135. Peterson, *et al.* See reference 139
136. Jaselskis, B., Huston, J. L. and Spittler, T. M. (1969). *J. Amer. Chem. Soc.*, **91,** 1874
137. Jaselskis, B., Spittler, T. M., Huston, J. L. (1967). *J. Amer. Chem. Soc.*, **89,** 2770
138. Hodgson, D. J. and Ibers, J. A. (1969). *Inorg. Chem.*, **8,** 326
139. Peterson, J. L., Claassen, H. H. and Appelman, E. H. (1970). *Inorg. Chem.*, **9,** 619
140. Zalkin, A., Forrester, J. D., Templeton, D. H., Williamson, S. M. and Koch, C. W. (1964). *J. Amer. Chem. Soc.*, **86,** 3569
141. Ibers, J. A., Hamilton, W. C. and MacKenzie, D. R. (1964). *Inorg. Chem.*, **3,** 1412
142. Bane, R. W. (1965), *Analyst*, **90,** 756
143. Meinert, H. (1966). *Z. Chem.*, **6,** 71
144. Nelson, L. Y. and Pimentel, G. C. (1967). *Inorg. Chem.*, **6,** 1758
145. Appelman, E. H. and Malm, J. G. (1965). *Preparative Inorganic Reactions, II,* 349. (N.Y.: Interscience)
146. Gunn, S. R. (1963). *Noble-Gas Compounds,* 149. (Chicago: University of Chicago Press)
147. O'Hare, P. A. G., Johnson, G. K. and Appelman, E. H. (1970). *Inorg. Chem.*, **9,** 332
148. Templeton, D. H., Zalkin, A., Forrester, J. D. and Williamson, S. M. (1963a). *J. Amer. Chem. Soc.*, **85,** 817
149. Cleveland, J. M. (1967). *Inorg. Chem.*, **6,** 1302
150. Cleveland, J. M. and Werkema, G. J. (1967). *Nature,* **215,** 732
151. Jaselskis, B. and Warriner, J. P. (1966). *Anal Chem.*. **38,** 563
152. Jaselskis, B. and Krueger, R. H. (1966). *Talanta,* **13,** 945
153. Jaselskis, B. and Warriner, J. P. (1969). *J. Amer. Chem. Soc.*, **91,** 201
154. Gunn, S. R. (1965). *J. Amer. Chem. Soc.*, **87,** 2290
155. Huston, J. L. and Claassen, H. H. (1970). *J. Chem. Phys.*, **52,** 5646
156. Ogden, J. S. and Turner, J. J. (1966). *Chem. Comm.*, 693
157. Gundersen, G., Hedberg, K. and Huston, J. L. (1970). *J. Chem. Phys.*, **52,** 812
158. Smith, D. F. (1963). *Science,* **140,** 899
159. Shamir, J., Selig, H., Samuel, D. and Reuben, J. (1965). *J. Amer. Chem. Soc.*, **87,** 2359
160. Selig, H. (1966). *Inorg. Chem.*, **5,** 183
161. Not allocated
162. Martins, J. F. and Wilson, E. B. (1968). *J. Mol. Spectr.*, **26,** 410
163. Huston, J. L. (1967). *J. Phys. Chem.*, **71,** 3339
164. Frame, H. D., Huston, J. L. and Sheft, I. (1969). *Inorg. Chem.*, **8,** 1549
165. Begun, G. M., Fletcher, W. H. and Smith, D. F. (1963). *J. Chem. Phys.*, **42,** 2236
166. Yeranos, W. A. (1965). *Bull. Soc. Chim. Belges.*, **74,** 407
167. Claassen, H. H., Gasner, E.L., Kim, H. and Huston, J. L. (1968). *J. Chem. Phys.*, **49,** 253
168. Huston, J. L. (1968). *Inorg. Nucl. Chem. Letters,* **4,** 29
169. Schreiner, F., Malm, J. G. and Hindermann, J. C. (1965). *J. Amer. Chem. Soc.*, **87,** 25
170. Claassen, H. H., Goodman, G. L., Malm, J. G. and Schreiner, F. (1965). *J. Chem. Phys.*, **42,** 1229
171. Not allocated
172. Murchinson, C., Reichmann, S., Anderson, D., Overend, J. and Schreiner, F. (1968). *J. Amer. Chem. Soc.*, **90,** 5690
173. Coulson, C. A. (1966). *J. Chem. Phys.*, **44,** 468
174. Collins, G. A. D., Cruickshank, D. W. J. and Breeze, A. (1970). *Chem. Comm.*, 884
175. Ruby, S. L. and Selig, H. (1966). *Phys. Rev.*, **147,** 348

176. Selig, H. and Peacock, R. D. (1964). *J. Amer. Chem. Soc.*, **86,** 3895
177. Fields, P. R., Stein, L. and Zirin, M. H. (1962). *J. Amer. Chem. Soc.*, **84,** 4164
178. Hazeltine, M. W. and Moser, H. C. (1967). *J. Amer. Chem. Soc.*, **89,** 2497
179. Flohr, K. and Appelman, E. H. (1968). *J. Amer. Chem. Soc.*, **90,** 3584
180. Stein, L. (1970). *Yale Sci.*, **44,** 2
181. Stein, L. (1970). *Science*, **168,** 362
182. Musher, J. I. (1968). *J. Amer. Chem. Soc.*, **90,** 7371
183. Selig, H., Claassen, H. H., Chernick, C. L., Malm, J. G. and Huston, J. L. (1964). *Science*, **143,** 1322
184. Eisenberg, M. and Des Marteau, D. D. (1970). *Inorg. Nucl. Chem. Letters*, **6,** 29
185. Appelman, E. H., Sloth, E. N. and Studier, M. H. (1966). *Inorg. Chem.*, **5,** 766
186. Holloway, J. H. (1966). *Chem. Comm.*, **22,**
187. Chernick, C. L., Claassen, H. H., Malm, J. G. and Plurien, P. L. (1963). *Noble Gas Compounds* (Chicago: University of Chicago Press)
188. Gunn, S. R. (1967). *J. Phys. Chem.*, **71,** 2934
189. Sessa, P. A. and McGee, H. A. (1969). *J. Phys. Chem.*, **73,** 2078
190. Stein, L. (1969). *J. Amer. Chem. Soc.*, **91,** 5396
191. Bartlett, N. and Sladky, F. (1968). *2nd European Symposium on Fluorine Chemistry*, (Göttingen)
192. Maricic, S., Veksli, Z., Slivnik, J. and Volavsek, B. (1963). *Croat. Chem. Acta*, **35,** 77
193. Meinert, H. and Kauschka, G. (1968). *Z. Chem.*, **9,** 114
194. Shaw, M. J., Hyman, H. H. and Filler, R. (1970). *J. Amer. Chem. Soc.*, **92,** 6498
195. Zalkin, A., Forrester, J. D. and Templeton, D. H. (1964). *Inorg. Chem.*, **3,** 1417
196. Selig, H. (1967). *Halogen Chemistry*, Ed. V. Gutmann, Vol. 1, 403 (London: Academic Press)
197. Shich, T. C., Feit, E. D., Chernick, C. L. and Yang, N. C. (1970). *J. Org. Chem.*, **35,** 4020

2
Interhalogen Compounds and Polyhalide Anions

A. I. POPOV
Michigan State University

2.1 INTRODUCTION

This review attempts to cover advances in the chemistry of interhalogen compounds and of polyhalide anions from 1965 through to most of 1970.

During this period the attention of the investigators centred primarily on three topics: the behaviour and the stabilities of polyhalide ions in non-aqueous solvents, spectroscopic investigations of the structure of polyhalide ions and of halogen complexes and the chemistry of halogen fluorides.

There is really no good chemical reason for separating the chemistry of halogen fluorides from that of the other interhalogens. This division is arbitrary and reflects only the greater chemical reactivity of the former compounds and, consequently, the greater difficulties of handling and manipulation.

Because of the difficulties inherent in making physico-chemical measurements on solutions of halogen fluorides, very few equilibria studies have been attempted in these media although the advent of new techniques of studying such solutions holds interesting promises for the future. On the other hand, the basic chemistry of interhalogen compounds containing only chlorine, bromine and iodine, has been quite thoroughly investigated during the first half of this century and the present research interests in this field are mostly concerned with the equilibria involving these species in aqueous and non-aqueous solvents as well as on the structures and bonding of the complexes formed by these Lewis acids with a large variety of electron pair donors. Because of this pragmatic (albeit temporary) difference of approach, it seems justifiable to divide accordingly this review into two parts.

2.2 CHEMISTRY OF NON-FLUORINE-CONTAINING INTERHALOGENS AND POLYHALIDES

2.2.1 Infrared and Raman spectra of polyhalide anions

Recent developments in the far infrared instrumentation made it possible to study the vibrational spectra of halogen complexes and of polyhalide ions in solid compounds as well as in solutions.

Vibrational spectra of the trichloride ion in solid salts as well as in benzene and acetonitrile–benzene solutions are reported by Evans and Lo[1]. The spectra of the solid compounds show a strong Raman-active band at $268\ cm^{-1}$ (ν_1) and a strong infrared-active band at $242\ cm^{-1}$ (ν_3). The ν_2 vibration was not observed but was estimated to be at $\sim 165\ cm^{-1}$. The spectrum is consistent with a linear, centrosymmetric structure of the ion.

In solutions, ν_1 and ν_3 bands are shifted to 281 and $232\ cm^{-1}$ respectively, indicating some loss of symmetry, and a new band appeared at $93 \pm 3\ cm^{-1}$ which is attributed to vibrations of ion-pairs or of higher ionic aggregates.

Raman spectra of acetonitrile solutions containing Cl_2 to Cl^- ratios in excess of 1:1 show a broad intense band at $482\ cm^{-1}$ which the authors attribute to the pentachloride ion, Cl_5^-. At 3:1 mole ratio of Cl_2/Cl^- the $275\ cm^{-1}$ band of Cl_3^- ion is no longer observable which seemingly indicates a complete conversion of Cl_3^- to Cl_5^-. This conclusion implies that the formation constant of Cl_5^- in acetonitrile solutions is surprisingly large. Further work is needed to elucidate this point.

Similar studies[2] on the salts of dichlorobromate(I) ion $BrCl_2^-$, support linear centrosymmetric structure Cl—Br—Cl. Upon addition of chlorine to $BrCl_2^-$ solution in acetonitrile a broad Raman band at 513 cm^{-1} was observed which is assigned to the $BrCl_4^-$ ion.

Evans and Lo[3] also studied the Raman spectra of solutions containing bromine and the bromide ion, at various mole ratios, in water, acetonitrile and n-propanol solutions. The spectra are fairly complex and change with the changing Br_2/Br^- ratios. Two bands at 160–170 cm^{-1} and 200–210 cm^{-1} can be assigned to Br_3^- vibrations but as Br_2/Br^- ratio is increased to about 2, a new band becomes prominent (in water and acetonitrile solutions) at about 250 cm^{-1} which the authors assign to Br_5^- species. It should be noted, however, that this interpretation contradicts a number of previous results which show that the formation constant of Br_5^- is at best very small[4].

A detailed infrared and Raman spectra of several trihalide ions is reported by Maki and Forneris[5]. Spectra of solid tetra-alkylammonium as well as of alkali metal salts were obtained. The infrared measurements were made down to 40 cm^{-1} which allowed the observation of the bending mode of the trihalide species. The results are shown in Table 2.1.

In general, the alkali metal salts gave more complex spectra than the tetra-alkylammonium salts. For example, caesium tri-iodide and rubidium dichloroiodate(I) show three infrared-active bands, i.e., the symmetric stretch is infrared-active and, therefore, the ions cannot be centro-symmetric. Similar results were obtained with caesium and rubidium dibromodate(I).

Far infrared spectra of the tri-iodide ion in the form of $Bu_4^nNI_3$ was investigated by Ginn and Wood in benzene, nitrobenzene, and pyridine solutions as well as in Nujol mulls[6]. The I_3^- asymmetric stretching frequency was found to be at 140, 138, 139 and 133 cm^{-1} respectively. The spectral data support linear centrosymmetric structure of the I_3^- ion rather than the bent structure proposed by Robin[7].

Hayward and Henda[8] studied the far infrared and Raman spectra of solid tetra-alkylammonium and caesium dibromoiodates(I) and tri-iodides. Confirming the results of other investigators, the results for the I_3^- salts were dependent on the cation. For example, the far infrared spectrum of the caesium tri-iodide has only one band at 150 cm^{-1} while that of the tetra-methylammonium tri-iodide has two bands located at 143 and 123 cm^{-1} respectively. The far infrared and Raman spectra of tetramethylammonium dichloroiodate are relatively simple and permit to assign 160 cm^{-1} band to the ν_1 mode (Raman-active symmetric stretch) and 171 cm^{-1} band to the ν_3 mode (infrared-active asymmetric stretch). The spectra of the caesium salts are more complex. The authors postulate that while IBr_2^- and I_3^- ions are linear and centrosymmetric in the tetra-alkylammonium salts, they are distorted in the caesium salts.

Far infrared and Raman spectra of solid tetra-alkylammonium poly-halides containing I_3^-, I_5^-, I_7^-, I_9^-, I_4Cl^- and I_2Br^- anions were studied by Parrett and Taylor[9]. The tri-iodides show a Raman active band at 104–108 cm^{-1} (ν_1) and two infrared active bands at about 130 cm^{-1} (ν_3) and 70 cm^{-1} (ν_2).

The Raman and infrared bands for the other polyhalides are tabulated without assignments.

Table 2.1 Infrared* and Raman† spectra of trihalide compounds. Unless otherwise indicated all spectra are for solid samples[5].

(From Uaki and Formeris[5] by courtesy of Spectrochimica Acta)

Cation	*Anion*	*Mode of observation*	ν_1 (cm^{-1})	ν_2 (cm^{-1})	ν_3 (cm^{-1})	*Other bands*	*Remarks*
Cs	I_3	i.r.	103 (s)	69 (m)	149 (s)	—	—
$(CH_3)_4N$	I_3	Raman	111 (10)	—	143 (3)	—	0.25 M solution in nitrobenzene
$(CH_3)_4N$	I_3	Raman	109 (10)	—	152 (2)	—	0.5 M solution in C_2H_5OH
K	I_3	Raman	114 (†)	—	148	—	0.8 M solution in H_2O
Cs	I_2Br	i.r.	117 (m)	84 (m)	168 (s)	—	—
Cs	IBr_2	i.r.	143 (m)	98 (m)	—	192 (m)	—
Cs	IBr_2	Raman	144 (5)	—	178 (10)	186 (6), 197 (1), 129 (1), 70 (1), 40 (1)	—
Rb	IBr_2	i.r.	144 (m)	97 (m)	—	195 (m)	—
Rb	IBr_2	Raman	146 (6)	—	180 (10)	190 (6), 130 (1)	—
$(CH_3)_4N$	IBr_2	i.r.	—	98 (w)	171 (s)	70 (w)	—
$(CH_3)_4N$	IBr_2	Raman	160 ± 1	—	—	—	—
$(CH_3)_4N$	IBr_2	Raman	157 (10)	—	173 (7)	—	0.3 M solution in nitrobenzene (173 cm^{-1} shift is superimposed in a nitrobenzene shift)
$(CH_3)_4N$	IBr_2	Raman	157 (10)	—	173 (5)	—	0.15 M solution in CH_3OH
$(Bu)_4N$	IBr_2	Raman	163 ± 1	—	—	—	—
$(Bu)_4N$	IBr_2	Raman	158 (10)	—	174 (7)	—	0.4 M solution in $CHCl_3$

K	IBr_2	Raman	158	—	—	—	2 M solution in H_2O
Cs	ICl_2	i.r.	—	129 (w)	218 (s)	64 (w)	—
Cs	ICl_2	Raman	268	—	—	—	—
Rb	ICl_2	i.r.	282 (m)	140 (w)	218 (s)	—	—
Rb	ICl_2	Raman	278	—	—	—	—
$(CH_3)_4N$	ICl_2	i.r.	—	138 (w)	226 (s)	73 (w)	—
$(CH_3)_4N$	ICl_2	Raman	254	—	—	—	(‡)
Cs	BrICl	i.r.	198 (s)	125–145 (m)	—	256 (m), 275 (m), 55 (m), 77 (m)	—
Cs	BrICl	Raman	200 (20)	—	180 (2)	247 (00), 271 (1), 146 (1), 174 (1)	—
Rb	BrICl	i.r.	202 (s)	125–145 (m)	—	259 (m), 279 (m), 55 (m), 77 (m)	—
Rb	BrICl	Raman	203 (20)	—	180 (2)	258 (00), 278 (0), 145 (0), 160 (1)	—
$(CH_3)_4N$	BrICl	i.r.	229 (s)	128 (w)	175 (m)	70 (m)	—
$(CH_3)_4N$	BrICl	Raman	230 (4)	—	174 (10)	—	Also bands observed at 156 and 254 cm^{-1} due to IBr_2^- and ICl_2^- impurities
$(CH_3)_4N$	BrICl	Raman	232 (1)	—	177 (5)	—	0.7 M solution in $(CH_2Cl)_2$–IBr_2^- and ICl_2^- impurities
$(Et)_4N$	BrICl	Raman	230 (2)	—	173 (5†)	—	0.8 M solution in dioxane; IBr_2^- and ICl_2^- impurities
K	BrICl	Raman	205 (5)	—	174 (10†)	—	0.8 M solution in H_2O; IBr_2^- and ICl_2^- impurities

* Infrared intensities are indicated by the abbreviations s = strong, m = medium, and w = weak
† Raman intensities are approximately proportional to the numbers given in parentheses
‡W. B. Person, G. R. Anderson, J. N. Fordemwalt, H. Stammreich, and R. Forneris (1961). *J. Chem. Phys.*, **35,** 908

2.2.2 Spectroscopic studies of interhalogens and of interhalogen complexes

The Raman spectrum of solid dimer I_2Cl_6 showed nine normal vibrational modes[10] expected for a planar bridged X_2Y_6 molecule with D_{2h} symmetry. A later work[11] reports both the Raman and the infrared spectra of the solid compound. The Raman results are in essential agreement with the earlier publication. Attempts to study solution spectra of I_2Cl_6 were unsuccessful due to its complete dissociation to iodine monochloride and chlorine in all solvents tried.

Nelson and Pimentel[12] passed a mixture of chlorine, bromine and an inert gas through a microwave discharge, condensed the products on a caesium iodide window at 20 K and obtained infrared spectra of the matrix. Analysis of the spectral data showed that several new bromine–chlorine compounds were formed among which they identified a T-shaped Br_2Cl_2 molecule with a trivalent–bromine atom. Less definitive evidence was obtained for the existence of the Br_3Cl species.

There are several reports on the infrared and far infrared studies of interhalogen complexes with heterocyclic amines. Wood and co-workers[13,14] studied the far infrared spectra of pyridine complexes with iodine monochloride and iodine monobromide (as well as iodine) in benzene and in pyridine solutions as well as of the solid compounds. The results are shown in Table 2.2. In pyridine solutions of Py·ICl and Py·IBr complexes, the

Table 2.2 Halogen-halogen and halogen-nitrogen stretching frequencies in interhalogen complexes

Donor	*Interhalogen (XY)*	*Solvent*	ν_{X-Y}	ν_{X-N}	*References*
Pyridine	ICl	Benzene	290	147	13, 14
Pyridine	ICl	Benzene	292	140	15
Pyridine	ICl	Pyridine	277	160	13, 14
Pyridine	ICl	Pyridine	278	151	15
Pyridine	ICl	(Solid)	285, 278	180, 172	15
Pyridine	IBr	Benzene	204	134	13, 14
Pyridine	IBr	Pyridine	195	144	13, 14
2-Methylpyridine	ICl	Benzene	290	133	15
3-Methylpyridine	ICl	Pyridine	278	—	15
3-Methylpyridine	ICl	(Solid)	282	165, 155	15
4-Methylpyridine	ICl	Benzene	289	135	15a
4-Methylpyridine	IBr	Benzene	201	124	15a
2,6-Dimethylpyridine	ICl	Benzene	290	117	15
2,6-Dimethylpyridine	ICl	Pyridine	278	—	15
2,6-Dimethylpyridine	ICl	(Solid)	272	140	15

authors observed some slow formation of the dichloroiodate(III) and dibromoiodate(I) ions. It has been observed previously[14a] that in relatively polar solvents, halogen complexes may ionise to give a halide ion and a corresponding pyridinium-type cation

$$Py \cdot YX \rightarrow PyY^{+} + X^{-} \tag{2.1}$$

The anion reacts with the complex to give the corresponding halide ion

$$2\text{Py}\cdot\text{YX} + \text{X}^- \rightarrow \text{Py}_2\text{Y} + \text{YX}_2^- \qquad (2.2)$$

Similar studies were carried out by Yarwood and Person[15] on the far infrared spectra of pyridine, 3-methylpyridine and 2,6-dimethylpyridine complexes of iodine monochloride in benzene and pyridine solutions as well as in Nujol mulls. The results are shown in Table 2.2. Plots of the integrated intensities of the 290 (ν_{I-Cl}) and 140 (ν_{N-I}) cm^{-1} bands for Py·ICl complex in benzene gave reasonable straight lines. The ν_{I-Cl} frequencies for the three complexes are essentially identical but the ν_{N-I} frequency in the 2,6-dimethylpyridine complex is lower than in the other two complexes which may reflect some steric hindrance of the two methyl groups in the 2,6-dimePy·ICl complex.

Infrared spectra of 4-methylpyridine complexes with iodine monochloride and iodine monobromide in non-polar and polar solvents show that in the former solvents (carbon disulphide, benzene and dichloromethane)

Table 2.3 Fundamental vibration band of iodine monobromide in various solvents[17]

Solvent	$\nu_{I-Br}(cm^{-1})$	*Solvent*	$\nu_{I-Br}(cm^{-1})$
$\frac{1}{n}$	267	Carbon disulphide	253
Carbon tetrachloride	261	1,2-Dibromoethane	249
n-Heptane	261	Benzene	249
Cyclohexane	261	Toluene	247
1,2-Dichloroethane*	259	*o*-Xylene*	247
Nitrobenzene*	254.5	Acetonitrile	244.5
Nitromethane	253	Pyridine†	205

* Iodine monobromide was dissolved in a 10–90% mixture of the respective solvent and carbon tetrachloride
† Solid Py·IBr complex dissolved in benzene

Table 2.4 Vibrational frequencies in solid IBr and ICl complexes[17]

Complex	$\nu_{I-X}(cm^{-1})$	$\nu_{N-I}(cm^{-1})$
Pyridine·IBr	199, 190	158.5
Pyridine·ICl	274, 266	178, 169.5(s)
4,4′-Bipyridine·2IBr	194.5, 187	138
4,4′-Bipyridine·2ICl	274	154.5
2,2′-Bipyridine·2IBr	211, 203	135
2,2′-Bipyridine·2ICl	284.5, 268	151(s), 130.5(w)

reactions (2.1) and (2.2) do not occur since there is no band for the IX_2^- ion. If 4-methylpyridine itself is used as solvent, a band at 168 cm^{-1} appears in the 4MePy·IBr system which may be due to the IBr_2^- or $(4MePy)_2I^+$ ions and a band at 222 cm^{-1} in the 4MePy·ICl system which can be assigned to an ICl_2^- ion vibration. Reactions (2.1) and (2.2) occur extensively in acetonitrile–benzene mixtures and Augdahl and Klaboe[43] likewise found quantitative formation of the I_3^- ion in acetonitrile solutions of triphenylarsine–iodine complexes.

Infrared spectra of pyridine and of pyridine ICl complexes in chloroform and benzene solutions were also studied by Yarwood[16]. No evidence for ionic reactions (2.1) and (2.2) was found.

Infrared absorption spectra of iodine monobromide in various solvents as well as in solid complexes were reported by Yagi *et al.*[17]. The results are shown in Tables 2.3 and 2.4. The solution spectra show a progressive shift of the ν_{I-Br} band to lower frequencies as the donor strength of the solvent is increased. The results are similar to those previously obtained for iodine monochloride solutions[18] except that in this case the magnitude of the shift is smaller. Several of the bands are split into doublets probably due to correlation splitting.

Far infrared spectra of solid pyridine complexes with iodine monochloride and iodine monobromide were also studied by Watari[19]. While he did not observe the splitting of the ν_{I-X} band, the band frequencies are in good agreement with the results of Yagi *et al.*[17]. It was noticed that upon complexation, the 405 cm^{-1} band of pyridine shifts to higher frequency by about 20 cm^{-1}. Similar behaviour of this band has been previously observed in pyridine–metal ion complexes[20].

Mössbauer spectroscopy was also used to study iodine monohalide complexes with pyridine, 2,2′- and 4,4′-bipyridines and pentamethylenetetrazole[21]. The iodine monobromide complexes exhibit mainly σ bonding while those of iodine monochloride show a substantial amount of π bonding. The data also indicate that 2,2′-bipyridine·2IX complex can exist in two different configurations, the IBr complex being in the *cis*-configuration while the ICl complex has the *trans*-configuration.

2.2.3 X-ray studies of polyhalide ions

The structures of several polyhalide ions had been quite thoroughly studied prior to 1965 by numerous investigators[4]. In recent years the structures of $CsIBr_2$, NH_4BrICl and $(C_2H_5)_4NI_3$ have been reported. It was confirmed[22] that IBr_2^- ion is essentially linear (178.0 degrees) and centrosymmetric. The interatomic distances were not reported. Crystal structure of ammonium bromochloroiodate(III) was difficult to determine exactly since the ions are statistically distributed in the crystal. The approximate values for the bond lengths were found to be 2.84 Å for I—Cl and 2.54 Å for I—Br.

Tetraethylammonium tri-iodide exists in two crystalline modifications. Crystal structures of the two forms is reported by Migchelsen and Vos[23]. Form I is orthorhombic with a space group *Cmca*. The cell parameters are: $a = 14.07$, $b = 15.220$ and $c = 14.061$ Å. Tri-iodide ion is symmetrical with I—I bond distances of 2.928 and 2.943 Å. Form II of the compound is also orthorhombic with a space group *Pnma* and cell parameters, $a = 14.552$, $b = 13.893$ and $c = 15.156$ Å. The crystal contains two distinct asymmetric tri-iodide ions with bond distances being 2.921, 2.961 and 2.892, 2.981 Å respectively.

2.2.4 Stability constants of polyhalide ions

A considerable amount of effort has been expended during the last few years on the determination of stability constants of polyhalide ions in various solvents by a variety of physico-chemical techniques. It is generally agreed that in aprotic solvents the stability of these species is several orders of magnitude higher than in aqueous solutions. Even among protic solvents water seems to be a rather exceptional solvent in this regard. It seems that there is only one known case where the stability of polyhalide ion is less than in water, namely, trifluoroacetic acid. The formation constant of Br_3^- ion was measured spectrophotometrically in this solvent[24] and was found to be less than 0.9 (M^{-1}) or one order of magnitude smaller than in water.

Recent work in aqueous solutions seems to be restricted to that of Mironov and Lastovkina[25, 26] who investigated the thermodynamics of the trihalide ion formation in solutions of high acidity (pH $\approx$ 1) and at ionic strength of 3.0. The values were obtained from calorimetric, spectrophotometric, extraction and solubility studies. Enthalpies of formation of X_3^- ions were determined either calorimetrically or from the temperature dependence of the formation constant. The results are summarised in Table 2.5. It is seen

Table 2.5 Thermodynamic constants for the reaction $X_2(aq)+Y^-(aq) \rightleftarrows X_2Y^-(aq)$ at 25°C in 0.1 M acid and ionic strength of 3.0[25, 26]

Ions	*Method*	K_f(M^{-1})	ΔG(kcal mol^{-1})	ΔH(kcal mol^{-1})	ΔS (cal deg^{-1} mol^{-1})
$ClBr_2^-$	calor.	1.4±0.3	−0.20±0.1	−2.2±0.2	−7.0±1.0
ClI_2^-	temp. coeff.	2.0±0.1	−0.40±0.1	1.7±0.1	7.0±1.0
Br_3^-	extrac.	11.3±1.4	1.44±0.19	1.66±0.14	0.74±1.60
Br_3^-	calor.	11.1±0.3	1.43±0.07	1.64±0.14	0.71±0.70
Br_3^-	spectr.	12.7±0.3	1.50±0.09	1.63±0.15	0.44±0.80
BrI_2^-	temp. coeff.	13.3±0.6	−1.5±0.02	−1.8±0.3	−1.0±0.8
BrI_2^-	calor.	13.3±0.6	−1.5±0.02	−2.2±0.1	−2.2±0.4
I_3^-	temp. coeff.	385±20	−3.5±0.02	−3.1±0.5	1.4±2.0
I_3^-	calor.	385±20	−3.5±0.002	−3.3±0.2	0.7±0.9

that there is good agreement between results obtained by different methods. In general, the formation constants are lower than the previously reported values[4] but previous measurements were carried at low ionic strengths and/or were extrapolated to infinite dilution.

Parker[27] studied the formation constants of the I_3^- and Br_3^- ion in dimethylformamide solutions by means of I^-–I_2 and Br^-–Br_2 concentration cells at 25 °C. The values obtained were 6.5×10^6 for the tri-iodide ion and 2.0×10^6 for the tribromide ion. It is interesting to note the close proximity of these values in dimethylformamide since they differ by nearly two orders of magnitude in aqueous solutions. It should be noted, however, that Povarov and Barbasheva[28], using the same technique and the same solvents, found the formation constant of the I_3^- ion to be 7.8×10^4. A higher value of 2.2×10^7 is reported by other investigators[29]. A polarographic study of bromine solutions in DMF by Siniki and Breant[30], give a much lower value of about 105 for the formation constant of Br_3^- ion in this solvent. In view of the previous results, the validity of the last value is doubtful.

Oxidation-reduction potentials of Cl^-–Cl_2, Br^-–Br_2 and I^-–I_2 couples were studied by Marchon in nitromethane and acetonitrile solutions[31]. The bromide and the iodide ions showed a two-step oxidation, first to the corresponding trihalide ion and then to the halogen. Chloride ion is oxidised in one step directly to the chlorine. These results contradict the earlier data of Nelson and Iwamoto[32] concerning the stability of Cl_3^- ion in acetonitrile. The pK's for the dissociation of I_3^- and Br_3^- were found to be 7.4 and 7.0 respectively and were identical in both solvents. The stability order, for the three polyhalogens is, therefore, $I_3^- > Br_3^- \gg Cl_3^-$.

The stability constants of the tri-iodide ion were reported in acetonitrile[33]

Table 2.6 Formation constants of the I_3^- and I_5^- ions in ethanol–water and acetonitrile–water mixtures at 25 °C[38]

Solvent (% by wt)	$\log K_{I_3^-} = \log(I_3^-)/(I_2)(I^-)$	$\log K_{I_5^-} = \log(I_5^-)/(I_2)^2(I^-)$
Ethanol		
50	3.68	2.4
70	3.77	1.8
90	4.05	1.7
100	4.13	1.8
Acetonitrile		
40	4.26	1.7
60	5.01	1.7
80	5.16	1.55
90	6.14	2.1
100	6.85	—

($K_f = 2 \times 10^7$) in dimethylsulphoxide[34] ($K_f = 1.6 \times 10^6$) and in ehtanol[35] ($>10^5$) solutions. The formation constant of Br_3^- in the latter solvent[35] is 182 ± 9. The values were determined spectrophotometrically and the measurements were complicated by the reaction of the halogens with the solvent. Voltametric study of the reduction of iodine in glacial acetic acid solutions containing sodium perchlorate or a mixture of sodium perchlorate and sodium acetate gave the formation constant of I_3^- ion of 1.7×10^5 in this

Table 2.7 Formation constant of Br_3^- ion in water–methanol mixtures[39]

% MeOH (wt)	K_f
0	16.3
10	17.5
20	22
30	31
40	43
50	58
60	78
70	100
80	125
90	150
100	176

solvent[36]. Formation constants of I_3^- and I_5^- ions were also measured potentiometrically in propylene carbonate solutions at constant ionic strength of 0.1[37]. The respective values reported are 8.3×10^7 and 83.

Formation constants of I_3^- and I_5^- ions in ethanol–water and acetonitrile–water mixtures have been determined potentiometrically by Tremillon and co-workers[38]. The results are shown in Table 2.6. It is seen that, in agreement with previous results, the stability of the tri-iodide ion increases sharply with increasing concentration of the organic solvent. It is somewhat more difficult to interpret the constancy of the I_5^- formation constant in solutions with different ratios of organic solvent and water.

Similar studies in water–methanol mixtures on Br_2–Br^- equilibrium were carried out by Dubois and Garnier[39]. Their results are shown in Table 2.7.

Benoit and co-workers[40, 41] used anhydrous sulpholane as solvent for the studies of halogen-halide ion equilibria by potentiometric and spectrophotometric techniques. The earlier work was carried out at 22 °C and in 0.1 M $LiClO_4$ while the later studies were at 30 °C and in solutions of very low ionic strength. The I_2Br^- and I_2Cl^- disproportionate in solutions

$$2I_2Br^- \rightleftarrows IBr_2^- + I_3^-$$
$$2I_2Cl^- \rightleftarrows ICl_2^- + I_3^-$$

The results, shown in Table 2.8, indicate that the stabilities of the trihalide

Table 2.8 Equilibrium constants of trihalide ions in sulpholane solutions

Ion	log K_f	log K_{disp}	λ_{max}(nm)	ε_{max}	*Reference*
I_3^-	7.49±0.01	—	367 295	26100	41
I_3^-	7.5	—	—	—	40
Br_3^-	7.33±0.02	—	—	—	41
Br_3^-	6.4	—	—	—	40
Cl_3^-	4.11±0.02	—	—	—	41
I_2Br^-	7.23±0.02	−1.55±0.15	275	46300	41
I_2Br^-	6.5	—	—	—	40
I_2Cl^-	7.23±0.03	−0.64±0.1	263	36500	41
Br_2Cl^-	4.8	—	—	—	40
ICl_2^-	—	—	232	66000	41

ions vary in the order $I_3^- > Br_3^- \gg Cl_3^-$ confirming the results of Marchon[31] in nitromethane and acetonitrile. Here again there is only a slight difference in stability between I_3^- and Br_3^-.

Topol[42] obtained thermodynamic data for the formation of alkali metal and ammonium polyiodides from e.m.f. measurements of solid state cells of

Table 2.9 Values of $\Delta G°$, $\Delta H°$ and $\Delta S°$ for the reaction $MI_x(s) + yI_2(s) \rightarrow MI_x + 2_y(s)$ at 25 °C[42]

Complex	$-\Delta G^0$, (kcal)	$-\Delta H^0$ (kcal)	$-\Delta S^0$ (cal deg^{-1} mol^{-1})
RbI–RbI_3	2.4±0.1	3.1	2.3
NH_4I–NH_4I_3	1.8±0.1	2.1	1.1
CsI–CsI_3	3.5±0.1	3.7	0.9
CsI_3–CsI_4	0.63±0.03	0.80	0.6

the type Ag/AgI/C, $MI_x - MI_{x+2y}$ and Ag/AgI/C, I_2(s) in the 20–150 °C temperature interval.

The overall reaction is

$$MI_x(s) + yI_2(s) \rightarrow MI_{x+2y}(s)$$

The systems investigated were RbI–RbI_3, NH_4I–NH_4I_3, CsI–CsI_3 and CsI_3–CsI_4. Values for the standard free energy, enthalpy and entropy for the above reaction were calculated from the experimental data and are given in Table 2.9.

2.2.5 Studies of molecular complexes of halogens

Relatively few studies report the preparation of new molecular complexes containing interhalogens*. Most of the interest during the last few years seems to have been centred on spectroscopic studies on the structure and bonding of some typical interhalogen complexes.

Donor–acceptor complexes of iodine monochloride and of iodine bromide (as well as those of iodine) with dimethylcyanamide were studied spectrophotometrically by Augdahl and Klaeboe in 1965 [43]. The Job's plots showed that the complexes were formed in 1:1 ratio but there was some variation

Table 2.10 **Formation constants and thermodynamic functions for the dimethylcyanamide complexes with ICl and IBr [43]**

Interhalogen	*Temp.* °C	K_f (M^{-1})	$-\Delta G^0$ (kcal mol^{-1})	$-\Delta H^0$ (kcal mol^{-1})	$-\Delta S^0$ (cal deg^{-1} mol^{-1})
ICl	26	120	2.9	7.3	15.2
IBr	20	18.8	1.64	5.6	13.3

in the values of the formation constants obtained at different wavelengths which may indicate the formation of a higher complex. The complexity constants were determined in the 10–40 °C temperature region and the standard entropy and enthalpy changes for the complex formation were calculated from the temperature dependence of log K_f. The results are shown in Table 2.10.

While bromine chloride is an extremely unstable species, it does form

Table 2.11 **Formation constants of pyridine–BrCl complexes in carbon tetrachloride solutions at 25 °C [44]**

Base	10^{-3} K_f (M^{-1})	*Base*	10^{-3} K_f (M^{-1})
Pyridine	1.2 ± 0.1	2,6-Dimethylpyridine	1.5 ± 0.1
4-Methylpyridine	2.1 ± 0.2	3,4-Dimethylpyridine	4.2 ± 0.4
2,4-Dimethylpyridine	6.7 ± 0.7	3,5-Dimethylpyridine	6.8 ± 0.8
2,5-Dimethylpyridine	5.2 ± 0.4		

rather stable 1:1 complexes with heterocyclic amines[44] which can be prepared by mixing equimolar amounts of the two components in 1,1,2-trichlorotrifluoroethane at ~ -25 °C. Addition compounds were prepared with pyridine, 4-methylpyridine, 2,4-, 2,5-, 2,6-, 3,4- and 3,5-dimethylpyridine and 2,3,6-trimethylpyridine. Attempts to prepare compounds with halogen-substituted pyridines were unsuccessful probably due to the inductive effect of the halogens. Formation constants of several complexes were determined

*There are, however, many papers on the preparation and stability of iodine complexes but these are reviewed elsewhere.

spectrophotometrically in carbon tetrachloride solutions and are given in Table 2.11. Comparison of these data with those for the iodine monochloride and iodine monobromide complexes shows that the Lewis acid strength of the interhalogen varies in the order ICl > IBr > BrCl.

The crystal structure of iodine monochloride complex with pentamethylene tetrazole, $C_6H_{10}N_4$, has been reported by Baenziger *et al.*[45]. The N—I—Cl group is linear and the N—I and I—Cl distances are 2.34 and 2.44 Å respectively. The comparison of the N—I distance in the above complex with that in pyridine–ICl complex of 2.26 Å[46] shows that the bonding is somewhat weaker in the pentamethylenetetrazole complex.

2.2.6 Miscellaneous studies

2.2.6.1 Bonding

Bonding in interhalogen compounds and polyhalogen anions and cations is thoroughly discussed by Wiebenga and Kracht[47] in terms of a modified Hückel theory. Only linear combinations of p orbitals were used in calculating the molecular orbitals. The calculations show that the stability of the interhalogen compounds is due to the electrostatic interaction energy rather than to the covalent energy. There is a good agreement between the calculated and observed energies of formation. The energy values for the trihalide ions were calculated for the free state and, therefore, are not directly comparable with the experimental value obtained in solutions or in crystalline solid.

The properties of crystalline IBr, ICl and I_2Cl_6 have been studied by Mössbauer spectroscopy[48]. The authors conclude that iodine–chlorine and iodine–bromine bonds are chiefly formed by the p electrons without sp hybridisation. It is interesting to note that the authors report preparation of a new interhalogen compound $I_2Cl_4Br_2$ by the oxidation of sodium iodide with $KBrO_3$ and the treatment of the resulting mixture with chlorine. Neither the properties of the compound nor the analytical data are given, but the Mössbauer spectra seem to indicate that there are two different iodine states in the molecule and that the two bromine atoms are situated at one end of the molecule. This seems to be the first report of the existence of a ternary interhalogen compound, contradicting the results of Campbell and Shemilt[49] who studied solid–liquid equilibria in the I–Br–Cl system at 29.8 °C and found no evidence for the formation of a ternary compound.

Molecular orbital calculations on the electronic structure of the tri-iodide ion showed that the existence of symmetrical and asymmetrical structures of I_3^- in solid salts may be explained by the lattice effects[50].

Nuclear quadrupole resonance studies on I_3^- and IBr_2^- ions showed that the formation of these species involves a transfer of charge from the I^- ion to the interhalogen molecule[51].

2.2.6.2 Electrochemistry

Several groups of investigators studied the poloragraphic behaviour of halogens in various solvents. Matsui and Date[52] investigated poloragraphically Br_2–Br^- equilibria in dimethylformamide using dropping mercury electrode as the indicator electrode. They found that the bromine solutions in dimethylformamide are unstable. Sinicki and Breant[30] postulate the reduction of bromine by formic acid formed by hydrolysis of the solvent. In the present work, polarograms were obtained in the dark as a function of time. There is a slow reaction between bromine and the solvent with the formation of HBr_3. The authors propose the following mechanism

$$Br_2 \rightleftarrows Br^+ + Br^-$$
$$Br^+ + 2DMF \rightarrow Br^- + 2DMF^+$$
$$2DMF^+ \rightarrow 2H^+ + 2R$$
$$2Br_2 + 2Br^- \rightleftarrows 2Br_3^-$$

The nature of the radical R remains unclear.

Electrochemical reactions at the DME are complicated by the reaction of bromine with mercury resulting in the formation of tribromomercurate ion $HgBr_3^-$. No evidence was found for the presence of formic acid in DMF solutions.

Kinetics of $I^- - I_3^- - I_2$ equilibria on a rotating platinum electrode in acetonitrile solutions was studied by Macagao *et al.*[53]. Barbasheva and co-workers also investigated anodic oxidation of iodine species on a rotating platinum electrode in dimethylformamide[54,55]. Two waves were observed corresponding to the reactions, $3I^- \rightarrow I_3^- + 2e^-$ and $2I_3^- \rightarrow 3I_2 + 2e^-$.

2.2.6.3 *Chemical reactions*

Bromine chloride hydrate was prepared by Glew and Hames[56], in the form of a dark red crystalline solid. Analysis of the product showed it to have the composition $BrCl \cdot 7.34H_2O \cdot$ instead of $BrCl \cdot 4H_2O$ reported in the earlier literature.

The rate of dissolution of metallic bismuth by the tri-iodide ions was studied by Macdonald and Wright[57]. Solutions contained an excess of potassium iodide so that $Bi^{(III)}$ existed in solutions primarily as the hexaiodo complex BiI_6^{3-}. The overall dissolution reaction is

$$2Bi + 3I_3^- + 3I^- \rightarrow 2BiI_6^{3-}$$

Reaction of iodine trichloride with phosphorus pentachloride[58] in tetrachloroethane solutions produces an orange coloured adduct $PCl_5 \cdot ICl_3$. The ^{31}P nuclear magnetic resonance spectra indicate a salt structure $PCl_4^+ICl_4^-$. The compound is soluble in dichloroethane and nitromethane. It decomposes at 221 °C. It is also reported that the reaction of iodine bromide with PCl_5 leads to the formation of $PCl_4^+ICl_2^-$.

A new series of polyhalogen complex anions containing five halogen atoms was prepared by reacting organic salts of polyhalide acids, such as $B \cdot HICl_2$, with additional amounts of halogen or interhalogen species[59,60]. The organic base B was a heterocyclic amine such as pyridine, bipyridine, phenanthroline or biquinoline, or a tetra-alkylammonium cation R_4N^+. Salts of the following anions were prepared: $I_2Cl_3^-$, $I_2Cl_2Br^-$, $I_2Br_3^-$, and

$I_2Br_2Cl^-$. Far infrared spectra of these salts were obtained. Pentahalide anions do not seem to show any isomerism since the compound $(CH_3)_4NI_2Cl_2Br$ was prepared by the addition of iodine monochloride to the bromochloroiodate as well as by the addition of iodine monobromide to the dichloroiodate.

2.3 HALOGEN FLUORIDES

2.3.1 Introduction

While most of the halogen fluorides have been known since the beginning of this century, their high chemical reactivity was an important deterrent to their study. In recent years, however, experimental techniques were developed which allowed the handling of these compounds with a reasonable degree of safety. Recent studies have been concerned mainly with the preparation and the structure of complex compounds of halogen fluorides as well as some initial studies on equilibria which exist in the liquid species. Earlier literature has been thoroughly reviewed in an excellent review by Stein[61].

2.3.2 Chlorine monofluoride, ClF

Christe and Sawodny prepared a number of new compounds of chlorine monofluoride[62] which reacts in 2:1 mole ratio with arsenic pentafluoride and boron trifluoride to give, respectively, the adducts $2FCl{\cdot}AsF_5$ and $2FCl{\cdot}BF_3$. The compounds are white solids, and are stable at $-78°$ and $-127\,°C$ respectively. They are completely dissociated in the gas phase at room temperature and do not appear to have a stable liquid phase. The stoichiometries of the salts were independent of the relative amounts of the reagents and 1:1 complexes, $FCl{\cdot}AsF_5$ and $FCl{\cdot}BF_3$, could not be isolated. The reaction with antimony pentafluoride gave an ill-defined product with stoichiometry close to 1:1 but the properties suggest some sort of polymeric species rather than a true 1:1 adduct.

The reaction of nitryl fluoride with $FCl_2^+AsF_6$ results in the displacement of two molecules of chlorine monofluoride

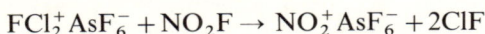

$$FCl_2^+AsF_6^- + NO_2F \rightarrow NO_2^+AsF_6^- + 2ClF$$

further supporting the proposed 2:1 stoichiometry.

The infrared spectra of the two solid compounds clearly show the vibrational bands characteristic of the anions SbF_6^- and BF_4^- respectively, indicating salt-like structures $FCl_2^+SbF_6^-$ and $FCl_2^+BF_4^-$. The authors interpreted the spectral data to indicate that the cation has a symmetric bent structure

which would make it unique in having the most electronegative element as the central atom. This interpretation, however, is challenged by Gillespie and Morton[63] who examined low temperature Raman spectra of the two complexes. The data support the ionic structure of these complexes $Cl_2F^+AsF_6^-$ and $Cl_2F^+BF_4^-$ as proposed above. However, contrary to the conclusions of Christe and Sawodny, the authors find that the cation has an asymmetric structure Cl—Cl—F. The vibrational frequencies for the cation were assigned as follows: ν_1 (Cl—Cl stretch)—doublet at 516 and 540 cm^{-1}, ν_2 (Cl—F stretch) at 743 cm^{-1} and ν_3 (bend) at 296 cm^{-1}.

The nitrosyl salt of difluorochlorate(I) anion $NO^+ClF_2^-$ was synthesised by Christe and Guertin[64] by reacting nitrosyl fluoride with chlorine monofluoride under vacuum at −196 °C. Thermochemical properties, infrared spectrum and electrical conductance of the compound in liquid NOF solutions have been measured. The compound is stable at −78 °C but dissociates completely into its component parts at room temperature. The infrared spectrum as well as appreciable conductance of the solutions of the compound in liquid NOF ($\Lambda = 4.98\ \Omega^{-1}\ cm^2$) are interpreted to indicate that the compound has the ionic structure indicated above.

Christe and Guerton[65] also prepared caesium, rubidium and potassium difluorochlorates(I) either by direct combination of the alkali fluorides with chlorine monofluoride or by a metathetical reaction.

$$NO^+ClF_2^- + M^+F^- \rightarrow M^+ClF_2^- + NOF$$

Attempts to prepare difluorochlorates of lithium and sodium or of alkaline earth metals were unsuccessful.

The three salts obtained are white solids and are quite stable at room temperature. The decompose at ~250 °C and the stability decreases with decreasing cationic radius $CsClF_2 > RbClF_2 > KClF_2$.

Raman and infrared spectra of the four difluorochlorate(I) salts were measured[66]. The data indicate that in $NOClF_2$ the anion is centrosymmetric and linear but that the crystal field effects distort the ion and lower its symmetry in the alkali metal salts.

2.3.3 Chlorine trifluoride, ClF_3

The studies of the physical and chemical properties of chlorine trifluoride have been quite numerous. Liquid–vapour equilibrium of the uranium hexafluoride–chlorine trifluoride system was studied at 2600 Torr[67] and of dichlorotetrafluorethane chlorine trifluoride mixture at 1600 and 2600 Torr[68]. In the first system the liquids are completely miscible and the solutions deviate considerably from ideality. The second system forms minimum boiling points azeotropes with the respective boiling points (at the two pressures studied) of 15.9 °C and 31.3 °C and containing 0.352 and 0.448 mole fraction of ClF_3 respectively. Activity coefficients of the components of the two systems were calculated from the vapour pressure data.

Infrared and Raman spectra of gaseous chlorine trifluoride have been recently investigated by Selig *et al.*[69]. The observed bands and their assignments are given in Table 2.12.

While the reaction of chlorine trifluoride with water most usually occurs with explosive violence, its study under controlled conditions and at various temperatures permitted Bougon *et al.*[70], to determine its mechanism and identify the reaction products.

The initial step in the reaction results in the formation of HF, ClO_2F and ClF

$$2ClF_3 + 2H_2O \rightarrow 4HF + ClO_2F + ClF$$

The reaction products, in turn, can react with water to give

$$5ClF + 2H_2O \rightarrow 2HF + ClO_2F + 2Cl$$

$$2ClO_2F + H_2O \rightarrow 2HF + 2ClO_2 + \tfrac{1}{2}O_2$$

Small amounts of ClO_3F were also detected; this compound is probably formed by a reaction of ClO_2F with oxygen

$$ClO_2F + \tfrac{1}{2}O_2 \rightarrow ClO_3F$$

While the hydrolysis of chlorine trifluoride thus gives a variety of products, no evidence was found for the formation of the fluorine oxide F_2O.

The reaction of chlorine trifluoride with uranyl fluoride, originally studied by Ellis and Forrest[71], has been recently reinvestigated by Shrewsberry and

Table 2.12 Observed frequencies and assignments of fundamentals of ClF_3 gas
(From Selig *et al.*[69] by courtesy of the Journal of Chemical Physics)

Assignment	*Raman*	*Infrared*
$\nu_1(a_1)$	752.1(^{35}Cl), 744.7(^{37}Cl) } p,s	742, 760 } s
$\nu_2(a_1)$	529.3 p, vs	522, 538 } m
$\nu_3(a_1)$	321, 337 } p, w	328 s
$\nu_4(b_1)$	not observed	702 vs
$\nu_5(b_1)$	431 dp, w	442 w
$\nu_6(b_2)$	not observed	328 s

Williamson[72] and by Luce and Hartmanshenn[73]. It appears that at 50 °C the reaction is,

$$2ClF_3 + UO_2F_2 \rightarrow UF_6 + ClF + ClO_2F$$

At higher temperatures, chlorine monofluoride reacts with chlorine oxyfluoride and uranyl fluoride to give,

$$2ClO_2F + 2ClF + UO_2F_2 \rightarrow UF_6 + 3O_2 + 2Cl_2$$

and, therefore, the overall reactions at higher temperatures can be represented by

$$4ClF_3 + 3UO_2F_2 \rightarrow 3O_2 + 2Cl_2 + 3UF_6$$

Chlorine trifluoride can act either as a fluoride-ion donor or a fluoride-ion acceptor. In the former case it reacts with strong fluorine-containing Lewis acids to give the corresponding salts $ClF_2^+MF_n^-$. Addition compounds of ClF_3 with boron trifluoride and phosphorus, arsenic and antimony pentafluorides were prepared by Christe and Pavlath[74]. Cryoscopic measurements, the infrared spectra and the electrical conductance studies of these compounds in liquid chlorine trifluoride were found to be consistent with

Table 2.13 Infrared and Raman spectra of ClF_2^+ AsF_6^-

(Gillespie and Morton)[76] — (Christe and Sawodny)[75]

(by courtesy of *Inorganic Chemistry*)

Rel. intens.	*Raman freq. shift,* Δν, cm^{-1}	*Assignment*	*Raman freq. shift,* Δν, cm^{-1}	*I.R. freq.* cm^{-1}	*Assignment*
14	373	$\nu_5(AsF_6^-)$	375 mw		$\nu_5(AsF_6^-)$
14	384	$\nu_2(ClF_2^+)$		406 m	$\nu_3(AsF_6^-)$
				520 w }	$\nu_2(ClF_2^+)$
26	544 }	$\nu_2(AsF_6^-)$	544 m	558 m }	
6	602 }		603	609 w	$\nu_2(AsF_6^-)$
77	693	$\nu_1(AsF_6^-)$	693 s		$\nu_1(AsF_6^-)$
				703 vs	$\nu_4(AsF_6^-)$
100	806 }	$\nu_1(ClF_2^+)$	811 vs	810 sh	$\nu_1(ClF_2^+)$
90	809 }				
51	821	$\nu_3(ClF_2^+)$		818 s	$\nu_3(ClF_2^+)$

Table 2.14 Infrared and Raman spectra of ClF_2^+ BF_4^-

Raman (Gillespie and Morton)[76] 110 °C — I.R. (Christe and Sawodny)[75]

(by courtesy of *Inorganic Chemistry*)

Rel. intens.	*Freq.*, cm^{-1}	*Assignment*	*Freq.*, cm^{-1}	*Assignment*
15	63			
10	91			
15	124			
15	134			
3	355	$\nu_2(BF_4^-)$		
6	373 }	$\nu_2(ClF_2^+)$		
12	396 }			
			519 m }	ν_4 ($^{10}BF_4^-$)
4	526	$\nu_4(BF_4^-)$	529 sh }	ν_4 ($^{11}BF_4^-$)
			537 ms	$\nu_2(ClF_2^+)$
54	762	$\nu_1(BF_4)$	766 m	$\nu_1(BF_4^-)$
76	788 }	$\nu_1(ClF_2^+)$	798 s	$\nu_1(ClF_2^+)$
100	798 }			
31	808	$\nu_3(ClF_2^+)$	813 sh	$\nu_3(ClF_2^+)$
4	930	$\nu_3(BF_4^-)$	978–1145 s	$\nu_3(BF_4^-)$

the proposed ionic structure of these species. The stability of the ClF_2^+ salts is given by the order $SbF_6^- > AsF_6^- > BF_4^- > PF_6^-$ which agrees with the known Lewis acid strength of the four fluorides.

Vibrational spectrum of the adducts $ClF_3 \cdot AsF_5$ (infrared and Raman) and $ClF_3 \cdot BF_3$ (infrared) were measured by Christe and Sawodny[75]. In the later work it was found that the assignments given in an earlier paper[74] were incorrect. Complete set of the fundamentals was observed and the force constants were calculated. The ionic character of the complexes as well as the bent structure of the ClF_2^+ (bond angle 90–120 degrees) cation were confirmed. The three fundamental vibrations for the cation occur at ν_1 798 cm^{-1}, ν_2 537 cm^{-1} and ν_3 813 cm^{-1}.

Gillespie and Morton recently redetermined the Raman spectra of the compounds $ClF_2^+ \cdot SbF_6^-$, $ClF_2^+AsF_6^-$ and $ClF_2^+BF_4^-$ [76]. The results and the vibrational assignments agree well with the results of Christe and Sawodny[75] with the exception of the 384 cm^{-1} band which is assigned to the ν_2 vibration of the ClF_2^+ ion. The data are shown in Tables 2.13 and 2.14. Since the assignments of the ν_2 vibrations differ, the calculated force constant of Gillespie and Morton (0.60 mdyn $Å^{-1}$) is about half of that calculated by Christe and Sawodny (1.216 mdyn $Å^{-1}$), both calculations were made assuming F—Cl—F angle of 100 degrees.

The authors postulate that the structure of the ClF_2^+ compound is similar to that of the $BrF_2^+SbF_6^-$ with some fluorine bridging between the metal atom and chlorine as shown below.

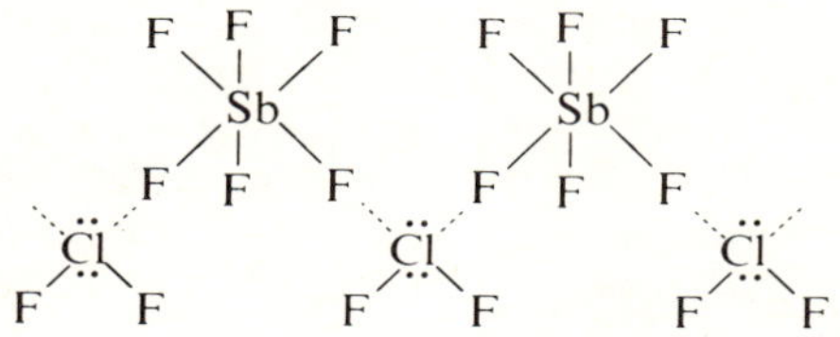

The vibrational spectrum of $ClF_3 \cdot SbF_5$ complex measured by Carter and Aubke[77] shows good agreement with the above results.

Chlorine trifluoride also acts as a fluoride-ion acceptor to give a series of tetrafluorochlorates(III). Christe and Guertin[78] studied electrical conductance of nitrosyltetrafluorochlorate(III) in liquid chlorine trifluoride at -23 °C. The salt is sparsely soluble in ClF_3 and a saturated solution was 0.079 M in concentration. The specific conductance of the solution was $2.1 \times 10^{-7}\ \Omega^{-1}\ cm^{-1}$ as compared with the specific conductance of the pure solvent of $9.2 \times 10^{-9}\ \Omega^{-1}\ cm^{-1}$. The authors conclude that this increase in conductance is indicative of the ionic structure of the complex $NO^+ClF_4^-$.

Vibrational spectra of the tetrafluorochlorates were measured[78–80]. Results obtained in the earlier work were unreliable due to the reactivity and hygroscopicity of the samples. The corrected results[80] did not alter the previous conclusions that the ion is square planar but new vibrational assignments were made and new values for the force constants were calculated.

Chlorine trifluoride was reacted with platinum hexafluoride by condensing the two reactants in a nickel vessel at -196 °C and slowly warming the mixture to room temperature[81]. Chlorine pentafluoride and a bright yellow solid with the composition ClF_3PtF_5 were obtained. The reaction, therefore, is given by,

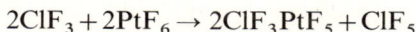

$$2ClF_3 + 2PtF_6 \rightarrow 2ClF_3PtF_5 + ClF_5$$

The ClF_3PtF_5 complex is very hygroscopic and reacts violently with water. Upon heating above 100 °C it decomposes according to the reaction,

$$ClF_3PtF_5 \rightarrow PtF_4 + ClF_3 + \tfrac{1}{2}F_2$$

The composition of the solid does not depend on the relative amounts of the two reagents. However, with an excess of chlorine trifluoride there is evidence for some formation of an unstable higher adduct of ClF_3.

Infrared evidence indicates that the platinum complex may have an ionic structure $ClF_2^+PtF_6^-$.

2.3.4 Chlorine pentafluoride ClF_5

Chlorine pentafluoride is the newest member of the halogen fluoride family and was first reported by Smith in 1963 [82]. Infrared and Raman spectra of this compound in the liquid and gaseous states were observed by Begun *et al.*[83]. The molecule was found to have a square pyramidal structure similar to that of BrF_5, IF_5 and $XeOF_4$.

While the original preparation by Smith involved heating a 14:1 mixture of fluorine and chlorine to 350 °C at 250 atm, Gatti *et al.*[84] showed that it can be very conveniently prepared by reacting 2:1 mixture of fluorine and chlorine trifluoride at 30 °C and atmospheric pressure. It can also be prepared by the fluorination of an alkali tetrafluorochlorates(III)[85]. As was seen above, chlorine pentafluoride is also formed in the reaction of chlorine trifluoride with platinum hexafluoride.

Only the most important physical properties of chlorine pentafluoride have been determined. The data are given in Table 2.15. The exact value of

Table 2.15 Physical constants of chlorine pentafluoride

Property	*Value*	*Reference*
Melting point	−103 °C±4 °C	85
Melting point	−93 °C	84
Boiling point	−14 °C	85
Boiling point	−12.9 °C	84
Heat of vapourisation	5.74 kcal mol^{-1}	84
Trouton's constant	22 cal deg^{-1} mol^{-1}	84
Trouton's constant	21.8 cal deg^{-1} mol^{-1}	85
Heat of formation	−56.0 kcal mol^{-1}	86

the melting point seems to be in doubt. Trouton's constant of 22 shows that the liquid is not associated. The variation of the vapour pressure with temperature is given by the equation,

$$\log P_{(mm)} = 7.2683 - \frac{1137.16}{T}$$

The compound is monomeric in the vapour state.

Bauer and Sheenan[86] studied the equilibrium

$$ClF_3(g) + F_2(g) \rightleftarrows ClF_5(g)$$

between 211 and 271 °C. The values of the equilibrium constant are listed in Table 2.16. The least squares treatment of the data gives

$$\ln K_p\,(\text{atm}^{-1}) = \frac{9175}{T} - 21.30$$

The enthalpy and entropy values for the reaction in the temperature range near 240 °C were found to be, $\Delta H = -18.2 \pm 0.9$ kcal and $\Delta S = -42.3 \pm 1.8$ cal deg^{-1} mol^{-1}. The heat of formation of gaseous ClF_5 was calculated to be -56.0 kcal mol^{-1}

Kinetics of dissociation of chlorine pentafluoride in the gaseous phase was studied by Blauer *et al.* in the 520–1120 K temperature interval[87].

Infrared spectrum of ClF_5 was observed in the sodium chloride region[88] but vibrational assignments have not been made.

Chlorine pentafluoride is a vigorous fluorinating agent but somewhat less corrosive than the trifluoride[85]. It is soluble in fluorinated plastics such as Kel-F and Teflon, and it reacts vigorously with water according to the equation

$$ClF_5 + 2H_2O \rightarrow FClO_2 + 4HF$$

Chlorine pentafluoride forms 1:1 addition compounds with arsenic pentafluoride and antimony pentafluoride[89]. Attempts to prepare an adduct with boron trifluoride were unsuccessful, probably due to relative weakness of

Table 2.16 Equilibrium constant for the reaction $ClF_3(g) + F_2(g) \rightleftarrows ClF_5(g)$ [86]

Temperature K	$K_p\,(\text{atm}^{-1}) \times 10^2$
484	8.52
486	7.80
493	7.40
497	7.06
516	3.10
518	2.58
542	1.39
544	1.04

the Lewis acid character in the latter compound. Chlorine pentafluoride does not appear to react with Lewis bases, indicating the difficulty of expanding the coordination number of chlorine beyond five.

Complexes of chlorine pentafluoride with chlorine trifluoride and antimony pentafluoride were studied by ^{19}F nuclear magnetic resonance by Bantov *et al.*[90]. The n.m.r. spectrum of pure ClF_5 confirms the tetragonal-pyramid structure of the molecule. The chlorine pentafluoride and chlorine trifluoride mixtures gave only the additive spectra of the two compounds indicating absence of interaction.

Spectra of binary mixtures with antimony pentafluoride containing 12.7, 16.5 and 25 mole per cent chlorine pentafluoride at 20 °C gave three resonance lines whose intensities were proportional to ClF_5 concentration. The spectrum can be interpreted as indicating a fluorine-bridged structure of the complex. Likewise, the very low electrical conductance of the solutions of complex in antimony pentafluoride ($10^{-6}\Omega^{-1}$ cm^{-1}) which does not increase with increasing temperature, seems to be additional evidence for the non-ionic configuration of the complex.

2.3.5 Bromine trifluoride, BrF_3

Bromine trifluoride is a highly self-ionised liquid as evidenced by its high specific conductance which has been recently re-examined in the 10–50 °C range using a very carefully purified product[91]. The values ranged from $8.34 \times 10^{-3}\ \Omega^{-1}\ cm^{-1}$ at 10 °C to $7.31 \times 10^{-3}\ \Omega^{-1}\ cm^{-1}$ at 50 °C with the 25 °C value being $8.03 \times 10^{-3}\ \Omega^{-1}\ cm^{-1}$. It seems that the lower values reported earlier[92] were due either to impurities or to errors in measurements[93].

Recent results on the infrared and Raman spectra of gaseous bromine trifluoride[69] are shown in Table 2.17.

Chretien and Martin[94,95] studied reactions of alkali metal fluorides with uranium fluorides in bromine trifluoride by cryoscopic and conductometric

Table 2.17 Observed frequencies and assignments for BrF_3 (from Selig *et al.* by courtesy of *The Journal of Chemical Physics*)

Assignment	*Raman*	*Infrared*
$\nu_1(a_1)$	675 p, s	668, 682 } s
$\nu_2(a_1)$	552 p, vs	547, 557 } w
$\nu_3(a_1)$	233* p, w	242 s
$\nu_4(b_1)$	612 vvw	604, 614, 621 } vs
$\nu_5(b_1)$	not observed	342, 350, 359 } vw
$\nu_6(b_2)$	not observed	242 s
$\nu_2+\nu_4 (= 1166)$		1162 w
$\nu_1+\nu_4 (=1289)$		1287 vw
$2\nu_1 (= 1350)$		1340 vw

* For liquid BrF_3

techniques. With the exception of lithium fluoride, the alkali metal fluorides are quite soluble in BrF_3. In general, electrical conductance of alkali fluorides increases in the order NaF > CsF > KF > RbF. The results are interpreted as indicating a progressive decrease in cation solvation. It seems somewhat unusual, however, that the caesium ion would be solvated in bromine trifluoride to a greater extent than the potassium ion.

Sodium fluoride reacts with uranium tetrafluoride to give $NaUF_5$ which precipitates out of solutions. Authors postulate similar reaction between rubidium fluoride and uranium tetrafluoride, but the product was not isolated. With caesium fluoride the authors postulate formation of complexes CsU_2F_9 and $Cs_2U_5F_{22}$, but the existence of these compounds has not been unambiguously characterised.

Martin also reports on electrical conductance studies in bromine trifluoride solutions containing xenon tetrafluoride[96]. Addition of varying amounts of phosphorus pentabromide, arsenic(III) oxide, and antimony(III) oxide produced an increase in conductance which is interpreted as evidence for the formation of xenon tetrafluoride complexes. Breaks in the conductance —mole ratios plots were observed and from these the stoichiometry of the complexes were deduced. In some cases, however, the breaks are rather minimal. Possible increase in the conductance of bromine trifluoride solution upon addition of PBr_5, As_2O_3 or Sb_3O_2 is not discussed.

Solutions of xenon difluoride and xenon tetrafluoride in bromine trifluoride do not show any enhancement of electrical conductance indicating the absence of a fluoride ion transfer reaction[97].

Potassium, rubidium and caesium trichlorotrifluoroplatinates(IV), $M_2PtCl_3F_3$, were prepared by a drop-wise addition of bromine trifluoride to solid hexachloroplatinates(IV) at room temperature[98]. Under the same conditions, reaction with the hexabromo- or heptaiodoplatinates was vigorous and the products appeared to be nonstoichiometric. Analysis showed that they contained a few per cent of bromine or iodine. When chloro-, bromo- or iodoplatinates were treated with boiling bromine trifluoride (127 °C), hexafluoroplatinates were formed.

Brown *et al.*[99] prepared addition compounds, $2BrF_3{\cdot}GeF_4$, $BrF_3{\cdot}2BF_3$ and $BrF_3{\cdot}BF_3$ by direct combination of the constituent fluorides. The infrared spectrum of the first adduct was obtained and the 600 cm^{-1} band, characteristic of the GeF_6^{2-} anion, was not found. It seems, therefore, that the ionic structure $(BrF_2^+)_2(GeF_6^{2-})$ for the complex appears to be ruled out.

The crystal structure of $BrF_3{\cdot}SbF_5$ complex[100, 101] was determined by Edwards and Jones. The structure is in agreement with the ionic formula $BrF_2^+SbF_6^-$, but, there is some fluorine bridging between the bent BrF_2^+ ion and the distorted octahedral SbF_6^- ion. While the bromine to fluorine bridge bond is considerably longer than the terminal bond (2.29 Å *v.* 1.70 Å), the authors consider that the former still represents a considerable degree of interaction and that the solid is formed of endless chains of BrF_2^+ and SbF_6^- units linked by fluorine bridges.

On the basis of the vibrational spectra of bromine trifluoride adducts with arsenic pentafluoride, germanium tetrafluoride and antimony pentafluoride, Christe and Schack conclude[102] that the complex $BrF_3{\cdot}SbF_5$ is ionic and consists of discrete BrF_2^+ and SbF_6^- ions. The anion, however, does not have O_h symmetry. The vibrational spectra of $BrF_3{\cdot}AsF_5$ and of $2BrF_3{\cdot}GeF_4$ do not yield unambiguous information on the structures of the two complexes. However, in the former case, and possibly in the second, bands attributed to the BrF_2^+ cation were detected. Thus ionic structures of these compounds $BrF_2^+AsF_6^-$ and $(BrF_2^+)_2GeF_6^{2-}$ with distorted octahedral anions cannot be ruled out.

A similar study of Carter and Aubke[77] is less positive about the ionic structure of $BrF_3{\cdot}SbF_5$ adduct. While BrF_2^+ and SbF_6^- species may be present in the lattice, the evidence indicates a strong cation–anion interaction.

Acid–base equilibria in liquid bromine trifluoride was studied spectroscopically by Surles *et al.*[104]. Infrared and Raman spectra of liquid bromine trifluoride as well as alkali metal fluorides, antimony pentafluoride and arsenic pentafluoride solutions in BrF_3 were obtained in the 700–200 cm^{-1}

region, and compared with the spectra of the solid complexes $MBrF_4$, $BrF_2^+SbF_6^-$ and $BrF_2^+AsF_6^-$. Solutions of potassium, rubidium and caesium fluorides showed the characteristic Raman bands of the BrF_4^- anion at 528 cm^{-1} (ν_1), 249 cm^{-1} (ν_3) and 455 cm^{-1} (ν_5). In addition, infrared bands were observed at 302 cm^{-1} (ν_2) and 570 cm^{-1} (ν_6). The spectra of pure liquid bromine trifluoride also show the ν_1, ν_5 and ν_3 bands of BrF_4^- ion, although with much smaller intensity than for the alkali fluoride solutions. Raman spectra of AsF_5 and SbF_5 in liquid bromine trifluoride show a sharp increase in the intensity of the 625 cm^{-1} solvent band and a decrease in the intensity of the 528 cm^{-1} band of BrF_4^-. It is assumed that the 625 cm^{-1} band is due to the solvated cation $BrF_2^+ (BrF_3)_n$.

Bromine trifluoride mixtures with hydrogen fluoride[105] and with chlorine trifluoride[106] were also studied by Surles and co-workers. Concentrations of BrF_4^- and BrF_2^+ ions in these solutions were obtained from the integrated intensities of their vibrational bands. Spectral evidence indicates that in this system hydrogen fluoride acts as a fluoride-ion acceptor according to the reaction

$$BrF_3 + HF \rightleftarrows HF_2^- + BrF_2^+$$

The equilibrium constant for this reaction at 25 °C was estimated to be $\sim 3 \times 10^{-3}$. It was also determined that the self-ionisation constant for bromine trifluoride at 25 °C $K_s = (BrF_2^+)(BrF_4^-)$ is ~ 0.8. Thus it appears that bromine trifluoride is very extensively self-ionised.

Conductance of BrF_3–HF mixture throughout the entire concentration range showed that initial addition of HF to BrF_3 causes a decrease in conductance, with a minimum occurring at ~ 70 mol per cent BrF_3. Further addition of HF produces an increase in its specific conductance until a maximum is reached at 30 mole per cent BrF_3 after which the conductance decreases rapidly to the specific conductance of pure hydrogen fluoride.

A similar study on the BrF_3–ClF_3 system indicates that in this case chlorine trifluoride is a stronger fluoride-ion donor than bromine trifluoride. The equilibrium in the mixture is

$$ClF_3 + BrF_3 \rightleftarrows ClF_2^+ + BrF_4^-$$

and the equilibrium constant at 25 °C is estimated to be $\sim 10^{-4}$.

An electrical conductance study of the BrF_3–ClF_3 system has been carried out. The results in chlorine trifluoride-rich mixtures agree with the data of Toy and Cannon[92]. However, below 32 mol per cent ClF_3 there is a sharper increase in the conductance than predicted by the extrapolation of Toy and Cannon.

2.3.6 Bromine pentafluoride, BrF_5

Chemical properties of bromine pentafluoride have been recently studied by Meinert and Gross[107]. Solutions of BrF_5 in acetonitrile and in trichlorofluoromethane are reported to be stable at room temperature provided that the preparation of these solutions is carried out with chilled solvents.

While direct reaction of bromine pentafluoride with water occurs with explosive violence, it can be moderated in acetonitrile solutions. Under these conditions, the hydrolysis reaction is,

$$BrF_5 + 3H_2O \rightarrow HBrO_3 + 5HF$$

Bromine pentafluoride reacts with pyridine in trichlorofluoromethane solutions at −40 °C to give a sparsely soluble colourless adduct. The adduct is very unstable and attempts to isolate it were unsuccessful. Under similar conditions bromine pentafluoride reacts with dioxane but complex formation was not observed and the reaction products were not identified. Caesium fluoride reacts with BrF_5 in acetonitrile solutions to give the salt $CsBrF_6$. Xenon difluoride is soluble in liquid bromine pentafluoride without reaction and can be recrystallised from the solvent.

Meinert *et al.*[108] studied the infrared spectrum of $BrF_5 \cdot 2SbF_5$ complex dispersed in Kel-F oil, as well as the ^{19}F nuclear magnetic resonance spectrum of the solid compounds and of the melt. Attempts to study the n.m.r. spectrum in solutions were unsuccessful due to the rectivity of the compound. The n.m.r. spectra of the solid and the melt were quite different, indicating a change in the structure of the complex upon melting.

Reactions of the complex with caesium hexafluorobromate(V) (at 65 °C in the melt) and with potassium fluoride gave bromine pentafluoride and the corresponding hexafluorostannate(V),

$$BrF_4Sb_2F_{11} + 2CsBrF_6 \rightarrow 3BrF_5 + 2CsSbF_6$$
$$BrF_4Sb_2F_{11} + 2KF \rightarrow BrF_5 + 2KSbF_6$$

Attempts to prepare BrF_5 addition compounds with boron trifluoride, phosphorus pentafluoride and titanium tetrafluoride were unsuccessful.

Tin(IV) fluoride reacts with bromine pentafluoride to give a solid white addition product $2BrF_5 \cdot SnF_4$ [103]. While the compound could have an ionic structure $(BrF_4^+)_2(SnF_6^{2-})$, its Mössbauer spectrum shows the absence of O_h symmetry of the SnF_6^{2-} anion. The present evidence, therefore, is in favour of a bridge structure.

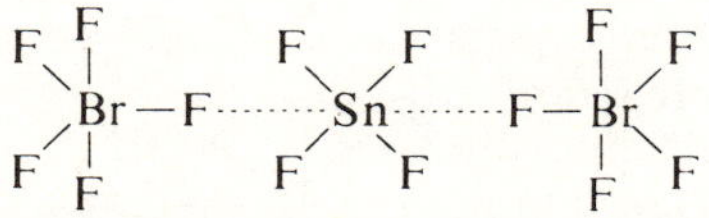

Reaction of platinum hexafluoride with bromium pentafluoride gave a black viscous liquid which analysed to $BrPt_2F_{14}$ [81]. Continuous pumping gave a red-brown product with composition close to that of PtF_4.

When bromine pentafluoride is reacted with antimony pentafluoride in the presence of bromine, a scarlet solid melting at 69 °C is obtained[109]. Crystallographic studies show that in this case the structure is ionic (Br_2^+) $(Sb_3F_{16}^-)$ and, therefore, in contrast to other halogen fluoride complexes, the cation does not include any fluorine.

2.3.7 Iodine monofluoride, IF

Recent work on iodine monofluoride is rather sparse. Schmeisser *et al.*[110] succeeded in preparing pure iodine monofluoride by bubbling fluorine through a solution of iodine in trichlorofluoromethane at −45 °C. A colourless solid compound was obtained which disproportionates to I_2 and IF_5 at −14 °C.

Iodine monofluoride dissolves to some extent in 20% oleum to give deep blue solutions which show absorption bands at 15 750, 20 000 and 24 500 cm^{-1} indicative of the I_2^+ cation.

Solid 1:1 complexes of iodine monofluoride with pyridine, quinoline and 2,2′-bipyridine were prepared at −78 °C in trichlorofluoromethane solutions. The iododipyridinium fluoride, IPy_2F previously reported by Schmidt and Meinert[111] could not be isolated but rather a 1:1 compound Py·IF with a m.p. of 110 °C was prepared.

Attempts to prepare caesium iododifluoride, $CsIF_2$, by reacting equimolar amounts of caesium fluoride and iodine monofluoride in acetonitrile solutions at −40 °C were unsuccessful. Solid compound identified by analysis as Cs_3IF_6 was isolated. On the other hand, reaction of IF with caesium chloride, under identical conditions led to the formation of $CsIF_4$ and $CsICl_2$. Since the existence of IF_6^{3-} anion is somewhat dubious, it may be possible that Cs_3IF_6 is a 'double salt', $CsIF_4{\cdot}2CsF$.

2.3.8 Iodine trifluoride, IF_3

Schmeisser *et al.*[112] postulate that the solid iodine trifluoride is not monomeric, but exists either in the form of a salt $IF_2^+IF_4^-$ or of a fluorine bridged polymer.

The reactivity of iodine trifluoride towards organic compounds is somewhat smaller than that of other halogen fluorides and 1:1 solid addition compounds have been prepared in trichlorofluoromethane solutions at −78 °C with quinoline, acetonitrile and pyridine[112]. In the latter case also a solid 2:1 compound $2Py{\cdot}IF_3$ was obtained. With the exception of acetonitrile adduct the addition compounds are stable at room temperature.

The reaction of iodine trifluoride with alkali fluorides in acetonitrile solutions at −50 °C gave potassium, rubidium and caesium tetrafluoroiodates(III). Similar reaction with nitrosylfluoride in CCl_3F at −78 °C yielded nitrosyliumtetrafluoroiodate(III), $NOIF_4$.

Difluoroiodonium(III) compounds IF_2AsF_6 and IF_2SbF_6 were obtained by reacting iodine trifluoride with arsenic and antimony pentafluoride. The ^{19}F resonance of the latter compound showed two resonance lines with relative intensities of 2.6:1.

Reactions of iodine trifluoride with caesium chloride in acetonitrile solutions appears to be quite complex and led to the formation of several polyhalide salts. The overall reaction is given by,

$$6IF_3 + 6CsCl \rightarrow 3CsIF_4 + CsIF_6 + CsICl_2 + CsICl_4$$

On the other hand, the reaction of iodine trifluoride with alkali metal fluorides in acetonitrile at −45 °C gave the colourless addition products $KI{\cdot}IF_3$, $RbF{\cdot}IF_3$ and $CsF{\cdot}IF_3$ [113]. Addition compounds with arsenic pentafluoride and antimony pentafluoride were prepared in liquid arsenic pentafluoride at −78 °C. The fact that the $IF_3{\cdot}SbF_5$ complex can be formed in arsenic pentafluoride solution is indicative of the vastly superior Lewis acid strength of antimony pentafluoride.

Iodine trifluoride forms 1:1 complexes with pyrazine, 2,2′-bipyridine and tetrafluorophthalic anhydride[114]. The first two compounds are quite stable and decompose respectively at 138 °C and 142 °C. They are slightly soluble in acetonitrile, acetone and pyridine. The tetrafluorophthalic anhydride begins to decompose at −12 °C. It is interesting to note that this is the first complex of iodine trifluoride where the donor atom is oxygen. The reaction with caesium fluoride leads to the formation of a complex salt Cs_3IF_6.

2.3.9 Iodine pentafluoride, IF_5

Recent work on iodine pentafluoride is mainly concerned with the complex compounds of this species and with the structure of these complexes.

At room temperature iodine pentafluoride does not react with chlorine-containing interhalogens (ICl, ICl_3, BrCl) or with chlorine[115]. However, upon addition of pyridine to a solution of iodine pentafluoride and a chlorine halide in trifluorotrichloroethane, yellow adducts are precipitated. The reaction is given by the equation,

$$XCl + 2Py + IF_5 \rightarrow XPy_2ClIF_5$$

with X = I, ICl_2, Br or Cl. It appears that the complex is ionic with the structure $XPy_2^+ IClF_5^-$.

Iodine pentafluoride also reacts under similar conditions with tetra-alkyammonium chlorides in 1:1 mol ratios. The authors postulate that the first step involves the formation of a tetra-alkylammonium fluoride and an unstable, and hitherto unknown, interhalogen IF_4Cl

$$R_4NCl + IF_5 \rightarrow R_4NF + IF_4Cl$$

The tetra-alkylammonium fluoride reacts with iodine pentafluoride to give the corresponding hexafluoroiodate(V) salt

$$R_4NF + IF_5 \rightarrow R_4NIF_6$$

while the postulated IF_4Cl presumably reacts with the chloride to give tetrafluorodichloroiodate(V)

$$R_4NCl + IF_4Cl \rightarrow R_4NIF_4Cl_2$$

Meinert *et al.*[116] obtained an addition compound of xenon difluoride with iodine pentafluoride, $XeF_2 \cdot 2IF_5$ by dissolving XeF_2 in IF_5 and concentrating the solution at 5 °C. The authors also claim that xenon tetrafluoride dissolved in liquid IF_5 converts the latter to iodine heptafluoride. However, those results are disputed by Nikolaev and co-workers who studied solutions of xenon tetrafluoride in iodine pentafluoride[117, 118]. Xenon tetrafluoride easily dissolves in iodine pentafluoride at room temperature. When excess liquid is removed under vacuum, a solid compound remains which analyses to $XeF_4 \cdot IF_5$. Thermogravimetric analysis shows that this compound is stable to 92 °C. At higher temperatures it decomposes to IF_5 and XeF_4. The ^{19}F nuclear magnetic resonance studies of the adduct in acetonitrile as well as

XeF_4 solutions in IF_5 show that while there is some interaction between XeF_4 and IF_5 in solution, the resulting species is a molecular compound rather than ionic. No evidence was found for the presence of iodine heptafluoride in the solution.

Study of I_2–IF_5 system in Freon 11[119] showed that between −80 °C and −30 °C a product is formed in solution which shows an absorption band at 460 nm. The authors postulate that the compound is an addition product $I_2 \cdot 3IF_5$ although no direct evidence for this stoichiometry is given. The addition of pyridine or of an alkali metal fluoride to the above solution results in the following reactions

$$I_2 \cdot 3IF_5 + Py \rightarrow IPy_2^+ IF_6$$

$$I_2 \cdot 3IF_5 + 5MF \rightarrow 5MIF_4$$

The possibility that the addition compound has a 1:1 stoichiometry, therefore, cannot be excluded.

The addition of iodine to an iodine pentafluoride solution in trifluorotrichloroethane (Freon 113) at temperatures above −30 °C, apparently produces $I^+IF_6^-$ [120]. Addition of pyridine results in the precipitation of dipyridineiodine(I) hexafluoroiodate(V), $IPy_2^+IF_6^-$. The compound is stable and melts at 166 °C. It is fairly soluble in acetonitrile and can be recrystallised from this solvent. The conductance data and the infrared spectra both indicate the ionic structure given above rather than the stoichiometrically equivalent $IF_3 \cdot Py$. The compound can also be prepared in the reactions

$$IPy_2F + IF_5 \rightarrow IPy_2IF_6 \qquad \text{(in Freon 113)}$$

$$KIF_6 + IPy_2NO_3 \rightarrow IPy_2IF_6 + KNO_3 \qquad \text{(in acetonitrile)}$$

Chemical properties of the hexafluoroiodate(V) anion, in the form of its tetraethylammonium salt, was studied by Klamm and Meinert[121]. The salt is soluble in acetonitrile, but in moist acetonitrile it is hydrolysed, presumably according to the reaction

$$Et_4NIF_6 + H_2O \rightarrow Et_4NIOF_4 + 2HF$$

The tetrafluoro-oxyiodate salt, however, has not been isolated. Infrared spectra of the reaction products show characteristic bands of $IO_2F_2^-$ anion as well as a band at 880–890 cm^{-1} which is ascribed to the IOF_4^- anion.

Hexafluoroiodate reacts with HCl, $SOCl_2$ and SO_2Cl_2 to give ICl_4^- anions. Colorimetric and conductrometric titrations of Et_4NIF_6 with Et_4NCl in dilute acetonitrile solutions show a 1:2 reaction and the authors postulate formation of a new ternary polyhalide anion $IF_4Cl_2^-$ according to the reaction

$$IF_6^- + 2Cl^- \rightarrow IF_4Cl_2^- + 2F^-$$

Vibrational spectrum of IF_6^- was studied by Christe and co-workers[122–124] and by Klamm *et al.*[125]. The latter authors conclude that the ion has C_{2v} symmetry with four fluorine atoms arranged in a square around the iodine atom and the other two fluoride atoms above the plane and bonded less strongly to the iodine.

2.3.10 Iodine heptafluoride, IF_7

Vibrational spectrum of iodine heptafluoride was re-examined by Claassen *et al.* and compared with the spectrum of rhenium heptafluoride[126]. Analysis of the data favours D_{5h} symmetry for both heptafluorides.

Deflection of molecular beams of iodine heptafluoride in a heterogeneous electric field showed that the molecule does not have a rigid polar structure and that its dipole moment is less than 0.3 D[127,128].

An adduct of iodine heptafluoride and arsenic pentafluoride was prepared by mixing the two components at −196 °C, slowly raising the temperature to 0 °C and pumping off the unreacted IF_7 and AsF_5[129]. A solid white 1:1 complex was obtained which shows a tendency to dissociate into component parts at room temperature. Dissociation pressure measured in the 25.2–41.5 °C temperature range is given by the equation:

$$\log P\,(\text{mm}) = 16.418 - \frac{4800}{T}$$

The solid has a face-centred cubic lattice with $a = 9.49$ Å. Unit cell contains four molecules. The enthalpy of formation of the complex is estimated to be −43.9 kcal mol^{-1}. Vibrational spectrum indicates ionic structure $(IF_6^+)(AsF_6^-)$. The broad line nuclear magnetic resonance spectrum of the complex was studied[130]. The chemical shifts observed for ^{129}I and ^{19}F confirm the ionic structure of the complex. Mössbauer spectra of the complex[131] also indicate highly ordered arrangement of fluorine atoms around the iodine atom which is consistent with the proposed octahedral structure of the cation.

2.3.11 Radon solutions in halogen fluorides

Radon solutions in several halogen fluorides (ClF, ClF_3, ClF_5, BrF_3, BrF_5, IF_5, and IF_7) as well as in some binary mixtures of these compounds have been carefully investigated by Stein[132,133]. With the exception of iodine pentafluoride, the other halogen fluorides converted radon into an involatile species. Evaporation of the resulting solutions to dryness yielded very small amounts of a solid residue (~0.2 μg). Attempts to analyse this residue by mass spectrometry thus far have been unsuccessful.

Electromigration of the oxidised radon species was studied in bromine trifluoride solutions as well as in HF–BrF_3 mixtures. It was established that the oxidised species migrates towards the cathode. No appreciable migration was observed in a solution of 5 mole per cent of BrF_3 in IF_5. However, if the solution was saturated with caesium fluoride in order to increase the electrical conductivity of the medium, radon is again found to be migrating to the cathode. It is evident that the oxidised species must be a radon fluoride cation but its exact nature is still under investigation.

References

1. Evans, J. C. and Lo, G. Y-S. (1966). *J. Chem. Phys.*, **44,** 3638
2. Evans, J. C. and Lo, G. Y-S. (1966). *J. Chem. Phys.*, **44,** 4356
3. Evans, J. C. and Lo, G. Y-S. (1967). *Inorg. Chem.*, **6,** 1483
4. Popov, A. I. (1967). *Polyhalogen Complex Ions,* in *Halogen Chemistry.* Gutmann, V. ed., Vol. I (New York: Academic Press)
5. Maki, A. G. and Forneris, R. (1967). *Spectrochim. Acta,* **23A,** 867
6. Ginn, S. G. W. and Wood, J. L. (1965). *Chem. Commun.*, 262
7. Robin, M. B. (1964). *J. Chem. Phys.*, **40,** 3369
8. Hayward, G. C. and Hendra, P. J. (1967). *Spectrochim. Acta,* **23A,** 2309
9. Parrett, F. W. and Taylor, N. J. (1970). *J. Inorg. Nucl. Chem.*, **32,** 2458
10. Stammreich, H. and Kawano, Y. (1968). *Spectrochim. Acta,* **24A,** 899
11. Forneris, R., Hiraishi, J., Miller, F. and Uehara, M. (1970). *Spectrochim. Acta,* **26A,** 581
12. Nelson, L. Y. and Pimentel, G. C. (1968). *Inorg. Chem.*, **7,** 1695
13. Ginn, S. G. W. and Wood, J. L. (1966). *Trans. Faraday Soc.*, **62(4),** 777
14. Haque, I. and Wood, J. L. (1967). *Spectrochim. Acta,* **23A,** 959

14a. Popov, A. I. and Pflaum, R. T. (1957). *J. Amer. Chem. Soc.*, **79,** 570

15. Yarwood, J. and Person, W. B. (1968). *J. Amer. Chem. Soc.*, **90,** 3930

15a. Haque, I. and Wood, J. L. (1967). *Spectrochim. Acta,* **23A,** 2523

16. Yarwood, J. (1969). *Trans. Faraday Soc.*, **65,** 934
17. Yagi, Y., Popov, A. I. and Person, W. B. (1967). *J. Phys. Chem.*, **71,** 2439
18. Person, W. B., Humphrey, R. E., Deskin, W. A. and Popov, A. I. (1958). *J. Amer. Chem. Soc.*, **80,** 2049
19. Watari, F. (1967). *Spectrochim. Acta,* **23A,** 1917
20. Clark, R. J. H. and Williams, C. S. (1965). *Inorg. Chem.*, **4,** 350
21. Wynter, C. I., Hill, J., Bledsoe, W., Shenoy, G. K. and Ruby, S. L. (1969). *J. Chem. Phys.*, **50,** 3872
22. Carpenter, G. B. (1966). *Acta Crystallogr.*, **20,** 330
23. Migchelsen, T. and Vos, A. (1967). *Acta. Crystallogr.*, **23,** 796
24. Alcais, P., Rothenberg, F. and Dubois, J. E. (1967). *J. Chim. Phys. Physicochim. Biol.*, **64,** 1818
25. Mironov, V. E. and Lastovkina, N. P. (1966). *Zh. Neorg. Khim.*, **11,** 580
26. Mironov, V. E. and Lastorkina, N. P. (1967). *Zh. Fiz. Khim.*, **41,** 1850
27. Parker, A. J. (1966)(A). *J. Chem. Soc.*, 220
28. Povarov, Yu. M. and Barbasheva, I. E. (1967). *Elektrokhimiya,* **3,** 745
29. Breant, M. and Sinicki, C. (1965). *Compt. Rend.*, **260,** 5016
30. Siniki, C. and Breant, M. (1967). *Bull. Soc. Chim. Fr.*, 3080
31. Marchon, J. C. (1968). *C. R. Acad. Sci. Ser. C,* **267,** 1123
32. Nelson, I. V. and Iwamoto, R. T. (1964). *J. Electroanal. Chem.*, **7,** 218
33. Guidelli, R. and Piccardi, G. (1967). *Electrochim. Acta,* **12,** 1085
34. Courtot-Coupez, J., Madec, C. and Le Demezet, M. (1969). *C. R. Acad. Sci. Ser. C,* **268,** 1856
35. Lormeau, S. and Mannebach, M. H. (1966). *Bull. Soc. Chim. Fr.*, 2576
36. Durand, G., Tremillon, B. (1970). *Anal. Chim. Acta,* **49,** 135
37. Courtot-Coupez, J. and L'Her, M. (1968). *C. R. Acad. Sci. Ser. C,* **266,** 1286
38. Barraque, C., Vedel, J. and Tremillon, B. (1969). *Anal. Chim. Acta,* **46,** 263
39. Dubois, J. E. and Garnier, F. (1965). *Bull. Soc. Chim. Fr.*, **6,** 1715
40. Benoit, R. L. and Guay, M. (1968). *Inorg. Nucl. Chem. Lett.*, **4,** 215
41. Deneux, M. and Benoit, R. L. (1970). *Can. J. Chem.*, **48,** 674
42. Topol, L. E. (1968). *Inorg. Chem.*, **7,** 451
43. Augdahl, E. and Klaeboe, P. (1965). *Acta. Chem. Scand.*, **19,** 807
44. Surles, T. and Popov, A. I. (1969). *Inorg. Chem.*, **8,** 2049
45. Baenziger, N. C., Nelson, A. D., Tulinsky, A., Bloor, J. H. and Popov, A. I. (1967). *J. Amer. Chem. Soc.*, **89,** 6463
46. Hassel, O. and Roemming, C. (1956). *Acta Chem. Scand.*, **10,** 696
47. Wiebenga, E. H. and Kracht, D. (1969). *Inorg. Chem.*, **8,** 738
48. Pasternak, M. and Sonnino, T. (1968). *J. Chem. Phys.*, **48,** 1997
49. Campbell, A. N. and Shemilt, L. W. (1946). *Trans. Roy. Soc. Can.*, **40,** 17
50. Brown, R. D. and Nunn, E. K. (1966). *Aust. J. Chem.*, **19,** 1567

51. Bowmaker, G. A. and Hacobian, S. (1968). *Aust. J. Chem.*, **21,** 551
52. Matsui, Y. and Date, Y. (1970). *Bull. Chem. Soc. Jap.*, **43,** 2828
53. Macagno, V. A., Giordano, M. C. and Arvia, A. J. (1969). *Electrochim. Acta*, **14,** 335
54. Barbasheva, I. E., Povarov, Yu. M. and Lukovtsev, P. D. (1967). *Elektrokhimiya*, **3,** 1149
55. Povarov, Yu. M., Barbasheva, I. E., and Lukovtsev, P. D. (1967). *Elektrokhimiya*, **3,** 1202
56. Glew, D. N. and Hames, D. A. (1969). *Can. J. Chem.*, **47,** 4651
57. Macdonald, D. D. and Wright, G. A. (1970). *Can. J. Chem.*, **48,** 2847
58. Becke-Goehring, M. and Hormuth, P. B. (1969). *Z. Anorg. Allg. Chem.*, **369,** 105
59. Yagi, Y. and Popov, A. I. (1965). *Inorg. Nucl. Chem. Lett.*, **1,** 21
60. Yagi, Y. and Popov, A. I. (1967). *J. Inorg. Nucl. Chem.*, **29,** 2223
61. Stein, L. (1967). *Physical and Chemical Properties of Halogen Fluorides*, in *Halogen Chemistry*. Gutmann, V., ed. Vol. I (New York: Academic Press)
62. Christe, K. O. and Sawodny, W. (1969). *Inorg. Chem.*, **8,** 212
63. Gillespie, R. J. and Morton, M. J. (1970). *Inorg. Chem.*, **9,** 811
64. Christe, K. O. and Guertin, J. P. (1965). *Inorg. Chem.*, **4,** 905
65. Christe, K. O. and Guertin, J. P. (1965). *Inorg. Chem.*, **4,** 1785
66. Christe, K. O., Sawodny, W. and Guertin, J. P. (1967). *Inorg. Chem.*, **6,** 1159
67. Aubert, J., Carles, M. and Bethuel, L. (1967). *Chim. et Ind.*, **98,** 661
68. Reynes, J., Carles, M. and Aubert, J. (1970). *J. Chim. Phys.*, **67,** 680
69. Selig, H., Claassen, H. H. and Holloway, J. H. (1970). *J. Chem. Phys.*, **52,** 3517
70. Bougon, R., Carles, M. and Aubert, J. (1967). *C. R. Acad. Sci., Ser. C*, **265,** 179
71. Ellis, J. F. and Forrest, C. W. (1960). *J. Inorg. Nucl. Chem.*, **16,** 150
72. Shrewsberry, R. C. and Williamson, E. L. (1966). *J. Inorg. Nucl. Chem.*, **28,** 2535
73. Luce, M. and Hartmanshenn, O. (1967). *J. Inorg. Nucl. Chem.*, **29,** 2823
74. Christe, K. O. and Pavlath, A. E. (1965). *Z. Anorg. Allg. Chem.*, **335,** 210
75. Christe, K. O. and Sawodny, W. (1967). *Inorg. Chem.*, **6,** 313
76. Gillespie, R. J. and Morton, M. J. (1970). *Inorg. Chem.*, **9,** 616
77. Carter, H. A. and Aubke, F. (1970). *Can. J. Chem.*, **48,** 3456
78. Christe, K. O. and Guertin, J. P. (1966). *Inorg. Chem.*, **5,** 473
79. Christe, K. O. and Sawodny, W. (1968). *Z. Anorg. Allg. Chem.*, **357,** 125
80. Christe, K. O. and Sawodny, W. (1970). *Z. Anorg. Allg. Chem.*, **374,** 306
81. Gortsema, F. P. and Toeniskoetter, R. H. (1966). *Inorg. Chem.*, **5,** 1925
82. Smith, D. F. (1963). *Science*, **141,** 1039
83. Begun, G. M., Fletcher, W. H. and Smith, D. F. (1965). *J. Chem. Phys.*, **42,** 2236
84. Gatti, R., Krieger, R. L., Sicre, J. E. and Schumacher, H. J. (1966). *J. Inorg. Nucl. Chem.*, **28,** 655
85. Pilipovich, D., Maya, W., Lawton, E. A., Bauer, H. F., Sheehan, D. F., Ogimachi, N. N., Wilson, R. D., Gunderloy, F. C., Jr. and Bedwell, V. E. (1967). *Inorg. Chem.*, **6,** 1918
86. Bauer, H. F. and Sheehan, D. F. (1967). *Inorg. Chem.*, **6,** 1736
87. Blauer, J. A., McMath, H. G., Jaye, F. C. and Engleman, V. S. (1970). *J. Phys. Chem.*, **74,** 1183
88. Notley, J. M. and Spiro, M. (1966). *J. Chem. Soc., B*, 362
89. Christe, K. O. and Pilipovich, D. (1969). *Inorg. Chem.*, **8,** 391
90. Bantov, D. V., Dzevitskii, B. E., Konstantinov, Yu. S., Sukhoverkhov, V. F. and Ustynyuk, Yu. A. (1968). *Dokl. Akad. Nauk SSSR*, **180,** 491
91. Hyman, H. H., Surles, T., Quarterman, L. A. and Popov, A. I. (1970). *J. Phys. Chem.*, **74,** 2038
92. Toy, M. S. and Cannon, W. A. (1966). *J. Phys. Chem.*, **70,** 2241
93. Christe, K. O. (1970). *J. Phys. Chem.*, **74,** 2039
94. Chretien, A. and Martin, D. (1966). *C. R. Acad. Sci., Ser. C*, **263,** 235
95. Martin, D. (1967). *Rev. Chim. Miner.*, **4,** 367
96. Martin, D. (1969). *C. R. Acad. Sci., Ser. C*, **268,** 1145
97. Martin, D. (1967). *C. R. Acad. Sci., Ser. C*, **265,** 919
98. Brown, D. H., Dixon, K. R. and Sharp, D. W. A. (1966). *J. Chem. Soc.*, 1244
99. Brown, D. H., Dixon, K. R. and Sharp, D. W. A. (1966). *Chem. Commun.*, 654
100. Edwards, A. J. and Jones, G. R. (1967). *Chem. Commun.*, 1304
101. Edwards, A. J. and Jones, G. R. (1969) (A), *J. Chem. Soc.*, 1467
102. Christe, K. O. and Schack, C. J. (1970). *Inorg. Chem.*, **9,** 2296
103. Sukhoverkhov, V. F. and Dzevitskii, B. E. (1966). *Dokl. Akad. Nauk. SSSR*, **170,** 1099
104. Surles, T., Hyman, H. H., Quarterman, L. A. and Popov, A. I. (1970). *Inorg. Chem.*, **9,** 2726

105. Surles, T., Hyman, H. H., Quarterman, L. A. and Popov, A. I. (1971). *Inorg. Chem.*, **10,** 611
106. Surles, T., Hyman, H. H., Quarterman, I. A. and Popov, A. I. (1971). *Inorg. Chem.*, **10,** 913
107. Meinert, H. and Gross, U. (1969). *Z. Chem.*, **9,** 190
108. Meinert, H., Gross, U. and Grimmer, A. R. (1970). *Z. Chem.*, **10,** 226
109. Edwards, A. J., Jones, G. P. and Sills, R. J. C. (1968). *Chem. Commun.*, 1527
110. Schmeisser, M., Sartori, P. and Naumann, D. (1970). *Chem. Ber.*, **103,** 880
111. Schmidt, H. and Meinert, H. (1959). *Angew. Chem.*, **71,** 126
112. Schmeisser, M., Ludovici, W., Naumann, D., Sartori, P. and Scharf, E. (1968). *Chem. Ber.*, **101,** 4214
113. Schmeisser, M. and Ludovici, W. (1965). *Z. Naturforsch.*, **20b,** 602
114. Schmeisser, M., Sartori, P. and Naumann, D. (1970). *Chem. Ber.*, **103,** 590
115. Klamm, H. and Meinert, H. (1970). *Z. Chem.*, **10,** 270
116. Meinert, H., Kauschka, G. and Rudiger, S. (1967). *Z. Chem.*, **7,** 111
117. Nikolaev, A. V., Opalovskii, A. A., Nazarov, A. S. and Tret'yakov, G. V. (1969). *Dokl. Akad. Nauk. SSSR*, **189,** 1029
118. Nikolaev, A. V., Opalovskii, A. A., Nazarov, A. S. and Tret'yakov, G. V. (1970). *Dokl. Akad. Nauk SSSR*, **191,** 629
119. Meinert, H. and Gross, U. (1968). *Z. Chem.*, **8,** 306
120. Meinert, H. and Jahn, D. (1967). *Z. Chem.*, **7,** 195
121. Klamm, H. and Meinert, H. (1970). *Z. Chem.*, **10,** 227
122. Christe, K. O. and Sawodny, W. (1967). *Inorg. Chem.*, **6,** 1783
123. Christe, K. O. and Sawodny, W. (1968). *Inorg. Chem.*, **7,** 1685
124. Christe, K. O., Guertin, J. P. and Sawodny, W. (1968). *Inorg. Chem.*, **7,** 626
125. Klamm, H., Meinert, H., Reich, P. and Witke, K. (1968). *Z. Chem.*, **8,** 469
126. Claassen, H. H., Gasner, E. L. and Selig, H. (1968). *J. Chem. Phys.*, **49,** 1863
127. Kaiser, E. W., Muenter, J. S. and Klemperer, W. (1970). *J. Chem. Phys.*, **53,** 53
128. Falconer, W. E., Buchler, A., Stauffer, J. L. and Klemperer, W. (1968). *J. Chem. Phys.*, **48,** 312
129. Christe, K. O. and Sawodny, W. (1967). *Inorg. Chem.*, **6,** 1783
130. Hon, J. F. and Christe, K. O. (1970). *J. Chem. Phys.*, **52,** 1960
131. Bukshpan, S., Soriano, J. and Shamir, J. (1969). *Chem. Phys. Lett.*, **4,** 241
132. Stein, L. (1969). *J. Amer. Chem. Soc.*, **91,** 5396
133. Stein, L. (1970). *Science*, **168,** 362

3
The Halogen Hydrides

T. C. WADDINGTON
University of Durham

3.1 INTRODUCTION

This article, though concentrating mainly on the last two or three years, provides background information of rather earlier date than this for the sake of completeness. Work on HF, HCl, HBr and HI up to 1956 is summarised in the Supplement II, Part I to *Mellors Inorganic Chemistry*[1] and on HCl up to 1968 in – Teil 1 B Lieferung 1, the Chlor system-number 6 of *Gmelin's Handbook of Inorganic Chemistry*[2], and up to 1959 on HF in the Fluor volume system – number 5 in *Gmelin's Handbook of Inorganic Chemistry*[3].

3.2 PHYSICAL PROPERTIES

In this section we propose to discuss briefly the physical properties of the isolated molecules, and then discuss the physical properties of the halides in the three states of aggregation.

3.2.1 Properties of the isolated molecules

The hydrogen halides are all polar molecules, the polarity decreasing steadily from hydrogen fluoride to hydrogen iodide. The dipole moments, bond strengths, vibration frequencies and most other properties also change steadily down the group. These molecular properties are summarised in Table 3.1.

Table 3.1 Physical properties of the HX molecules

	HF	HCl	HBr	HI
Formula weight	20.0064	36.4610	80.9710	127.9124
Electronic ground state	$^1\Sigma^+$	$^1\Sigma^+$	$^1\Sigma^+$	$^1\Sigma^+$
Bond length, in pm	91.7	127.4	141.4	160.9
Dipole moment, μ, $\times 10^{18}$ c.s.u.	1.74	1.07	0.788	0.382
Mean polarisability $\alpha \times 10^{-3}$m^3	2.46	2.63	3.61	5.45
Polarisability along the bond $\alpha_{\parallel}$, $\times 10^{-30}$ m^3		3.13	4.22	6.58
Polarisability perpendicular to the bond, $\alpha_{\perp}$, $\times 10^{-30}$ m^3		2.39	3.31	4.89
Bond dissociation energy $D°$(HX), in kJ	573.98	428.13	362.50	294.58

The bond lengths of the bonds in the hydrogen halides have been determined via the moments of inertia, obtained from the rotational fine structure of the infrared spectra of the molecules, microwave spectra and in one or two cases from the pure rotational infrared spectra. The results by the various methods are in very good agreement[4,5]. The best values are listed in Table 3.1. The values are accurate to about ± 0.05 pm. Isotopic substitution of H, D and T and ^{35}Cl, ^{37}Cl does not appear to affect the bond lengths within this order of accuracy.

Some information, from rotational fine structure in their u.v. spectra is also available for the short lived species HCl^+ and HBr^+. Their bond lengths are HCl^+, 131.5 pm and HBr^+ 144.8 pm [4,5].

The dipole moments of the molecules have been determined many times and a critical discussion of the best values has been given by Nelson, Lide and Maryott[6] (see also Robinette and Sanderson[7]). The values are given in Table 3.1.

The mean optical polarisabilities of the molecules α, are given by $\alpha = (\alpha_{\parallel} + 2\alpha_{\perp}/3)$ where $\alpha_{\parallel}$ and $\alpha_{\perp}$ are the polarisabilities parallel and perpendicular to the H—X bond axis. These have been critically reviewed in Landolt–Bornstein[8] and are in good agreement with values obtained later from intensity measurements on Raman spectra by Yoshimo and Bernstein[9]. The values are reported in Table 3.1.

The fundamental vibration frequencies and anharmonicity of the HX molecules have been determined with some accuracy. They are summarised for HF up to 1959 and for HCl up to 1968 in Gmelin. Values for HF, HCl and HBr are also reported in the JANAF[10] tables but these are average, weighted values to allow for the distribution of isotopes. Values for HI are quoted by Nakamoto[11]. Transitions in the rotation-vibration spectrum of the hydrogen halides can be quoted in the form

$$(E\text{-}Eo)/hc$$
$$= \{w_e - w_e x_e(v+\tfrac{1}{2}) + w_w y_e(v+\tfrac{1}{2})^2 - w_e z_e(v+\tfrac{1}{2})^3\}(v+\tfrac{1}{2})$$
$$+\{B_e - D_e J(J+1)\}J(J+1) - \{\alpha_e - \gamma_e(v+\tfrac{1}{2})\}(v+\tfrac{1}{2})J(J+1)$$

Some values of these constants are given in Table 3.2.

Table 3.2 Vibration and rotation constants for the HX molecules (cm^{-1})

	HF	HCl		HBr	HI
		HCl^{35}	HCl^{37}		
w_0					
w_e	4138.33	2990.94	2988.48	2649.65	2309.53
$w_e x_e$	89.652	52.82	52.51	45.57	
$w_e y_e$	0.980*	0.224			
$w_e z_e$	0.025*	−0.012			
B_e	20.9548	10.593	10.576	8.4665	
D_e	0.002_2*	5.32×10^{-4}	5.30×10^{-4}	3.53×10^{-4}	
α_e	0.7939	0.305	0.305	0.2325	
γ_e	0.005_0*	0.901×10^{-3}	0.970×10^{-3}		

* R. M. Tulley, H. M. Kaylor and A. H. Nielsen. (1967). *Phys. Rev.* [2], **77**, 529
HCl data from Gmelin
HBr data from JANAF
HI data from Nakamoto

The root mean square amplitudes of vibration of the hydrogen halides, calculated on the harmonic oscillation approximation[12] are as follows: $^1H^{19}F$, 6.529 pm; $^1H^{35}Cl$, 7.587 pm; ^1HBr, 7.995 pm; $^1H^{127}I$, 8.544 pm. Considerable attention has recently been focused on the infrared spectra of hydrogen halides in inert matrices, where both monomer and multimer vibrations have been detected[13, 14].

The bond dissociations energies of the HX molecules are fairly well established. In most cases the value depends on the determined values of the bond dissociation energies of hydrogen and of the gaseous halogen X_2. The values are listed in Table 3.1 and are taken from the $\Delta H_f^0$0K values in US Bureau of Standards Technical Note 270–273 [15].

On the Mulliken notation the ground states of the HX molecules are all $^1\Sigma^+$ with electronic energy levels represented as follows:

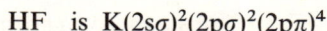

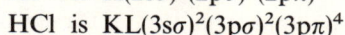

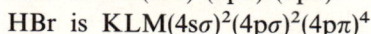

HF is $K(2s\sigma)^2(2p\sigma)^2(2p\pi)^4$
HCl is $KL(3s\sigma)^2(3p\sigma)^2(3p\pi)^4$
HBr is $KLM(4s\sigma)^2(4p\sigma)^2(4p\pi)^4$
HI is $KLMN(5s\sigma)^2(5p\sigma)^2(5p\pi)^4$

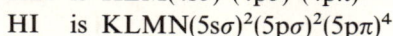

the representation is in general

$$\text{core } (ns\sigma)^2(np\sigma)^2(np\pi)^4$$

the first excited electronic configuration of the molecule is the doublet $^3\Pi$, $^1\Pi$

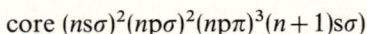

$$\text{core } (ns\sigma)^2(np\sigma)^2(np\pi)^3(n+1)s\sigma)$$

These two descriptions of the lowest electron configurations have been given using the united atom approximation. From this approximation it might appear that the transition between them is a Rydberg transition.

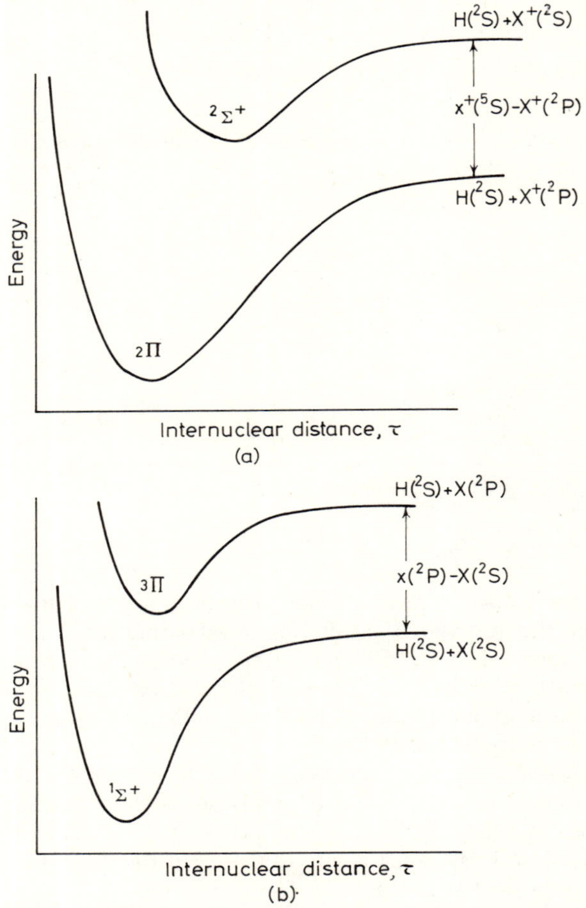

Figure 3.1 Potential energy diagrams for HX and HX^+ molecules (a) dissociation of HX^+; (b) Dissociation of HX

However, on dissociation the $^3\Pi$ and $^1\Pi$ states go to normal (configurationally unexcited) atoms. A better approximation to the electron configurations in the actual molecules would be

$$N:\sigma^2\pi_x^4,\ ^1\Sigma \text{ and } Q:\sigma^2\pi_x^3\sigma^*,\ ^3\Pi,\ ^1\Pi$$

where σ is a bonding, σ^* is the corresponding antibonding orbital resulting from $\sigma_n p_x$ and σ_{1SH} of the separated atoms and π_x is a non-bonding orbital which is essentially $\pi_n p_x$. Both in the case of HCl and HBr there appear to be a number of excited states of Π multiplicity. Recent studies of the excited states of the hydrogen halides have been made by Vanderslice and co-workers[16–18].

The HX molecules ionise to $^2\Pi_i$, the lowest electron configuration of the HX^+ molecular ion, which can be represented by core: $(ns\sigma)^2(np\sigma)^2\ (np\pi)^3$ and to $^2\Sigma^+$ core $(ns\sigma)^2(np\sigma)(np\pi)^4$. Both of these states have been observed in emission spectroscopy for HCl^+ and HBr^+. The reported bond lengths are HCl^+, $^2\Pi_i$, 131.5 pm and HCl^+, $^2\Sigma^+$ 151.4 pm [4,5]. The corresponding values for the bromide are HBr^+, $^2\Pi_i$, 145.9 pm and HBr^+, $^2\Sigma^+$, 166.6 pm [4,5]. Both these sets of bond lengths are longer than in the corresponding HX molecule and the electron in both cases has clearly been removed from a bonding orbital. The $(np\sigma)$ orbital is clearly more strongly bonding than the $(np\pi)$ orbital. The potential energy diagrams for HX and HX^+ are given in Figure 3.1.

The photoelectron spectra of all HX molecules has been studied by Lempka, Passmore and Price[19]. The results are shown in Figure 3.2. These yield the following values (in eV) for ionisation to the $^2\Pi_{\frac{3}{2},\frac{1}{2}}$ and $^2\Sigma^+$ states.

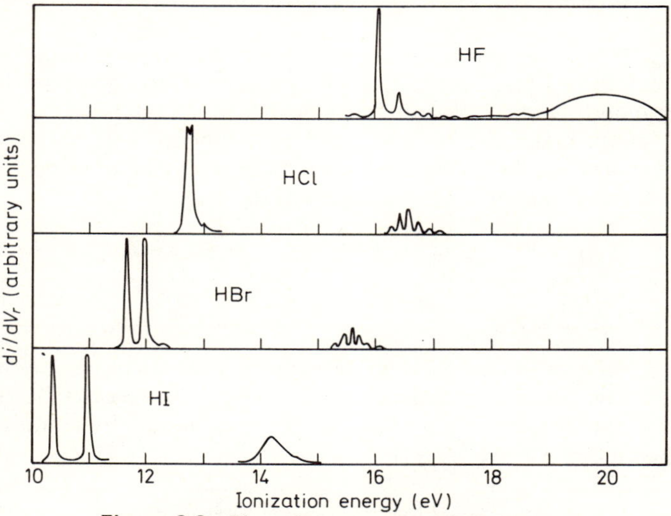

Figure 3.2 Photoelectron spectra of HX molecules (From Price *et al.*[19] by courtesy of The Royal Society)

HF, 16.05, 18.6; HCl 12.74, 12.82, 16.23; HBr, 11.67, 12.00, 15.29 and HI, 10.38, 11.05, 13.85, the experimental error being $\pm$ 0.01 eV in most cases.

It is worth noting that the $^2\Pi_{\frac{3}{2},\frac{1}{2}}$ doublet of HX^+ is resolved for HCl^+, HBr^+ and HI^+ but not for HF^+, and also that predissociation of the $^2\Sigma^+$ state occurs for HF^+ and HI^+.

The thermochemistry of the hydrogen halides and related species has been summarised most recently in the JANAF thermochemical tables and addenda[10] and in the US National Bureau of Standard Technical Note

270–273 [15] which revises and supplements the data in the US National Bureau of Standard Circular 500 [20]. The data selected in the two publications are in most cases in very good agreement and where not, the differences usually reflect real uncertainties in the best value to choose. The relevant data from these tables is summarised in Table 3.3.

For consistency the values from the US Bureau of Standard Technical Note 270–273 are generally adopted, an asterisk indicates that the value has been taken from the JANAF tables. The JANAF values for the enthalpies of F^-(g) and Cl^-(g) differ significantly from the Bureau of Standard values. They are based on Berry and Reimann's[21] values of the electron affinities, whereas the Bureau of Standards are based on Bailey's[22]. Since Bailey's value for the electron affinity of iodine, disagrees with that of Steiner, Seman and Branscomb[23], which is the same as Berry's and Reimann's, the JANAF values have been adopted.

The classical work of Schneider, Bernstein and Pople[24], who measured the proton chemical shifts of gaseous hydrogen halides relative to methane has not been superseded. Their values for δ were HF, $\delta = -0.25 \times 10^{-5}$; HCl, $\delta = 0.09 \times 10^{-5}$; HBr, $\delta = 0.44 \times 10^{-5}$ and HI, $\delta = 1.33 \times 10^{-5}$, where

$$\delta = \frac{H_{CH_4} - H_{HX}}{H_{CH_4}}$$

where H_{CH_4}–H_{HX} is the chemical shift difference between methane and the hydrogen halide and H_{CH_4} is the absolute position of the methane signal in the main magnetic field. These values can be regarded as those for the isolated molecules, shifts measured in the liquid phase being very different. The absolute shielding values for the hydrogen halides may be obtained by comparison with the value for methane of 3.04×10^{-5} to give

	HF	HCl	HBr	HI
σ	2.79×10^{-5}	3.08×10^{-5}	3.48×10^{-5}	4.37×10^{-5}

These shielding values are the sum of two terms, the diamagnetic shielding term σ_{dia} and the paramagnetic shielding term σ_{para}. Some calculations have been made of the shielding values for the hydrogen halides by Hameka[25] and these are given below:

	$\sigma_{dia} \times 10^5$	$\sigma_{para} \times 10^5$	$\sigma \times 10^5$
HF	3.468	−0.682	2.79
HCl	3.826	−0.420	3.41
HBr	4.034	−0.317	3.72
HI	4.343	−0.166	3.18

Considering the difficulty of calculations on systems as complex as molecules the agreement is not bad. The most important point that the calculation brings out is the relative size of the paramagnetic and diamagnetic terms.

The ^{19}F nucleus in HF also has magnetic moment and the fluorine chemical

Table 3.3 Thermochemical data for the HX molecules and related species

Formula and description	*State*	ΔH_f° 0 K kJ mol^{-1}	ΔH_f° 398.15 K kJ mol^{-1}	ΔG_f° 298.15 K kJ mol^{-1}	S° 298.15 K J deg^{-1} mol^{-1}
H_2	gas	0	0	0	130.57
H	gas	216.00	217.97	203.26	114.60
H^+	gas	1528.04	1536.31	—	108.83*
H^+	aq. standard state, m = 1		0	0	0
H^-	gas	143.93	139.70	—	108.85*
F_2	gas	0	0	0	202.67
F	gas	76.90	78.99	61.92	158.64
F^+	gas	1757.95	1766.23	—	—
F^-	gas	−255.64*	−260.24*	—	145.47
	aq. standard m = 1		−332.63	−278.82	−13.81
HF	liquid		−299.78	—	75.40
	gas	−271.08	−271.12	−273.22	173.67
HF	aq. standard undissociated state m = 1		−320.08	−295.85	88.70
HF	aq. ionised standard state m = 1		−332.63	−278.82	−13.81
HF_2^-	aq. standard state m = 1		−649.94	−578.15	92.47
Cl_2	gas	0	0	0	222.96
Cl	gas	120.00	121.68	105.70	165.09
Cl^+	gas	1375.95	1383.82	—	167.45*
Cl^-	gas	−229.28*	−233.89*	—	153.25*
	aq. standard state m = 1		−167.16	−131.26	56.48
HCl	gas	−92.13	−92.31	−95.30	186.80
	aq. standard state, m = 1		−167.16	−131.26	56.48
Br_2	liquid		0	0	152.23
	gas	45.70	30.91	3.14	245.35
Br	gas	117.94	111.88	82.43	174.91
Br^+	gas	1260.97	1261.10	—	
Br^-	gas	−221.75	−233.89	—	163.39*
	aq. standard state, m = 1		−121.55	−103.97	82.42
HBr	gas	−28.56	−36.40	−53.43	198.59
	aq. standard state, m = 1		−121.55	−103.97	82.42
HBr^+	gas	1092.44	1090.77		
I_2	crystal	0	0	0	116.14
	gas	65.52	62.44	19.36	260.58
I	gas	107.24	106.84	70.28	180.68
I^+	gas	1116.17	1121.98	—	—
I^-	gas	−189.95	−196.65	—	169.16*
I^-	aq. standard state, m = 1		−55.19	−51.59	111.29
HI	gas	28.66	26.48	1.72	206.48
	aq. standard state, m = 1		−55.19	−51.59	111.29

*Value taken from JANAF Thermochemical Tables[10]

shift in HF and the proton–fluorine coupling constant have also been measured.

The chemical shift of the fluorine resonance in HF relative to F_2 is 625×10^{-6}. The fluorine proton coupling constant is 615 H[26]. Though recent values[27] of the electron coupled spin–spin interaction J_{HF}, obtained by molecular beam radio frequency spectroscopy are +530 H_z.

Nuclear magnetic resonance signals from ^{35}Cl, ^{79}Br and ^{127}I are much more difficult to detect than from ^{19}F and there is no information available of gas phase values for these compounds. However, these nuclei possess nuclear electric quadrupole moments and the electric quadrupole coupling constants have been measured in the gas phase. The experimental results have been recently summarised by Lucken[28] and are given in Table 3.4.

Table 3.4 Halogen coupling constants in the hydrogen halides in the gas phase

Compound	*Nucleus*	*Coupling constant* (MH_z)	*Halogen σ population*
HCl	^{35}Cl	−67.51	1.39
HBr	^{79}Br	535.44	1.30
HI	^{127}I	−1831.07	1.20

The halogen σ orbital populations, calculated by the method of Townes and Dailey[29,30], assuming no sp hybridisation of the halogen σ-orbital, are also given in Table 3.4. The values indicate that the polarity of the bond decreases from chlorine to iodine, a result which would be expected in view of the electronegativities of the halogens, but in view of the approximations inherent in the Townes and Dailey treatment it is difficult to say more about these values.

3.2.2 Physical properties of the gases

The limiting densities, L_0, in kilogrammes per cubic metre (kg m^{-3}) are:

HF, 0.892 60; HCl, 1.626 74; HBr, 3.610 17; HI, 5.706 91

Actual standard Densities, L_1, in the same units are:

HCl, 1.639 14; HBr, 3.644 27

Values of the critical constants for the hydrogen halides are:

	HF	HCl	HBr	HI
T_c(K)	503.2	324.7	363.0	424.1
p_c(atm)		81.6	84	82
ρ_c(kg m^3)		424		

Hydrogen fluoride is associated in the gas phase; this is shown by its vapour density. The other hydrogen halides are not associated to any significant degree at normal pressures. The vapour density of the gas over liquid hydrogen fluoride reaches a maximum at −34 °C where it has a value of about 86.

It rapidly falls with increasing temperature, the vapour density at atmospheric pressure dropping from about 58 at 25 °C to about 20.6 at 80 °C. This association has been interpreted both in terms of the formation of a series of chain polymers and in terms of an equilibrium between monomers and hexamers

$$6HF \rightleftharpoons (HF)_6$$

It is difficult to make a choice between the two models on the basis of vapour density versus pressure data alone. However, both recent infrared measurements[31, 32] and electron diffraction studies[33] on the vapour appear to favour the monomer-hexamer equilibrium model, though the infrared work suggests that chain dimers may occur in high concentrations at some temperatures and pressures.

The n.m.r. spectra[34] of gaseous hydrogen fluoride shows that both ^1H and ^{19}F chemical shifts are strongly dependent on pressure and this behaviour has been interpreted in terms of the association of monomers to predominantly cyclic hexamers. Both nuclei in the molecule gave single lines instead of doublets, which means that rapid exchange takes place in the gas phase.

3.2.3 Physical properties of the liquids

The liquids fall into two classes. Liquid hydrogen fluoride is clearly a highly associated liquid with a high boiling point, long liquid range and high dielectric constant. The other three hydrogen halides have normal boiling points, short liquid ranges and relatively low dielectric constants. The physical properties of the liquids are summarised in Table 3.5.

Table 3.5 Physical properties of the liquid hydrogen halide

Property	HF	HCl	HBr	HI
M.p. (K)	189.79	158.5	184.6	222.2
B.p. (K)	292.66	189.0	206.1	238.1
Liquid range	102.87	30.5	21.5	15.9
Density of liquid (kg/dcm^3)	1.23 (−89.37 °C) 1.002 (0 °C)	1.187 (−114 °C)	2.603 (−84 °C)	2.85 (−47 °C)
Refractive index	1.1574 (0 °C)			
Viscosity/(centipoise)	0.256 (0 °C)	0.51 (−95 °C)	0.83 (−67 °C)	1.35 (−35.4 °C)
Surface tension (dynes/cm)	10.1 (0 °C)			
Dielectric constant	175 (−73 °C) 134 (−42 °C) 111 (−27 °C) 84 (0 °C)	9.28 (−95 °C)	7.0 (−85 °C)	3.39 (−50 °C)
Cryoscopic constant	1.309°			
Specificity conductivity	1.4×10^{-5} (−15 °C)			
Enthalpy of vaporisation ($kJ\ mol^{-1}$ at the boiling point)	7.485	16.15	17.61	19.77
Entropy of vaporisation at the boiling point	25.58	85.44	85.44	83.03

In spite of the strong hydrogen bonding that obviously occurs in liquid hydrogen fluoride, the surface tension and viscosity are a great deal lower than the values for water and this clearly rules out the presence of three-dimensional networks. The structure of liquid hydrogen fluoride is probably that of long linear hydrogen bonded polymers. The extremely low values for the enthalpy and entropy of vaporisation of liquid hydrogen fluoride at the boiling point (7.485 kJ mol^{-1} for ΔH(vap) and 25.58 J deg^{-1} mol^{-1} for ΔS(vap)compared to the corresponding values of 48.069 kJ mol^{-1} for ΔH(vap) and 118.80 for ΔS(vap) at 25 °C of H_2O) are to be attributed to the polymeric nature of HF vapour. Computed values for the vaporisation of HF liquid to the *monomeric* vapour at 25 °C are 28.66 kJ mol^{-1} for ΔH(vap) and 98.27 J deg^{-1} mol^{-1} for ΔS(vap).

The infrared and Raman spectra of the liquid hydrogen halides show the usual loss of vibrational structure associated with condensed phases. There is also, particularly in the case of liquid hydrogen fluoride, considerable shifts in the vibrational frequencies from the value for the monomers in the gas phase. Additionally, in the case of liquid hydrogen fluoride there is an extremely strong absorption in the infrared spectrum[36] between 400 and 1000 cm^{-1}.

The specific conductivities of the liquid hydrogen halides indicate that there must be self-ionisation in the pure liquids, presumably

$$HX \rightleftharpoons H^+(\text{solvated}) + X^-(\text{solvated})$$

Minimal solvation would imply species such as H_2X^+ and HX_2^-. Other evidence for the existence of species such as HX_2^- is readily obtained for all the hydrogen halides, but only in the case of H_2F^+ is there any firm evidence for the existence of the H_2X^+ species in solution or in solids[37]. Species such as H_2Cl^+ have been detected only in the mass spectrometer[38]. The proton affinity of this species has, however, been estimated[39] to be greater than 500 kJ mol^{-1}.

The vapour pressures over the liquid hydrogen halides are given by the equations[1,35].

$$\log_{10}p_{\text{mm}}(\text{HF, liquid}) = 7.3739 - \frac{1316.79}{T}$$

$$\log_{10}p_{\text{mm}}(\text{HCl liquid}) = 4.657\,39 - \frac{905.53}{T} + 1.75 \log_{10} T - 0.005\,007\,7\,T$$

$$\text{or less accurately} = 8.443 - \frac{1023.0}{T}$$

$$\log_{10}p_{\text{mm}}(\text{HBr, liquid}) = 7.465 - \frac{945.7}{T}$$

$$\log_{10}p_{\text{mm}}(\text{HI, liquid}) = 26.119 - \frac{1636}{T} - 7.111 \log_{10} T + 0.002293T$$

$$\text{or less accurately} = 7.630 - 1127/T$$

where T is the absolute temperature.

3.2.4 Physical properties of the solid hydrogen halides

Crystalline hydrogen fluoride appears to have one form throughout the temperature range studied[40]. The lattice is orthorhombic (Space Group *Bmmb*) with four molecules to the unit cell and $a_0 = 342$; $b_0 = 432$; $c_0 = 541$ pm. The structure consists of linear zig-zag chains of HF molecules, with an F—F distance of 249 pm and F—F—F bond angle of 120.1 degrees. Inelastic neutron scattering studies have confirmed the strong bonding along the chains and the weak bonding between them[41, 42].

Measurements of specific heat show that solid hydrogen chloride undergoes a first order phase transition at about 94.8 K. The crystal structure of the low-temperature α-phase is orthorhombic[43] with space group $Bb2_1m$, four molecules to the unit cell and $a_0 = 508.2$; $b_0 = 541.0$; $c_0 = 582.6$ pm. The structure is clearly very similar to that of crystalline HF and a recent neutron diffraction study on solid DCl has shown that the lattice contains hydrogen bonded zig-zag chains of DCl molecules with a Cl—Cl distance of 368.8 pm, a D—Cl distance of 125 pm and a Cl—Cl—Cl bond angle of 93.5 degrees[44, 45]. The structure is much more closely packed than that of HF.

The high temperature, β-phase of HCl is face centred cubic[44, 45] with $a_0 = 548.2$ pm at 118.5 K. This is clearly a close packed structure, but with relatively little change in Cl—Cl distance or in molecular volume from the α-phase. A neutron diffraction study of this phase in DCl has given a D—Cl bond length of 117 ± 4 pm.

Solid hydrogen bromide seems to show much the same behaviour as solid hydrogen chloride though the structures have not been so thoroughly studied. The low temperature form below 110 K has a face centred orthorhombic structure[46], with four molecules per unit cell and $a_0 = 555$; $b_0 = 564$; $c_0 = 606$ pm and there is a face-centred cubic form stable at higher temperatures with $a = 577$ pm. The structure of the low-temperature form probably consists of hydrogen bonded chains similar to those in the orthorhombic HF and HCl lattices.

Solid hydrogen iodide shows a number of transition points by calorimetry but the lattice parameters of only one phase appear to have been determined. This is face centred tetragonal[46] with four molecules to the unit cell and $a_0 = 619$; $c_0 = 668$ pm at -150 °C.

Some of the physical properties of the solid hydrogen halides are listed in Table 3.6.

Table 3.6 The physical properties of the solid hydrogen halides

Property	HF	HCl	HBr	HI
M.p. (K)	189.79	158.5	184.6	222.2
Enthalpy of sublimation (kJ mol^{-1})		19.07 (158.91 K)		
Enthalpy of fusion (kJ mol^{-1})	3.93	1.99		

3.3 CHEMICAL PROPERTIES

3.3.1 Formation

The formation and methods of manufacture of hydrogen fluoride and aqueous hydrofluoric acid have been reviewed by Gmelin[3] and Mellor[1], as have the manufacture and laboratory preparations of hydrogen chloride and aqueous hydrochloric acid. Laboratory scale preparations of gaseous HBr and gaseous HI from bromine and hydrogen and from iodine and hydrogen are described by Dodd and Robinson[35], as are preparations of deuterium fluoride, chloride, bromide and iodide. An alternative method of preparation of anhydrous hydrogen bromide and iodide in the laboratory, which may also be used conveniently for the preparation of deuterium bromide and iodide, is by dropping water (or D_2O) onto phosphorus tribromide or tri-iodide. The resulting gases are then purified by fractionation. Small quantities of anhydrous deuterium chloride may be very conveniently prepared in the laboratory by dropping D_2O onto phosphorus pentachloride.

The kinetics and mechanisms of the reactions of the elements X_2, with hydrogen to form the gaseous hydrogen halides has been very extensively studied[47–49]. With the exception of the reaction between iodine and hydrogen below 700 °C the reactions proceed by a chain mechanism. This mechanism involves some or all of the following elementary steps.

Initiation

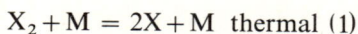

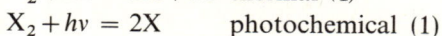

$$X_2 + M = 2X + M \quad \text{thermal (1)}$$
$$X_2 + h\nu = 2X \quad \text{photochemical (1)}$$

Propagation

$$X + H_2 = HX + H \tag{2}$$
$$H + X_2 = HX + X \tag{3}$$

Inhibition

$$H + HX = H_2 + X \tag{4}$$
$$X + HX = X_2 + H \tag{5}$$

Termination

$$X + X + M = X_2 + M \tag{6}$$
$$X + H = HX \tag{7}$$
$$H + H = H_2 \tag{8}$$

In the case of iodine and hydrogen the reaction below 700 °C mainly proceeds by a bi-molecular mechanism

$$H_2 + I_2 = 2HI \tag{9}$$
$$HI + HI = H_2 + I_2 \tag{10}$$

Detailed kinetic analysis has been carried out very successfully on the reactions between hydrogen and bromine and hydrogen and iodine, but the reactions of hydrogen with fluorine and with chlorine have proved much more difficult to interpret. This is probably due to the very great chain lengths encountered in these reactions, which means that termination steps on the walls of the containing vessel become important, as do the effects of small quantities of impurities and self-heating. Both reactions can lead to very vigorous explosions, and in the case of HCl, use of additives can sensitise the reaction.

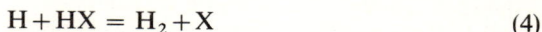

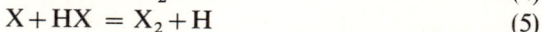

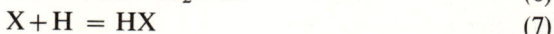

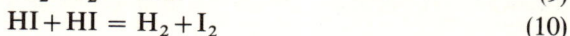

With the exception of fluorine, the reaction

$$H+X_2 = HX+X$$

is very exothermic, whereas the reaction

$$X+H_2 = HX+H$$

is endothermic. Table 3.7 gives some experimental activation energies for some of these reactions.

Table 3.7 Experimental activation energies (kJ mol^{-1}) for some of the steps in the hydrogen–halogen reactions

Reaction	F	Cl	Br	I
*$X+X+M = X_2+M$	0	0	0	0
†$X+H_2 = HX+H$		22.9 ± 0.58	82.4 ± 1.6	142.7 ± 2.9
‡$H+HX = H_2+X$	139.3	18.8	3.8	0
‡$H+X_2 = HX+X$		12.6	3.8	0

* S. W. Benson, *The Foundations of Chemical Kinetics*, McGraw-Hill.1960
† G. C. Fettis and J. H. Knox *The Rate Constants of Halogen Atom Reactions*; Progress in Reaction Kinetics Vol. 2. Pergamon Press. 1964
‡ B. A. Thrush, *The Reactions of Hydrogen Atoms*; Progress in Reaction Kinetics Vol. 3. Pergamon Press. 1965

For the H_2 and Br_2 reaction the following of the steps given are usually considered

$$Br_2+M \rightarrow 2Br+M$$
$$Br+H_2 \rightarrow HBr+H$$
$$H+Br_2 \rightarrow HBr+Br$$
$$HBr+H \rightarrow H_2+Br$$
$$2Br+M \rightarrow Br_2+M$$

which can be shown by applying the stationary state hypothesis, to lead to the rate equation

$$\frac{d[HBr]}{dt} = \frac{k_t[H_2][Br_2]^{\frac{1}{2}}}{1+k_i\{[HBr]/[Br_2]\}}$$

The formation reactions of the hydrogen halides have been suggested as the basis for fuel cell operation[50].

3.3.2 Chemical reactions of the gaseous hydrogen halides

Thermodynamically we may say that the reactions of the gaseous hydrogen halides with the elements:

$$M+nHX = MX_n+\tfrac{1}{2}nH_2$$

should occur if $\Delta G_r = \Delta G_f^\circ MX_n - n\Delta G_f^\circ HX(g)$, for the reaction is negative. This means that the reactions should proceed to any n-valent halide MX_n if:

for the fluoride $\Delta G_f^\circ MF_n$ is $< -274n$ kJ mol^{-1}
for the chloride $\Delta G_f^\circ MCl_n$ is $< -96n$ kJ mol^{-1}
for the bromide $\Delta G_f^\circ MBr_n$ is $< -54n$ kJ mol^{-1}
for the iodide $\Delta G_f^\circ MI_n$ is < 0 kJ mol^{-1}

These are, of course, necessary but not sufficient conditions for the reactions to proceed. The kinetics of the reaction may be very slow in spite of the favourable thermodynamics.

Using the free energies of formation from the Bureau of Standards circular 500 [20] and from Latimer[51], one can see that most of the metals will react with most of the hydrogen halides. The reactions are in many cases, e.g. with the alkali metals, the alkaline earths, zinc, aluminium and the lanthanides, extremely exothermic. Silver should react with hydrogen chloride, bromide and iodide but not with hydrogen fluoride. Copper should form CuF_2 with HF, but not $CuCl_2$ with HCl, and $CuBr_2$ and CuI_2 with HBr and HI. Iron should in theory react to give $FeCl_3$ with HCl but in practice the reaction only proceeds to $FeCl_2$. Thermodynamically the reactions of all the hydrogen halides with titanium should give the tetrahalides, but the reactions only proceed at elevated temperatures.

The reactions of HF, HCl and HBr with the non-metal silcon to form the tetrahalides

$$Si + 4HX = SiX_4 + 2H_2$$

are all thermodynamically very favourable, though only in the case of HF is there any reaction at room-temperature. In practice species such as $SiHCl_3$ etc. are formed. With arsenic the reaction

$$As + 3HF = AsF_3 + \tfrac{3}{2}H_2$$

is favourable but the reactions with the other hydrogen halides are not.

The thermodynamics of the reactions of oxygen with the hydrogen halides

$$O_2 + 4HX = 2H_2O + 2X_2$$

are favourable for all except HF, but do not proceed at room temperature even with HI in the absence of light.

The exchange reaction between the halogens and the hydrogen halides

$$2HX + Y_2 = 2HY + X_2$$

all go with F replacing Cl replacing Br replacing I.

These reactions between the halogens and the hydrogen halides appear to follow a consecutive bimolecular path[52, 53].

$$HX + Y_2 = HY + XY$$
$$HX + XY = HY + X_2$$

The reactions have been used as bases for chemical lasers, with both pulsed and continuous operation[54, 55].

The reactions of the hydrogen halides with the oxides are more complicated. The reaction

$$MO_{n/2} + nHX = MX_n + \tfrac{1}{2}nH_2O$$

will proceed if ΔG, given by

$$\Delta G = \Delta G_f^\circ MX_n - \Delta G_f^\circ MO_{n/2} - n(\Delta G_f^\circ HX - \tfrac{1}{2}\Delta G_f^\circ H_2O),$$

is negative. This corresponds to

$$\Delta G < \Delta G_f^\circ MF_n - \Delta G_f^\circ MO_{n/2} + 153n \text{ kJ mol}^{-1}$$

for the fluorides;

$$\Delta G < \Delta G_f^\circ MCl_n - \Delta G_f^\circ MO_{n/2} - 25n \text{ kJ mol}^{-1}$$

for the chlorides;

$$\Delta G < \Delta G_f^\circ MBr_n - \Delta G_f^\circ MO_{n/2} - 67 \text{ kJ mol}^{-1}$$

for the bromides and

$$\Delta G < \Delta G_f^\circ MI_n - \Delta G_f^\circ MO_{n/2} - 121 \text{ kJ mol}^{-1}$$

for the iodides.

The reactions of the gaseous hydrogen halides with the alkali metal oxides M_2O, the alkaline earth oxides, including MgO and zinc and cadmium oxides are all strongly exothermic. On the other hand, the conversion of Al_2O_3 to the aluminium trihalides is thermodynamically unfavourable even for the fluoride, and cupric oxide should not react with any of the hydrogen halides. With ZrO_2, HF should react to give ZrF_4 but the reactions of HCl, HBr and HI are thermodynamically unfavourable. SeO_2 should just give SeF_4 and $SeCl_4$. The reaction

$$SiO_2 + 4HX = SiX_4 + 2H_2O$$

is thermodynamically favourable only for HF, which does in fact etch silica. The reaction of HCl with Sb_2O_3 to form $SbCl_3$ is thermochemically favourable and is complete in 45 min at 300 °C [56].

The reactions of the alkali metal halides themselves with the hydrogen halides are of some interest

$$MX + HY = MY + HX$$

All the alkali metal fluorides except lithium should react with hydrogen chloride to give the chloride and hydrogen fluoride. Again, with the exception of lithium, they should all react with hydrogen bromide to give the bromide and hydrogen fluoride. KF, RbF and CsF should all react with hydrogen iodide to give the corresponding iodide and hydrogen fluoride.

The reactions

$$MCl + HBr = MBr + HCl$$

and

$$MBr + HI = MI + HBr$$

are exothermic and have favourable free energy changes for all salts except the lithium halides.

The reactions of the hydrides with the hydrogen halides show some interesting features. The thermochemistry of all the salt-like hydrides is such that the reactions

$$MH + HX = MX + H_2$$

are all favourable. So are the reactions

$$B_2H_6 + 6HX = 2BX_3 + 6H_2$$

which proceed smoothly in the gas-phase. Indeed by utilising the correct stoichiometry of HX, particularly of the higher hydrogen halides, it is possible to produce partly substituted species such as $B_2H_2Cl_4$.

The reactions

$$CH_4 + 4HX = CX_4 + 4H_2$$

are, of course, uniformly thermodynamically unfavourable but with the higher members of group IV such as silicon the reactions

$$SiH_4 + 4HX = SiX_4 + 4H_2$$

are all favourable. These reactions can be carried out. The reaction

$$PH_3 + 3HF = PF_3 + 3H_2$$

actually has a ΔG of $-96\ \mathrm{kJ\ mol^{-1}}$ though of course it does not proceed, the protonation of phosphine taking place preferentially. Similarly, the reaction of arsine with HF and HCl to give hydrogen and AsF_3 or $AsCl_3$ respectively is thermodynamically favourable, though the reaction does not proceed at room temperature.

Nitrides will in many cases react with the gaseous hydrogen halides to give the corresponding halides and ammonia, this is particularly true of the salt-like nitrides. The salt-like carbides will in many cases react to give the chloride and acetylene, e.g.

$$CaC_2 + 2HCl = CaCl_2 + C_2H_2$$

The reactions of magnesium boride and magnesium silicide with HCl, admittedly in solution, were the basis of the original preparations by Stock of the boron hydrides and the silicon hydrides.

3.3.3 Aqueous solutions of the hydrogen halides

All the acids are extremely soluble in water. Liquid hydrogen fluoride is miscible in all proportions. The other hydrogen halides all form constant boiling mixtures with water. The HCl, H_2O constant boiling mixture is used as a primary standard in volumetric analysis and its composition as a function of pressure is given below

P(mm Hg)	730	750	760	770	780
Wt% of HCl	20.293	20.245	20.221	20.197	20.173

The boiling point at 760 mm Hg is 108.6 °C. Vapour–liquid equilibrium for the HCl–H_2O system from −10 °C to +70 °C has recently been reported[57]. The HBr, H_2O constant boiling mixture contains 47.63% by wt of HBr and boils at 124.3 °C under a pressure of 760 mm.

HCl, HBr and HI are very strong acids in water whereas HF is a comparatively weak one. The ionisation of HF in water is complicated. In addition to the straight forward ionisation

$$HF + H_2O \rightleftharpoons H_3O^+(aq) + F^-(aq)$$

or

$$HF \rightleftharpoons H^+(aq) + F^-(aq)$$

we have

$$2HF+H_2O \rightleftharpoons H_3O^+(aq)+HF_2^-(aq)$$

or

$$F^-(aq)+HF \rightleftharpoons HF_2^-(aq)$$

The dissociation constant[58] for the first process is 1.20×10^{-3} corresponding to a pK_a value of 2.92 and for the second[58]

$$\frac{[HF_2^-]}{[HF][F^-]} = 2.57 \times 10^{-1}$$

The other hydrogen halides are such strong acids that values of the dissociation constants have to be obtained from $\Delta G_f^\circ X^-(aq)$ and estimated free energies of solution of the undissociated acids. The pK_a values are approximately[58]

HCl, −7; HBr, −9; HI, −9.5.

The enthalpies of formation, free energies of formation and absolute entropies of the dissociated hydrogen halides in water are given in Table 3.3. These free energy values[15] are probably the best source of the thermodynamic values for the electrode potentials for the reversible reactions

$$\tfrac{1}{2}X_2 \text{ (standard state)}+e = X^-aq.$$

The values are for acid solutions of unit activity

$$\tfrac{1}{2}F_2(g)+e = F^-(aq) \qquad E^\circ = +2.899V$$
$$\tfrac{1}{2}Cl_2(g)+e = Cl^-(aq) \qquad E^\circ = +1.3604V$$
$$\tfrac{1}{2}Br_2(l)+e = Br^-(aq) \qquad E^\circ = +1.0776V$$
$$\tfrac{1}{2}I_2(s)+e = I^-(aq) \qquad E^\circ = +0.5347V$$

These potentials mean that aqueous HI is quite an effective reducing agent for many systems. Aqueous HBr is also a reducing agent and even aqueous HCl can yield chlorine in solution with strong oxidising agents such as permanganate.

Using the thermodynamic data in Table 3.3 and the computed single ion solvation enthalpy of the proton of -1090.8 kJ mol^{-1}, due to Halliwell and Nyburg[59] we obtain the following values for the single ion solvation enthalpies of the halide ions in water,

$$H_{F^-} = -507.5; H_{Cl^-} = -366.5; H_{Br^-} = -333.0 \text{ and } H_{I^-} = -304.2 \text{ kJ mol}^{-1}$$

The situation with regard to single ion absolute hydration entropies is less satisfactory. The values of S^0_{abs} $(X^-(aq))$ relative to a value of $S^0_{abs}(H^+(aq))$ $= 0$ are accurately known as are the absolute values of $S^0_{abs}(X^-(gas))$ based on the Saekur–Tetrode equation. Probably $S^0_{abs}(H^+(aq))$ in absolute terms is around -20 J deg^{-1} mol^{-1} with an uncertainty of ± 8 J deg^{-1} mol^{-1} [60]. This value leads to the following absolute hydration entropies at 25 °C and the following absolute values of $S^0_{abs}(X^-(aq))$.

Absolute hydration entropies, ΔS_{X^-}, $F^- = -139$; $Cl^- = -77$; $Br^- = -61$; $I^- = -48$; $H^+ = -129$ J deg^{-1} mol^{-1}.

Absolute entropies of the ions in aqueous solution $S^0_{abs}(X^-(aq))$ values, $F^- = +6$; $Cl^- = 76$; $Br^- = 102$; $I^- = 121$; $H^+ = -20$ J deg^{-1} mol^{-1}

Combining these values of absolute hydration entropies with those for absolute enthalpies gives the following set of absolute free energies of hydration,

ΔG_{X^-}; $F^- = -466$; $Cl^- = -344$; $Br^- = -315$; $I^- = -290$ and

$H^+ = -1052$ kJ mol^{-1}

The limiting equivalent conductivities of the hydrogen ion and the halide ions in water at 25 °C are[61] (in cm^2 Int. ohm^{-1} equivalent^{-1}) H^+, 349.8; F^-, 55.4; Cl^-, 76.35; Br^-, 78.14 and I^-, 76.84. These values indicate that the fluoride ion is more heavily solvated than the other halides.

If it is assumed that one can distinguish certain solvent molecules near an ion, which have lost their translational degrees of freedom and move as one entity with the iron during its Brownian movement, then it is possible to define a 'primary solvation' number. If one further regards the entropy of hydration as arising entirely from the loss of translational degrees of freedom of water molecules in this primary hydration sheath, and assumes that this loss of entropy is equal to that associated with the formation of one molecule of water of crystallisation in the solid state (-25 J mol^{-1} deg^{-1}), then the primary hydration numbers of ions may be calculated.

The following values[60] are found for the halide ions: F^-, 5; Cl^-, 3; Br^-, 2; I^-, 1. Calculations of hydration numbers by other methods tend to give different figures but the *average* values[60] can be taken to be F^-, 4 ± 1; Cl^-, 2 ± 1; Br^-, 1 ± 1; I^-, 1 ± 1.

The solid–liquid diagrams of the solutions of the hydrogen halides with water all show compound formation. The phase diagram of the system HF/H_2O shows the existence of three compounds: $HF{\cdot}H_2O$; $2HF{\cdot}H_2O$ and $4HF{\cdot}H_2O$.

They are presumably $H_3O^+F^-$, $H_3O^+HF_2^-$ and $H_3O^+H_3F_4^-$.

The phase diagram of the HCl–H_2O system on the other hand shows the existence of the compounds $HCl{\cdot}6H_2O$; $HCl{\cdot}3H_2O$; $HCl{\cdot}2H_2O$ and $HCl{\cdot}H_2O$. The crystal structure of $HCl{\cdot}H_2O$ has been shown by Yoon and Carpenter[62] to be $H_3O^+Cl^-$ with the hydroxonium ion hydrogen bonded to chloride ions and this has been confirmed by study of the infrared spectrum[63]. The hydrate $HCl{\cdot}2H_2O$ has been shown to be $H_5O_2^+Cl^-$ by Lundgren and Olovsson[64].

The phase diagram of the system HBr–H_2O shows the existence of the definite hydrates $HBr{\cdot}H_2O$; $HBr{\cdot}2H_2O$; $HBr{\cdot}3H_2O$ and $HBr{\cdot}4H_2O$. Probably by analogy with the HCl–H_2O system the first two can be formulated as $H_3O^+Br^-$ and $H_5O_2^+Br^-$. A crystal structure determination[65] on $HBr{\cdot}4H_2O$ shows that it can be formulated as $(H_7O_3^+)(H_9O_4^+)\ 2Br^-{\cdot}H_2O$. The solid–liquid phase diagram of the HI–H_2O system shows the existence of the hydrates $HI{\cdot}2H_2O$; $HI{\cdot}3H_2O$ and $HI{\cdot}4H_2O$. The hydrate $H_2O{\cdot}HI$ does not appear to be stable.

3.3.4 The hydrogen halides as non-aqueous solvents

The solvent properties of liquid hydrogen fluoride have been fairly extensively studied. Its high dielectric constant and high acidity make it a particularly

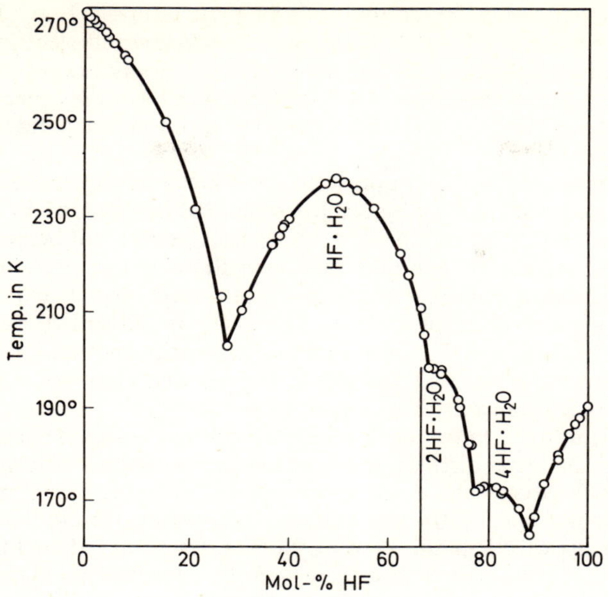

Figure 3.3 The HF—H_2O phase diagram
(From Gmelin's *Handbuch der Inorganische Chemie*, by courtesy of Verlag Chemie)

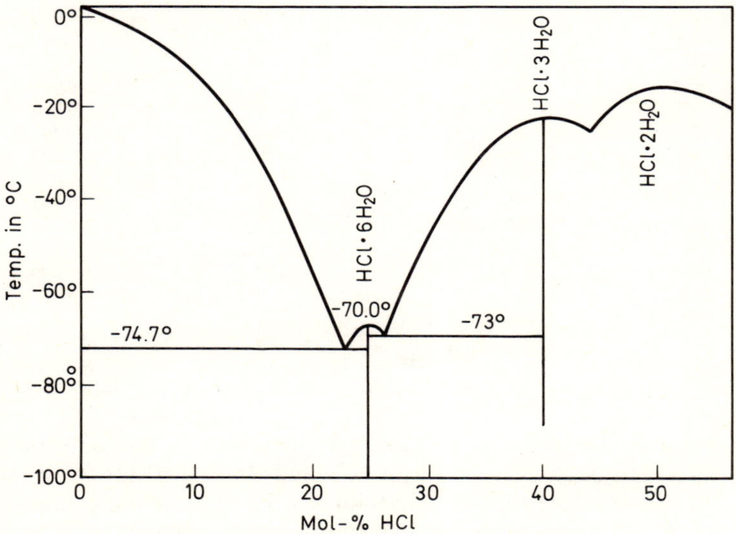

Figure 3.4 The HCl—H_2O phase diagram

powerful solvent, even for compounds of high lattice energy. The higher hydrogen halides on the other hand have much lower dielectric constants and small solvating properties and are therefore very poor solvents for salts of moderate lattice energy, such as the alkali halides. On the other hand, salts with low lattice energies like the tetra-alkyl ammonium halides are often quite soluble.

Work in liquid hydrogen fluoride to 1964 has been reviewed by Hyman and Katz[37] and work on the higher hydrogen halides to 1965 by Peach and Waddington[66]. Only an outline of work to that period will be given here.

The chief problem with liquid hydrogen fluoride has been the extreme reactivity of the solvent, even with glass and quartz, and its extraordinarily unpleasant physiological properties. However, the advent of fluorine containing plastics such as Teflon (polytetrafluoroethylene) and Kel-F (polychlorotrifluoroethylene) and the use of copper and stainless steel vacuum lines have made quantitative studies easier.

The chief impurity in liquid hydrogen fluoride is water. This can only be removed by fractional distillation of the solvent in either a platinum, copper or Teflon still. Most salts dissolve in the solvent with reaction and most non-ionic compounds with protonation. Even when simple fluorides such as NaF are dissolved in the solvent, bifluorides rather than fluorides are recovered on its removal. The solvent is most effective in solvating anions, but not particularly effective in solvating cations. The solubilities of a number of fluorides in liquid hydrogen fluoride are given in Table 3.8.

Table 3.8 The solubilities of fluorides in liquid hydrogen fluoride

Data are at 12°C (unless otherwise stated) and solubilities are in kg/100 kg of HF.

Fluoride	*Solubility*	*Fluoride*	*Solubility*
LiF	10.3	AgF_2	0.048
NaF	30.4	PbF_2	2.62
KF	36.5 (8 °C)	NiF_2	0.037
RbF	110.0 (20 °C)	FeF_2	0.006
CsF	199.0 (10 °C)	CrF_2	0.036
NH_4F	32.6 (17 °C)	HgF_2	0.54
AgF	83.2 (19 °C)	AlF_3	<0.002
TlF	580.0	CeF_3	0.043
BeF_2	0.015	FeF_3	0.081
MgF_2	0.025	SbF_3	0.536
CaF_2	0.87	CeF_4	0.10
SrF_2	14.83	ThF_4	<0.006
BaF_2	5.60	NbF_5	6.8
CuF_2	0.010	TaF_5	15.2
		SbF_5	Miscible in all proportions

Information on equivalent ionic conductivities in the solvent is very scanty but figures of $\Lambda_\infty = 120$ for K^+ and $\Lambda_\infty = 280$ for F^- have been reported. These suggest that the mobility of the fluoride ion is anomalously high but that the anomaly is not so pronounced as it is in water. The value of the Hammett acidity function, H_0, for the purest hydrogen fluoride is

-10.98, which is about the same as for sulphuric acid. This value drops drastically with the addition of even a small amount of water. The proposed ionisation mechanism is:

$$3HF_2 = H_2F^+ + HF_2^-$$

Only $HClO_4$ and HSO_3F appear to be sufficiently strong protonic acids to increase the concentration of H_2F^+ in the solvent. Other protonic acids are themselves protonated and may undergo reaction.

Thus

$$HNO_3 + 4HF = NO_2^+ + H_3O^+ + 2HF_2^-$$

and

$$CH_3{\cdot}COOH + 2HF = CH_3{\cdot}C(OH)_2^+ + HF_2^-$$

SbF_5 and AsF_5 appear to be the strongest Lewis acids in liquid hydrogen fluoride. The conductivities of the solutions are high and their Raman spectra show strong lines due to SbF_6^- and AsF_6^-. NbF_5 and TaF_5 have only limited solubility in liquid HF. They are somewhat weaker acids than SbF_5 and AsF_5. Boron trifluroide is the only other fluoride ion acceptor, apart from the Group V fluorides, to behave as an acid in the solvent and its low solubility and the low conductivity of its solutions indicate that it is rather a weak acid. Curiously enough, some metal ligand bonds are preserved in liquid hydrogen fluoride. Thus AgCN dissolves readily in the solvent to form a stable solution containing mono- and di-HCN complexes of Ag(I) [67].

Hydrazinium difluoride has been used as a reducing agent in liquid hydrogen fluoride with a number of metal hexafluorides to prepare anhydrous complex fluorides with the metal in a lower oxidation state. $N_2H_6UF_7$, and $N_2H_6UF_6$, $N_2H_6MoF_6$ and $N_2F_6ReF_6$ have been made this way[68].

The behaviour of nitrates in the solvent is similar to that of nitric acid. Sulphates are converted to fluorosulphates.

$$K_2SO_4 + 5HF = 2K^+ + SO_3F^- + H_3O^+ + 2HF_2^-$$

Some substances, such as SO_2 and $CF_3{\cdot}COOH$ dissolve without ionisation.

A number of fluorides dissolve in the solvent to give solutions of high conductivity and appear to be acting as fluoride ion donors. Examples of this type of behaviour are shown by chlorine trifluoride, bromine trifluoride, and xenon hexafluoride:

$$ClF_3 + HF \rightarrow ClF_2^+ + HF_2^-$$
$$BrF_3 + HF \rightarrow BrF_2^+ + HF_2^-$$
$$XeF_6 + HF \rightarrow XeF_5^+ + HF_2^-$$

The solubilities of organic, covalent compounds in liquid hydrogen fluoride are extremely high and in many cases the solutions have a high conductivity, indicating that the solute has been protonated. Solubilities are summarised in Table 3.9.

Many organic reactions have been carried out successfully in liquid hydrogen fluoride. Aromatic fluorides can be obtained by carrying out the diazotisation reaction in the solvent

$$ArNH_2 \xrightarrow[HF]{NaNO_2} ArN_2F \longrightarrow ArF + N_2$$

Diazotisation in liquid hydrogen fluoride has proved particularly useful for the preparation of ortho-substituted fluorine derivatives directly from the ortho-substituted amines.

Unsaturated hydrocarbons in liquid hydrogen fluoride add solvent molecules across the double bond smoothly. For example, propylene is converted to 2-fluoropropane and cyclohexene to fluorocyclohexane in yields of 61% and 70% respectively. Addition of hydrogen fluoride to substituted acetylenes,

Table 3.9 Solubilities of organic compounds in liquid hydrogen fluoride

Hydrocarbons	Saturated aliphatic hydrocarbons are insoluble. Aromatic hydrocarbons, such as benzene, toluene, the xylenes and mesitylene tend to be soluble and to be protonated. Aromatic and aliphatic compounds also tend to be soluble if they carry substituents that can be protonated, e.g. containing nitrogen, oxygen or sulphur atoms. In general, the presence of electron-withdrawing groups, such as halogen or nitro on an aromatic ring, has the effect of lowering solubility. Unsaturated compounds, such as butadiene, tend to polymerise.
Alcohols	Aliphatic alcohols are miscible in all proportions.
Amines	They are protonated to produce extremely soluble salts, as are heterocyclic nitrogen bases.
Esters and ethers	They are protonated and are very soluble.
Carboxylic acids	They are protonated and very soluble. Acetic acid is miscible in all proportions.

$HC{\equiv}C{-}R$, occurs in yields of 85–90% by the introduction of the acetylene into liquid hydrogen fluoride containing oxygenated molecules such as ether or acetone. Both fluorine atoms add to the carbon atom which was originally not attached to a proton. Thus $R{\cdot}CF_2{\cdot}CH_3$ is formed from $R{\cdot}C{\equiv}C{\cdot}H$.

Because of the very high potential needed for the anodic reaction in liquid hydrogen fluoride

$$F^- \rightleftharpoons \tfrac{1}{2}F_2 + e^-$$

the solvent is particularly suited to performance of anodic oxidation reactions. This, added to the fact that a wide variety of organic compounds form conducting solutions has made the electrochemical insertion of fluorine into organic molecules a technique of wide commercial application as well as one of great use in the research laboratory. On the inorganic side the production of NFH_2, NF_2H and NF_3 by the electrolysis of ammonium fluoride in liquid hydrogen fluoride represents the only convenient synthetic route to these difficulty available compounds. Trifluoracetic acid is most readily obtained by the electrolysis of acetic acid in liquid HF. When the organic starting material is insoluble (e.g. an aliphatic hydrocarbon), soluble organic or inorganic additives are introduced to increase the conductivity and speed electrolysis. Some of the products of anodic oxidation of a variety of compounds in liquid HF are summarised in Table 3.10.

Other reactions of industrial importance are the preparations of the 'freons' by the solvolysis of carbon tetrachloride dissolved in liquid hydrogen fluoride. Antimony(V)chloride, dissolved in and partly solvolysed by hydrogen hydrogen fluoride, acts as a 'transfer' catalyst

$$CCl_4 + HF \xrightarrow{SbF_nCl_{5-n}} CCl_3F + HCl$$

$$CCl_3F + HF \xrightarrow{SbF_nCl_{5-n}} CCl_2F_2 + HCl$$

$$CCl_2F_2 + HF \xrightarrow{SbF_nCl_{5-n}} CClF_3 + HCl$$

The various mixed chlorofluoromethanes are separated by fractional distillation. Fluorinated ethane derivatives may be prepared in the same manner.

Liquid hydrogen fluoride is currently employed very extensively in biologically related research[37,69]. Carbohydrates and proteins dissolve readily in the anhydrous solvent, frequently with only minor chemical consequences. Complex organic compounds potentially capable of eliminating the elements of water often dissolve without dehydration. It is important that solution be carried out in such a way that the heat released, which is often high, is dissipated without a rise in the temperature of the liquid. Cellulose dissolves freely to form conducting solutions and the material recovered from such

Table 3.10 Products of anodic oxidation in liquid hydrogen fluoride

Reactant	*Product*
NH_4F	NF_3, NHF_2, NH_2F
H_2O	OF_2
SCl_2, SF_4	SF_6
$NaClO_4$	ClO_4F
$(CH_3)_3N$	$(CF_3)_3N$
$(CH_3{\cdot}CO)_2O$	$CF_3{\cdot}COF$
$(CH_3)_2S$, CS_2	CF_3SF_5, $(CF_3)_2SF_4$
CH_3CN	CF_3CN, $C_2F_5{\cdot}NF_2$

solutions, designated a glucosan, yields glucose on mild hydrolysis. The behaviour of a variety of sugar esters in the solvent has also been examined; chemical attack is very slow indeed and may be largely due to the presence of traces of water in the solvent.

Anhydrous hydrogen fluoride is a powerful solvent for amino acids and proteins. All common amino acids and many proteins dissolve freely. Even tryptophan and cystine, amino acids which are practically insoluble in water, dissolve in hydrogen fluoride (to give 3% and 5% solutions respectively). All of the globular proteins that were examined dissolved in the solvent, and many fibrous proteins normally insoluble in water, such as silk fibroin, dissolved readily. Proteins are remarkably stable in liquid hydrogen fluoride, for example the hormone insulin and ACTH were recovered after 2h in the solvent at 0 °C with their biological activity substantially intact. Liquid hydrogen fluoride is now extensively used as a reagent for peptide synthesis. This stems from the discovery by Japanese workers in 1965 [70] that liquid hydrogen fluoride containing excess anisole releases a wide variety of protecting groups commonly used in peptide synthesis. Anisole was added to the solvent to combine with the benzyl fluoride released, thus pulling the reaction to completion and preventing side reactions. Thus if carbobenzoxy chloride ($C_6H_5{\cdot}CH_2{\cdot}O{\cdot}CO{\cdot}Cl$) is used to block an amino ($-NH_2$) group of an amino acid during a peptide synthesis treatment of the product with liquid hydrogen fluoride containing anisole gives the following reaction:

$$C_6H_5{\cdot}CH_2{\cdot}O{\cdot}CO{\cdot}NH\sim + HF$$
$$\longrightarrow C_6H_5CH_2{\cdot}O{\cdot}CO{\cdot}F + H_2N\sim$$
$$C_6H_5{\cdot}CH_2{\cdot}O{\cdot}COF \longrightarrow C_6H_5{\cdot}CH_2F + CO_2$$

Similarly protected S—H groups can be solvolysed.

$$\sim S{\cdot}CH_2{\cdot}C_6H_5 \xrightarrow{HF} \sim SH + C_6{\cdot}H_5{\cdot}CH_2F$$

Thus if we take oxytocin, protected by *S*-benzyl, *S*-(*p*-methoxy)benzyl and *N*-carbo-benzoxy groups, treatment for 1 h at room temperature with hydrogen fluoride containing anisole removes them[70]. The product has more than four times the biological activity of the same product deprotected by conventional means.

The above procedure is currently in wide use and has already aided in the synthesis of bradykinin and other polypeptides. The usefulness of liquid hydrogen fluoride in peptide synthesis in solution has suggested its extension to solid-phase peptide synthesis[71]. Treatment with an anisole solution in liquid hydrogen fluoride has been shown to remove not only the protecting groups from the peptide but also the synthesised peptide from the ion exchange resin on which it has been synthesised. Thus HF–anisole generates free bradykinin from its fully protected, resin bound precursor in a single step[72]. The action of hydrogen fluoride on nucleotides and other esters of phosphorus(V) acids shows that P—O rather than C—O bond cleavage takes place. The reaction is fast and has been suggested as a method for the base analysis of RNA[73].

The higher hydrogen halides, because of their low dielectric constants, dissolve only ionic materials of low lattice energy. Another result of this low dielectric constant is the complex variation of the equivalent conductivity of salts with concentration in these solvents. This variation of conductance with concentration of, for example, Me_2S in liquid hydrogen chloride is very much the same as that predicted by Fuoss and Kraus[74] for a solvent of the same dielectric constant on the assumption that two modes of ionisation of the ion pair A^+B^- are possible

$$A^+B^- \rightleftharpoons A^+ + B^-$$
$$A^+B^- + A^+ \rightleftharpoons A_2B^+$$
$$A^+B^- + B^- \rightleftharpoons AB_2^-$$

Ionisation of the non-conducting ion-pairs to single ions occurs in very dilute solutions, whereas their association to produce conducting ion-triplets takes place in more concentrated solutions. This theory implies a conductivity minimum in plots of conductivity against concentration and this is indeed found in these solutions. Because of the high acidities of the solvents it is difficult to find effective acids in these solvents. This is shown by the extremely low conductivities of both protonic acids and Lewis acids in the solvents. For example, boron(III)chloride which has probably been most extensively used as a Lewis acid in conductivity titrations in liquid hydrogen chloride, has a molar conductance of only 4.0×10^{-4} cm^2 ohm^{-1} mol^{-1} and the Raman spectrum of its solution shows no evidence of tetrachloroborate ion. Other Lewis acids which have been used successfully in conductometric titrations against bases are BF_3, B_2Cl_4, $SnCl_4$, ICl, and PF_5

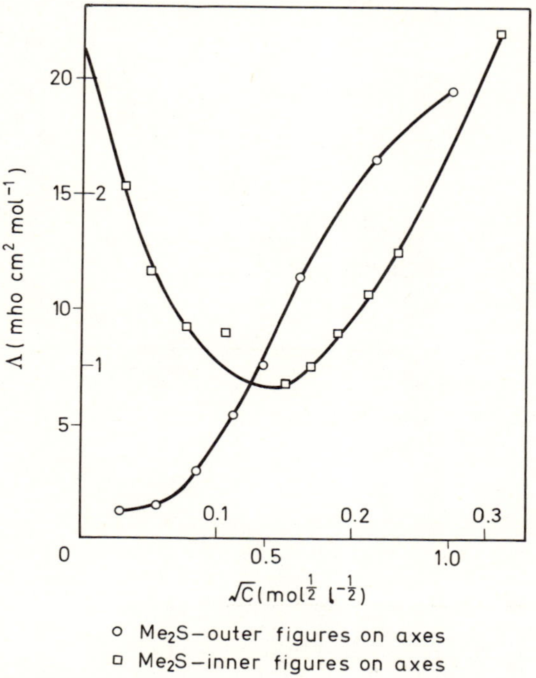

Figure 3.5 The variation of conductivity of dimethyl sulphide in liquid hydrogen chloride with concentration (From Waddington, by courtesy of The Chemical Society)

in liquid hydrogen chloride and BBr_3, $SnBr_4$ in liquid hydrogen bromide and BI_3 in liquid hydrogen iodide.

The reaction of PF_5 with chloride ions in liquid hydrogen chloride is of some interest; hexafluorophosphates are formed and the stoichiometry of the reaction is given by

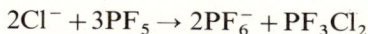

$$2Cl^- + 3PF_5 \rightarrow 2PF_6^- + PF_3Cl_2$$

Conductometric titrations show a break when the molar ratio of base to PF_5 is 2:3 (see Figure 3.6). This behaviour is quite different to that of a conventional Lewis acid such as BCl_3 which gives a conductometric end-point at a ratio of 1:1 (see Figure 3.6).

The only protonic acids which seem to function with any acidity in the solvents are chlorosulphuric acid, hydrogen bromide and hydrogen iodide in liquid hydrogen chloride. No titrations have been attempted with $HClSO_3$

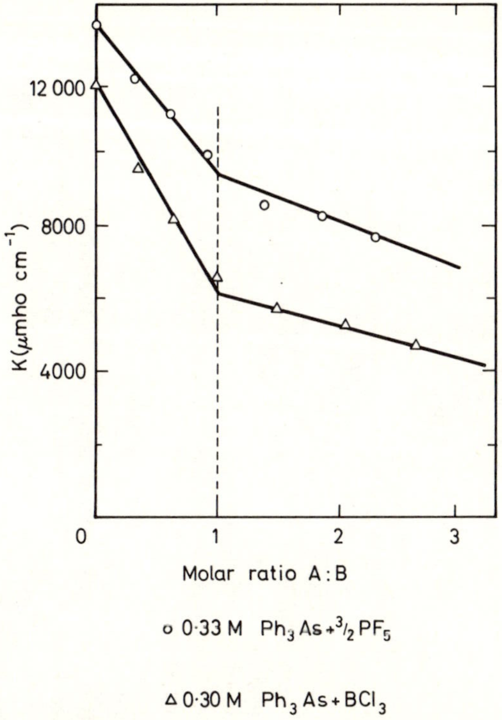

Figure 3.6 The conductivity titration of a base $(C_6H_5)_3As$ with PF_5 and BCl_3 in liquid hydrogen chloride (From Waddington, by courtesy of The Chemical Society)

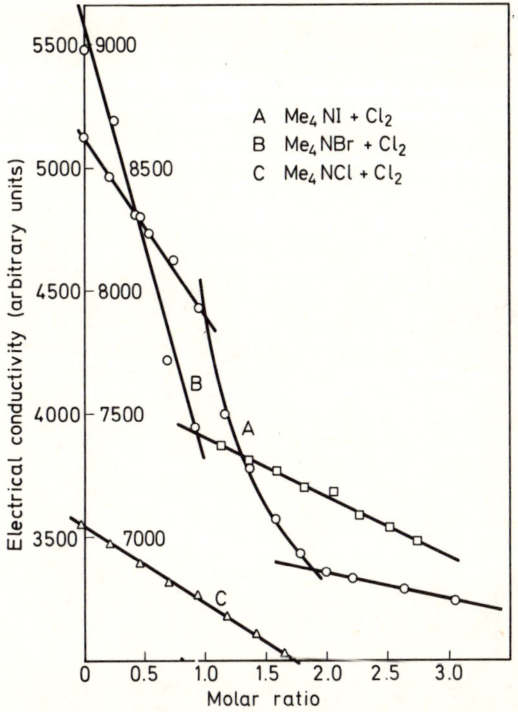

Figure 3.7 The oxidations of the halide ions by chlorine in liquid hydrogen chloride followed conductometrically. The outer scale on the vertical axis refers to titrations B and C the inner scale to titration A
(From Waddington[119], by courtesy of The Chemical Society)

but both HBr and HI failed to give end-points in titrations with chloride ions in liquid hydrogen chloride. However, on treatment of tetra-alkylammonium chlorides in liquid hydrogen chloride with the above three acids, R_4NSO_3Cl, $R_4NHClBr$ and R_4NHClI were respectively recovered on removing excess solvent and letting the systems warm to room temperature[75]. It thus appears that these protonic acids can function as acids in liquid hydrogen chloride and that reactions such as

$$HCl_2^- + HSO_3Cl \rightarrow SO_3Cl^- + 2HCl$$
$$HCl_2^- + HBr \rightarrow HClBr^- + HCl$$

and

$$HCl_2^- + HI \rightarrow HClI^- + HCl$$

do occur. Like some acids in water they are so weak that the end-points of titrations with them cannot be determined conductiometrically.

A solvolysis reaction can be said to occur in liquid hydrogen chloride when a ligand is replaced by a chlorine atom, e.g.

$$MX + HCl \rightarrow MCl + HX$$

This type of reaction has been observed in solutions when X is phenyl, hydroxyl or halogen, for example $Ph_3SnCl + HCl \rightarrow Ph_2SnCl_2 + C_6H_6$.

The only hydroxyl compound found to be solvolysed was triphenylcarbinol

$$Ph_3COH + 3HCl \rightarrow Ph_3C^+HCl_2^- + H_3O^+Cl^-$$

It is not possible to solvolyse either benzoic acid or ethyl benzoate in liquid hydrogen chloride. In liquid hydrogen chloride various fluorides are solvolysed (e.g. AsF_3, SbF_3), some are partially solvolysed (e.g. AsF_5, SbF_5, GeF_4, B_2F_4) and some are not solvolysed at all (SiF_4, BF_3, PF_3). Boron trifluoride acts as an acid, forming the BF_3Cl^- ion. Attempts to prepare the chlorotribromoborate ion, $BClBr_3^-$, from BBr_3 and Me_4NCl in liquid hydrogen chloride lead to solvolysis and formation of tetramethylammonium tetrachloroborate[76]

$$Me_4NCl + BBr_3 + 3HCl \xrightarrow[HCl]{liq} Me_4NBCl_4 + 3HBr$$

In liquid hydrogen bromide, BCl_3 has been reported to solvolyse, and various halides are solvolysed in liquid hydrogen iodide (Ph_3CCl, $GeCl_4$, BCl_3, BBr_3). A recent report has stated that at $-75\,°C$, XeF_2 is soluble in liquid HCl with neither oxidation nor exchange, about $-60\,°C$ Xe, Cl_2 and HF are produced[77].

Many compounds which are too weak to function as bases in aqueous systems are readily protonated in the liquid hydrogen halides and function as strong bases. Thus phosphine is protonated to phosphonium ions and the salts $PH_4^+BCl_4^-$, $PH_4BF_3Cl^-$ and $PH_4^+BBr_4^-$ can be prepared. Dimethyl sulphide is a strong base in liquid hydrogen chloride. Phosphorus pentachloride ionises as a strong base in liquid hydrogen chloride

$$PCl_5 + HCl \rightarrow PCl_4^+ + HCl_2^-$$

A wide range of π-bonded systems, such as ethylenes, acetylenes, C = O, P = O, C = N, N = N and similar systems are bases in the higher hydrogen halides and their protonation reactions have been studied in some detail, particularly in liquid hydrogen chloride. In addition, a number of transition metal systems in low oxidation states can be protonated. Thus ion pentacarbonyl, $Fe(CO)_5$, acts as a base in liquid hydrogen chloride and its protonation can be demonstrated by conductometric titration with BCl_3 [78]. The product of the reaction is $Fe(CO)_5H^+BCl_4^-$. Ferrocene, nickelocene and cobaltocene can also all be protoned in the solvent and the salts $(\pi\text{-}C_5H_5)_2FeH^+BCl_4^-$ and $(\pi\text{-}C_5H_5)_2NiH^+BCl_4^-$ isolated[79]. Similarly $[(\pi\text{-}C_5H_5)Fe(CO)_2]_2$ and $[(\pi\text{-}C_5H_5)Mo(CO)_3]_2$ can also be protonated and the compounds $[(\pi\text{-}C_5H_5)Fe(CO)_2]H^+\ BCl_4^-$ and $[(\pi\text{-}C_5H_5)Mo(CO)_3]_2H^+\ BCl_4^-$ isolated[80]. They contain M—H—M bonds (M = Fe or Mo) with no bridging carbonyl groups.

Oxidation-reduction reactions have been studied in liquid hydrogen chloride using chlorine, bromine and iodine monochloride as oxidising agents. The reactions of these oxidising agents with simple halide ions are instructive. Chlorine oxidises the iodide ion in two stages, to ICl_2^- and to ICl_4^-, and both stages can be detected conductometrically (see Figure 3.7). Bromine in liquid hydrogen chloride oxidises the iodide ion to IBr_2^- and in a conductometric a 1:1 end-point is found for Br_2 *v.* I^-. The reactions of iodine monochloride are more complicated. In titrations of ICl against I^-, two end-points are found, at 1:1 and 2:1 for ICl *v.* I^-.

Iodine is first liberated

$$I^- + ICl + HCl \rightarrow I_2 + HCl_2^- \qquad 1:1$$

and the ICl_2^- ion is then formed in an acid base reaction

$$HCl_2^- + ICl \rightarrow ICl_2^- + HCl \qquad 2:1$$

Oxidations of P^{III}, S^{II} and Se^{II} compounds in liquid hydrogen chloride have also been studied. The oxidation of PCl_3 by both Cl_2 and Br_2 can be followed conductometrically in the solvent. Both reactions have 1:1 end-points and formation of PCl_4^+ and PCl_3Br^+ ions respectively occurs. The latter ion can be isolated as the tetrachloborate, $PCl_3Br^+BCl_4^-$. The reaction of PCl_3 with ICl is difficult to follow conductometrically; the end-points are variable and iodine is precipitated during the reaction. The final product is $PCl_4^+ICl_2^-$ [81]. When dialkyl sulphides are dissolved in liquid hydrogen chloride, the R_2SH^+ ion is formed and this can be oxidised to R_2SCl^+ by chlorine[82]. Similarly, RSCl is protonated to $RSClH^+$ and this ion is again oxidised to give $RSCl_2^+$ by chlorine. Transition metal species in low oxidation numbers can also be oxidised in the solvent. Thus $Fe(CO)_5$ gives with Cl_2, Br_2 and NOCl the ions $Fe(CO)_5Cl^+$, $Fe(CO)_5Br^+$ and $Fe(CO)_5NO^+$ [78]. The first two have been isolated as their tetrachloroborates.

3.4 PROPERTIES OF THE HYDROGEN DIHALIDE ANIONS AND HIGHER SPECIES

These species can all be regarded as solvates obtained from halides dissolved in the liquid hydrogen halides but the simplest members of the series of

Table 3.11 Structural information on the alkali metal and ammonium difluorides

Compound	*Crystal class*	*Space group*	*z*	*Lattice parameter* (pm)	*Fluorine-fluorine distance* (pm)	*Reference*
$LiHF_2$	Rhombohedral	$R\bar{3}m$	1	a = 472.5 α = 37° 3′	227.0	L. K. Frevel and H. W. Rinn *Acta. Crystallogr.*, **15**, 256, 1962
$NaHF_2$	Rhombohedral	$R3m$	1	a = 500.4 α = 40° 33′	226.0	McGaw and Ibers, *J. Chem. Phys.*, **39**, 2677, 1963
KHF_2	Tetragonal	$P4/mnm$	4	a = 567; c = 681	225.8	S. W. Peterson and H. A. Levey *J. Chem. Phys.*, **20**, 704, 1952
$RbHF_2$	Tetragonal	$P4/mnm$	4	a = 590; c = 726	225.8	R. Kruh, K. Fueva and T. E. McEver *J. Amer. Chem. Soc.*, **78**, 4526, 1956
$CsHF_2$	Tetragonal	$P4/mnm$	4	a = 614; c = 784	225.8	R. Kruh, K. Fueva and T. E. McEver *J. Amer. Chem. Soc.*, **78**, 4526, 1956
NH_4HF_2	Orthorhombic	$Ibam$	4	a = 840.8; b = 816.3 c = 367.0	227.5 226.9	M. T. Rogers and L. Helmholz, *J. Amer. Chem. Soc.*, **62**, 1533, 1940

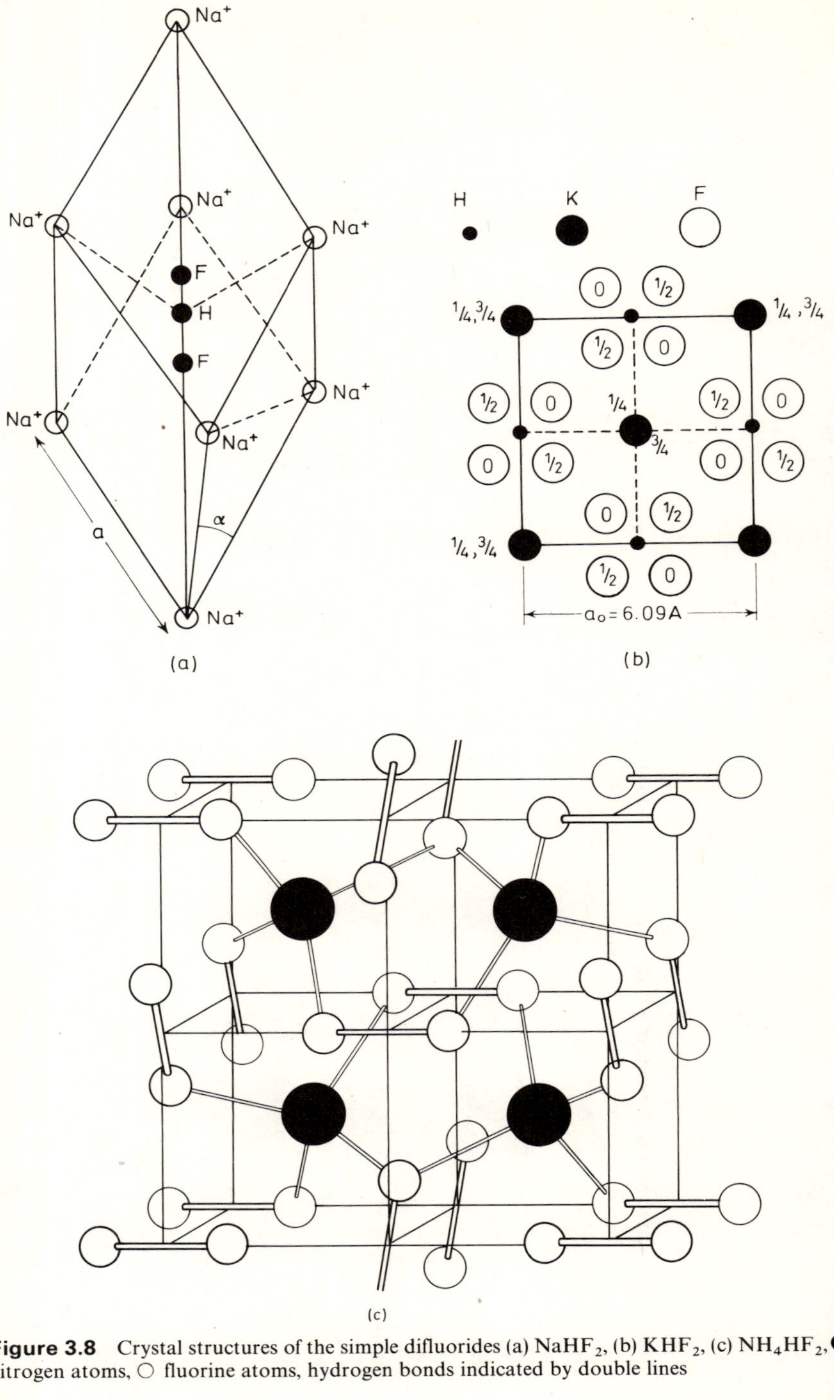

Figure 3.8 Crystal structures of the simple difluorides (a) $NaHF_2$, (b) KHF_2, (c) NH_4HF_2, ● nitrogen atoms, ○ fluorine atoms, hydrogen bonds indicated by double lines

solvates, the hydrogen dihalide anions are so important that they are best regarded as complex ions in their own right.

3.4.1 The hydrogen difluoride anion, HF_2

The acid fluorides have been known since the time of Berzelius and Fremy[83]. The simplest members of this series are the alkali metal and ammonium difluorides MHF_2. They are so stable that they may be crystallised from aqueous solutions containing hydrofluoric acid. The difluoride ion also occurs in the salts SrF_2, HF and BaF_2, HF, which may be formulated as $M^{2+}F^{-}HF_2^{-}$. The compounds are all isomorphous with the corresponding azides and clearly the difluoride ion, HF_2^{-}, is about the same size as the azide ion, N_3^{-}. Lithium and sodium difluorides have a rhombohedral unit cell containing one formula unit, potassium, rubidium and caesium difluorides are all tetragonal, with four formula units in the unit cell. Ammonium

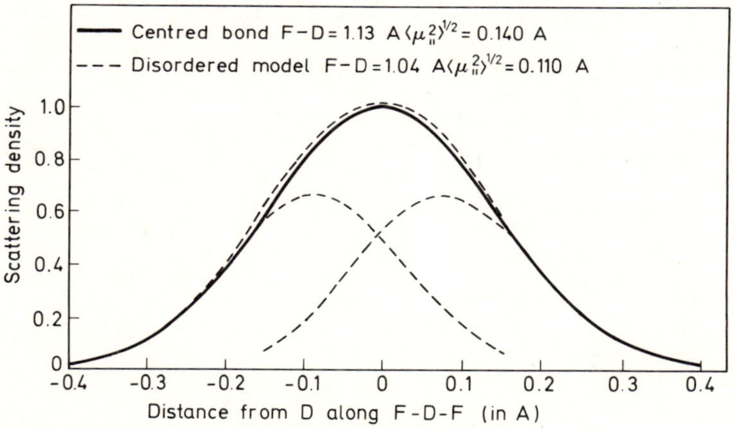

Figure 3.9 Proton density along the bond in HF_2^{-}; a demonstration of how two half-hydrogens of appropriate amplitudes of thermal vibration can give the same result as a centred hydrogen
(From Ibers and McGaw[85], by courtesy of The American Institute of Physics)

difluoride is orthorhombic with four formula units in the unit cell. The lattice parameters, space groups and F—F distances in the difluoride ion are given in Table 3.11 and some of the typical crystal structures are shown in Figure 3.8. From the data in the table it will be seen that the F—H—F bond distance is about 50 pm shorter than twice the Van der Waals radius of the fluoride ion. It is clearly an example of a very strong hydrogen bond and as a three atom system, it represents the simplest hydrogen bonding situation possible. One of the most important questions about the difluoride ion is whether the hydrogen is symmetrically placed in the ion or not. Quite early on Pitzer and Westrum[89] concluded from crystal entropy measurements that the salt KHF_2 possessed no residual entropy at absolute zero. They suggested that this meant that the HF_2^{-} ion was either symmetrical or if it did have a central potential barrier this could be easily tunnelled through by the proton. The hydrogen bond in KHF_2 was one of the first subjects of study using broad line n.m.r. techniques. Waugh, Humphrey and Yost determined the second moments of the hydrogen and fluorine magnetic resonance absorption lines in a polycrystalline sample of KHF_2. They concluded that the H atom is centred to within ± 0.06 Å. Because of various uncertainties in the corrections for the effects of thermal motion their estimate of the accuracy to which they had demonstrated that the proton was centrally placed in the ion was over optimistic.

Potassium hydrogen difluoride was one of the first materials to be studied by neutron diffraction measurements on a single crystal[86]. The data from these measurements has been re-analysed by Ibers[90] and McGaw and Ibers[85] have studied both $NaHF_2$ and $NaDF_2$ by neutron diffraction. The basic problem in both of these studies is uncertainty about zero-point motional amplitudes. McGaw and Ibers[90] were able to obtain equally good agreement with their data for a wide range of displacements of the proton from the centre of the F—F bond, provided that suitable adjustments were made in

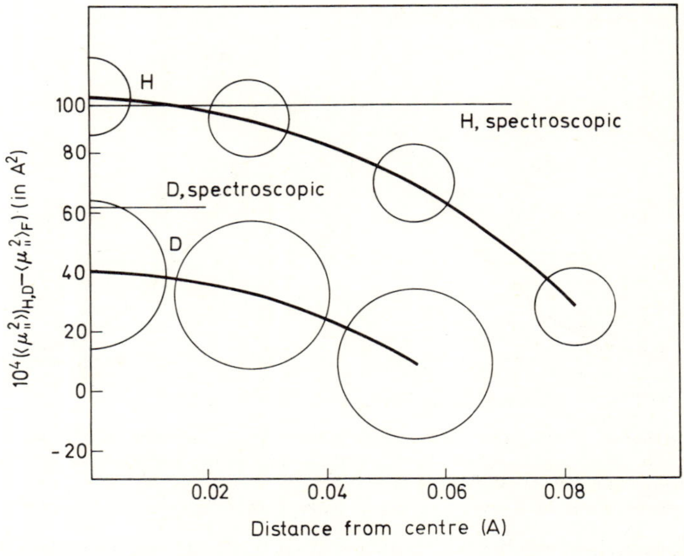

Figure 3.10 The combination of neutron diffraction and vibrational data, used to show that the H-bond is centred in the difluoride ion (From Ibers and McGaw[85], by courtesy of The American Institute of Physics)

the amplitude of vibration of the proton along the bond. This is illustrated in Figure 3.9. However, McGaw and Ibers[90] pointed out that a combination of neutron diffraction data and spectroscopic data could be used to demonstrate very powerfully that the F—H—F and F—D—F bonds are symmetrical. Essentially the neutron diffraction data enable one to calculate a value for $<x_H^2> - <x_F^2>$ for each assumed position of the proton in the bond and then the normal co-ordinate analysis of the vibrational spectra of the HF_2^- ion enables the true value of $<x_H^2> - <x_F^2>$ to be found. This is shown in Figure 3.10.

The first infrared study of the difluoride ion was due to Ketelaar[91]. He observed two bands in KHF_2 that he ascribed to fundamental modes; one at 1450 cm^{-1} and the other at 1225 cm^{-1}. He ascribed both fundamentals to the two components of the symmetrical stretch, suggesting that the ion must have a large double potential minimum. Later infrared studies by Ketelaar and Vedder[92] and polarised infrared studies on oriented crystals by Coté and Thompson[93] have shown that the band at 1450 cm^{-1} is the symmetrical stretching frequency and the band at 1225 cm^{-1} is the bending mode. The symmetrical stretching mode has been found in the Raman spectrum at 600 cm^{-1} [94]. The far infrared spectra of sodium and potassium difluorides have been studied by two groups of workers[95,96]. The inelastic neutron scattering spectrum of KHF_2 has been measured by Boutin, Safford and Brajovic[97].

Another important question in relation to the difluoride ion has been the hydrogen bond strength in the ion. This is best defined in terms of the process

$$HF_2^-(\text{gas}) \rightarrow HF(\text{gas}) + F^-(\text{gas}),\ \Delta H_1$$

though this is related to the process

$$HF_2^-(\text{gas}) \rightarrow 2F^-(\text{gas}) + H^+(\text{gas}),\ \Delta H_2$$

by the enthalpy of dissociation, ΔH_3, of HF(g) into H^+(g) and F^-(gas), which has a value of 1536.6 kJ mol^{-1}.

ΔH_1 can be obtained by means of the thermochemical cycle below

$$\begin{array}{ccc} MHF_2(c) & \xrightarrow{\Delta E(MHF_2)+2RT} & M^+(g) + HF_2^-(g) \\ \Delta H_D \uparrow & & \downarrow \Delta H_1 \\ MF(c) + HF(g) & \xleftarrow[-\Delta E(MF)-2RT]{} & M^+(g) + HF(g) + F^-(g) \end{array}$$

where $\Delta E(MHF_2)$ and $\Delta E(MF)$ are the lattice energies of the difluoride and corresponding fluoride and ΔH_D is the enthalpy of formation of the difluoride from the solid fluoride and gaseous HF, hence

$$\Delta H_1 = \Delta E(MF) - \Delta E(MHF_2) - \Delta H_D$$

Thus calculations of the lattice energies of the fluorides and difluoride coupled with a knowledge of the dissociation enthalpy of the difluorides can give a value for the hydrogen bond energy in the difluoride ion.

Such a set of calculations was carried out by Waddington[98] on the potassium, rubidium and caesium salts and led to a value of 243 ± 20 kJ mol^{-1} for the bond energy in the difluoride ion. Harrell and McDaniel[99] have measured the enthalpy of the reaction

$$(CH_3)_4NF(s) + HF(g) = (CH_3)_4NHF_2(s)$$

as 155 kJ mol^{-1}; but their claim that the true energy of the hydrogen bond in HF_2^- is within a few kilojoules or so of this value rests on the assumption that the lattice energies of $(CH_3)_4NF$ and $(CH_3)_4NHF_2$ are equal, an unproven and unlikely condition. *Ab initio* quantum mechanical calculations on the HF_2^- ion have been made by Bessis and Bratoz[100], Clementi and McLean[101] and McLean and Yoshimine[102]. They led respectively to values of 1727, 1252 and 1725 kJ mol^{-1} for ΔH_2 and hence to values of 191, −285 and 188 kJ mol^{-1} for ΔH_1.

3.4.2 The hydrogen dichloride anion, HCl_2^-

The range of stabilities of the hydrogen dichlorides is much more restricted than that of the hydrogen difluorides. The only alkali metal dichloride that has been well characterised is the caesium hydrogen dichloride, which has a dissociation vapour pressure of 4.4 atm. at room temperature[103]. The rubidium salt appears to exist, but at room temperature potassium chloride co-exists with liquid hydrogen chloride under pressure without the formation of the $KHCl_2$ salt[104]. Tetramethylammonium hydrogen dichloride has a dissociation vapour pressure of about 1 mm at room temperature and that of the other tetra-alkylammonium halides is even lower. All these salts are extremely hygroscopic and have to be studied in the absence of water. A rather curious compound is obtained when gaseous hydrogen chloride is passed into a saturated solution of caesium chloride in water[105]. A white solid is precipitated which is stable only in the presence of water vapour and gaseous hydrogen chloride. It has the composition $CsCl(\frac{1}{3}H_3O^+HCl_2^-)$ [106].

The infrared and Raman spectra of the dichlorides seem to indicate these fall into two classes[107].

Type I salts, which include those of caesium, tetramethylammonium and hexadecyltrimethylammonium are characterised by two very broad and strong bands in the infrared, one in the 1520–1670 cm^{-1} region and the other near 1200 cm^{-1}. There is also a very weak band at about 220 cm^{-1}. Detailed investigation of the infrared spectrum of $CsHCl_2$ and $CsDCl_2$ at liquid hydrogen temperatures by Nibler and Pimentel[108] has shown that an extremely weak feature at about 630 cm^{-1} in $CsHCl_2$ sharpens and increases in intensity on cooling. They ascribe this to the fundamental of the bending frequency, ν_2, the much more intense feature at 1200 cm^{-1} to its first harmonic, $2\nu_2$, the feature at 1520–1670 cm^{-1} to the asymmetrical stretching frequency, ν_3, and the weak band at 220 cm^{-1} to the symmetrical stretching frequency of the HCl_2^- ion, ν_1. These assignments are confirmed by the Raman and inelastic neutron scattering spectra of $CsHCl_2$ recorded by Ludman, Waddington and Stirling[109].

Type II dichloride salts, which include those of trimethylethylammonium, tetraethylammonium, tetrapropylammonium and $CsCl(\frac{1}{3}H_3O^+HCl_2^-)$, appear to be characterised by very broad, asymmetric absorptions extending between about 600 and 1300 cm^{-1}, which may consist of a main broad band centred in the 700–800 cm^{-1} region and another broad band in the 1000–1050 cm^{-1} region. Evans and Lo[107] concluded tentatively that the hydrogen dichloride ion in Type II salts was symmetrical and in Type I was asymmetrical. This is supported by the inelastic neutron-scattering studies which indicate that the band at about 200–220 cm^{-1} in $CsHCl_2$ involves the motion of the hydrogen atom, which can only be the case if the ion is asymmetrical[109].

The results of nuclear quadrupole resonance studies[110–112] on the ^{35}Cl nuclei in a number of solid hydrogen dichlorides support the division of the salts into two classes. $CsHCl_2$ and Me_4NHCl_2, both Type I salts, show a single ^{35}Cl n.q.r. resonance at about 21 MHz, which shifted by about $\frac{1}{2}$ MHz on deuteration of the HCl_2^-, whereas $CsCl(\frac{1}{3}H_3O^+HCl_2^-)$ and Et_4NHCl_2 both show a single ^{35}Cl n.r. peak at about 11 MHz which shifts hardly at all on deuteration of the HCl_2^-.

The crystal structures of a hydrogen dichloride salt of each class has now been determined by x-ray diffraction. The crystal structure of $CsCl(\frac{1}{3}H_3O^+ HCl_2^-)$, a Type II salt, was determined by Schroeder and Ibers[106] in 1966.

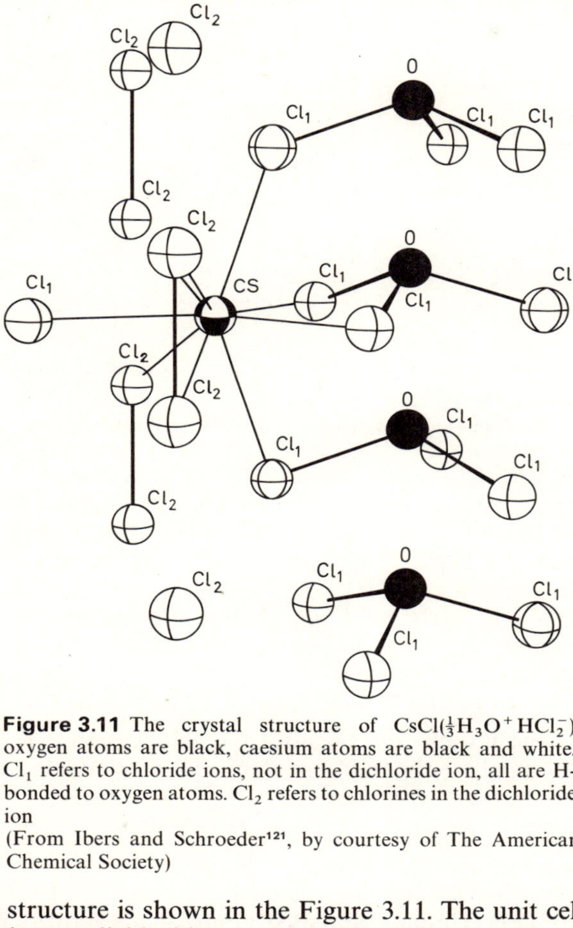

Figure 3.11 The crystal structure of $CsCl(\frac{1}{3}H_3O^+HCl_2^-)$ oxygen atoms are black, caesium atoms are black and white. Cl_1 refers to chloride ions, not in the dichloride ion, all are H-bonded to oxygen atoms. Cl_2 refers to chlorines in the dichloride ion
(From Ibers and Schroeder[121], by courtesy of The American Chemical Society)

Part of the structure is shown in the Figure 3.11. The unit cell is hexagonal and the hydrogen dichloride occur in strings Cl—H—Cl ... Cl—H—Cl ... Cl—H—Cl... parallel to the hexagonal axis. There is a plane of symmetry perpendicular to this axis through the middle of the hydrogen dichloride ions and the chlorines in the ion are crystallographically equivalent. No evidence for the proton positions was found from the x-ray study. However, because of the equivalence of the chlorine atoms, the HCl_2^- ion in these salts must possess a symmetrical single or double potential minimum. The Cl—H—Cl distance in the ion is 314 ± 2 pm, compared with a Cl...Cl distance between ions of 362 pm, which is just twice the ionic radius of a chloride ion.

The crystal structure of Me_4NHCl_2 has been reported by Swanson and Williams[113] from a single crystal x-ray diffraction study. The crystal is orthorhombic with $a = 927 \pm 1$ pm, $b = 773 \pm 1$ pm, $c = 1159 \pm 1$ pm, and four molecules to the unit cell. The space group is *Pnma* verified by the successful refinement. The structure, a view of which is shown in Figure 3.12 is

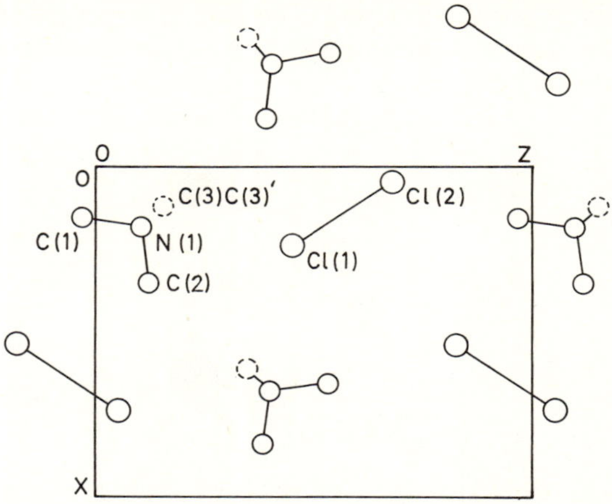

Figure 3.12 A projection of the crystal structure of Me_4NHCl_2 (From Williams and Swanson[113], by courtesy of Pergamon Press)

composed of two distinct entities: the Me_4N^+ ion and the HCl_2^- ion. The Cl—H—Cl distance is 322 ± 2 pm. There is no mirror plane through the ion perpendicular to the Cl—H—Cl bond. These two structures do confirm a difference between the Type I and Type II salts containing the hydrogen chloride ion.

The value of the hydrogen bond energy in the hydrogen dichloride ion has been estimated in two ways. Vallee and McDaniel[114] have argued that by plotting the enthalpies of dissociation,

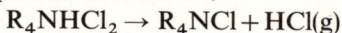

$$R_4NHCl_2 \rightarrow R_4NCl + HCl(g)$$

of a series of hydrogen dichlorides against the cation radii, one can extrapolate to a situation where the radius of the cation has no effect and hence determine ΔH for the isolated process

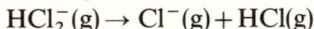

$$HCl_2^-(g) \rightarrow Cl^-(g) + HCl(g)$$

They obtain a value of about 75 kJ mol^{-1} this way.

Salthouse[115] has also attempted to calculate the energy of the hydrogen bond in the HCl_2^- ion. He has computed approximate lattice energies for Me_4NCl and Me_4NHCl_2 and combined these with the enthalpy of dissociation of Me_4NHCl_2 in a cycle similar to the one used for the bifluorides. A value of 67 kJ mol^{-1} was obtained.

3.4.3 The hydrogen dibromide, hydrogen di-iodide and mixed hydrogen dihalide ions

In addition to the HF_2^- and HCl_2^- ions, the HBr_2^- [116,117] and HI_2^- [118] ions have been prepared as have all the possible HXY^- species[119,120] with the exception of $HBrI^-$, which appears to be incapable of existence. The stability of salts of all these species always appears to be less than that of the corresponding HCl_2^- salts containing the same cation. Thus the only other caesium salt that has been prepared at room temperature at the moment is $CsHBr_2$ and this has a much higher dissociation energy than $CsHCl_2$. All these salts, like the hydrogen dichlorides, are extremely hygroscopic and have to be prepared in the absence of any moisture.

The only physical property that has been studied of most of them is their infrared spectra and a list of frequencies and assignments is given in Table 3.12.

Table 3.12

Species	ν_1	$2\nu_2$	ν_2	ν_3
CsClHBr	170	1100	508	1705
CsClDBr	170	820	410	1300
$Me_4NBrHBr$	126	1038		1420
$Me_4NBrDBr$	123	752		1070
CsClHI		990	485	2200
CsClDI		730	350	1640
$HFCl^-$	275	835		2710
$HFBr^-$	220	740		2900
HFI^-	180	635		3145
IHI^-		1165		1700
IDI^-		840		1300

Again the fundamental of the bending frequency of these ions, ν_2, is very weak in the infrared and the overtone $2\nu_2$ is much more intense.

Only one crystal structure has been reported and this is of the salt $CsBr(\frac{1}{3}H_3O^+HBr_2^-)$ [121]. The structure is completely analogous to that of $CsCl(\frac{1}{3}H_3O^+HCl_2^-)$ and indeed the compound is made in the same way, by passing gaseous hydrogen bromide into a saturated solution of caesium bromide in water. This salt has a mirror plane through the middle of the ion perpendicular to the Br—H—Br axis and may contain a symmetrical bond. Its infrared spectrum is very different from that of the other HBr_2^- salts.

Estimates of the hydrogen bond energy in the HBr_2^- and HI_2^- ion have been made by Vallee and McDaniel[122] using the same method they employed for HCl_2^-; they lie between 50 and 63 kJ mol^{-1}.

The decompositions of the mixed hydrogen dihalides, MHXY, is of some interest in that the nature of the products is determined by the bond energy of the hydrogen halide, HY, produced rather than the lattice energy of the accompanying simple salt.

Thus the $HClBr^-$ salts decompose to bromides and hydrogen chloride

$$MHClBr \rightarrow MBr + HCl$$

whilst

$$MHClI \rightarrow MI + HCl$$

3.4.4 Higher species

Though the systems KH_2F_3, KH_3F_4 and KH_4F_5 can be isolated[123] from the interaction of potassium fluoride and anhydrous hydrogen fluoride under varying conditions at low temperatures only the salt KH_2F_3 and its deuterated analogue KD_2F_3 have been studied in any detail. An x-ray diffraction study of the crystal structure[124] of KH_2F_3 has shown that the fluorine atoms occur in groups of three, forming a triangle with F—F distances of 2.33 pm and an apex angle of about 135 degrees. The nuclear magnetic resonance spectra[125] of crystalline KH_2F_3 and NaH_2F_3 are consistent with a rigid chain A—B—A—B—A system of five nuclei with spin $\frac{1}{2}$. Any structure with a 'normal' HF distance, as in pure hydrogen fluoride, is ruled out. Clearly the hydrogen bonds are strong but there is some evidence that, unlike the HF_2^- ion, the hydrogens are not situated symmetrically between the fluorines. The neutron inelastic scattering spectrum[97] of KH_2F_3 and NaH_2F_3, together with a detailed normal coordinate analysis of the observed infrared spectrum of KH_2F_3 between 4000 and 200 cm^{-1} followed by the determination of the Urey–Bradley force field[126], suggest that the hydrogen atoms are not exactly halfway between the fluorines.

Similar species can be isolated and stabilised, at least at low enough temperatures, with the other hydrogen halides provided the cation accompanying the complex anion is large enough. Thus Waddington and Salthouse[119] have isolated $Bu_4NH_2Cl_3$, $Bu_4NBr(HCl)_2$ and $Bu_4NI(HCl)_2$ at low temperatures from solutions of the tetrabutyl ammonium halides in liquid hydrogen chloride.

References

1. *Supplement II. Part I. Mellors Inorganic Chemistry (1957).* (London: Longmans)
2. *Teil, B. Lieferung* **1,** *Chlor, System—number,* **6.** *Gmelin's Handbuch der Inorganische Chemie* (1968). (Weinheim: Verlag Chemie)
3. *Fluor, System—number,* **5**, *Gmelin's Handbuch der Inorganische Chemie* (1959). (Weinheim: Verlag Chemie)
4. *Tables of Interatomic Distances and Configuration in Molecules and Ions* (1958). Special Publication No. 11. The Chemical Society—London
5. *Supplement to Special Publication No.* **11.** (1965) Special Publication No. 18. The Chemical Society—London
6. Nelson, Lide and Maryatt. (1967). *U.S. Dept. Comm. NSRDS-NBS10,* **10,** 1/49
7. Robinette, W. H. and Saunderson, R. B. (1969). *Appl. Opt.,* **8,** 711
8. *Landolt-Bernstein, Tabelen* (1951), 6. Auflage., Bd. 1. Tiel 3
9. Yoshimo, Y. and Bernstein, H. J. (1958). *J. Mol. Spectrosc.,* **2,** 237
10. JANAF (1965). *Thermochemical tables* PB 168370 and *First, Second and Third Addenda* (to 1968) PB 165370-1, -2, -3. Distributed by Clearinghouse for Federal Scientific and Technical Information. U. S. Department of Commerce, Springfield, Virginia. 22151
11. Nakamoto, K. (1970). *Infra-red Spectra of Inorganic and Co-ordination Compounds* 2nd edn. (New York: Wiley)
12. Cyvin, S. J. (1968). *Molecular Vibrations and Mean Square Amplitudes* (London: Elsevier)
13. Barnes, A. J., Hallam, H. F. and Scrimshaw, G. F. (1969). *Trans. Faraday Soc.,* **65,** 3150, 3159, 3172
14. Shurrell, H. F. and Harvey, K. B. (1969). *Canad. Spectrosc.,* **14,** 84
15. National Bureau of Standards. (1968). Technical Note 270-3. *Selected Values of Chemical Thermodynamic Properties.* (Washington: U.S. Government Printing Office)
16. Tilford, S. G., Ginter, M. L. and Vanderslice, J. T. (1970). *J. Mol. Spectrosc.,* **33,** 505
17. Tilford, S. G., Ginter, M. L. and Bass, A. M. (1970). *J. Mol. Spectrosc.,* **34,** 327
18. Ginter, M. L. and Tilford, S. G. (1970). *J. Mol. Spectrosc.,* **34,** 206
19. Lempka, H. J., Passmore, T. R. and Priçe, W. C. (1968). *Proc. Roy. Soc. A.,* **304,** 53
20. National Bureau of Standards (1949). Circular 500 *Selected Values of Chemical Thermodynamic Properties.* (Washington: U.S. Government Printing Office)

21. Berry, R. S. and Reimann, C. W. (1963). *J. Chem. Phys.*, **38**, 1540
22. Bailey, T. L. (1958). *J. Chem. Phys.*, **28**, 792
23. Steiner, B., Seman, M. L. and Branscomb, L. M. (1962). *J. Chem. Phys.*, **37**, 1200
24. Schneider, W. G., Bernstein, H. J. and Pople, J. A. (1958). *J. Chem. Phys.*, **28**, 601
25. Hameka, H. F. (1959). *Mol. Physics*, **2**, 64
26. Solomon, I. and Bloembergen, N. (1956). *J. Chem. Phys.*, **25**, 261
27. Menter, J. S, and Klemperer, W. (1970). *J. Chem. Phys.*, **52**, 6033
28. Lucken, E. A. (1969). *Nuclear Quadrupole Coupling Constants.* (New York: Academic Press)
29. Townes, C. H. and Dailey, B. P. (1949). *J. Chem. Phys.*, **17**, 782
30. Dailey, B. P. and Townes, C. H. (1953). *J. Chem. Phys.*, **23**, 118
31. Smith, D. F. (1958). *J. Chem. Phys.*, **28**, 1040
32. Pham Van Huong and Couzi, M. (1969). *J. Chim. Phys. Physiochim. Biol.*, **66**, 1309
33. Janzen, J. and Bartell, L. S. (1969). *J. Chem. Phys.*, **50**, 3611
34. Mackor, E. L., Maclean, C. and Hilbers, C. W. (1968). *Rec. Trav. Chim. Pays-Bas*, **87**, 655
35. Dodd, R. E. and Robinson, P. L. (1957). *Experimental Inorganic Chemistry.* (Amsterdam: Elsevier)
36. Maybury, R. H., Gordon, S. and Katz, J. J. (1955). *J. Chem. Phys.*, **23**, 1277
37. Hyman, H. H. and Katz, J. J. (1965). *Liquid Hydrogen Fluoride.* Chapter 2, of *Non-Aqueous Solvent Systems* Ed. by Waddington, T. C. (London: Academic Press)
38. Armstrong, D. A. and Back, R. A. (1967). *Canad. J. Chem.*, **45**, 3079
39. Field, F. H. and Lampe, F. W. (1958). *J. Amer. Chem. Soc.*, **80**, 5583
40. Atoji, M. and Lipscomb, W. N. (1954). *Acta. Crystallogr.*, **7**, 173
41. Boorsch, P. (1970). *Ber.* Kernforschungsanlage, Juelich
42. Axmann, A., Biem, W., Borsch, P., Hossfeld, F. and Stiller, H. (1969). *Discuss. Faraday Soc.*, **48**, 69
43. Sándor, E. and Farrow, R. F. C. (1967). *Nature*, **213**, 171
44. Sándor, E. and Farrow, R. F. C. (1967). *Nature*, **215**, 1265
45. Sándor, E. and Farrow, R. F. C. (1969). *Discuss. Faraday. Soc.*, **48**, 78
46. Natta, G. (1933). *Gazz*, **63**, 425
47. Benson, S. W. (1960). *The Foundations of Chemical Kinetics.* (New York: McGraw-Hill)
48. Fettis, G. C. and Knox, J. H. (1964). *The Rate Constants of Halogen Atom Reactions* from *Progress in Reaction Kinetics Vol* **2** (London: Pergamon Press)
49. Thrush, B. A. (1965). *The Reactions of Hydrogen Atoms* from *Progress in Reaction Kinetics Vol.* **3** (London: Pergamon Press)
50. Childs, W. V. (1968). *U.S. Pat.* 3 445 292
51. Latimer, W. M. (1952). *Oxidation Potentials*, 2nd ed. (New York: Prentice Hall)
52. Anlauf, K. G., Mayotte, D. H., Polanyi, J. C. and Bernsterin, R. B. (1969). *J. Chem. Phys.*, **51**, 5716
53. Goldfinger, P., Noyes, R. M. and Wen, W. Y. (1969). *J. Amer. Chem. Soc.*, **91**, 4003
54. Airey, J. R. (1970). *J. Chem. Phys.*, **52**, 156
55. Spencer, D. J., Mircls, H., Jacobs, T. A. and Gross, R. W. E. (1970). *Appl. Physics. Letters*, **16**, 235
56. Ivaskeutsev, I. Y., Kutakova, L. I. and Ketov, A. N. (1968). *Izvest. Vyssh. Ucheb. Zavebl, Khim., Khim. Technol.*, **11**, 388
57. Kao, J. T. (1970). *J. Chem. Eng. Data.*, **15**, 362
58. Sillen, L. G. and Martell, A. E. (1964). *Stability Constants of Metal-Ion Complexes. Special Publication No.* **17** (London: Chemical Society)
59. Halliwell, H. F. and Nyburg, S. C. (1963). *Trans. Faraday Soc.*, **59**, 1126
60. Conway, B. E. and Bockris, J. O'M. (1954). *Ionic Solvation in Modern Electrochemistry* Ed. by Bockris, J. O'M. (London: Butterworths)
61. Robinson, R. A. and Stokes, R. H. (1959). *Electrolyte Solutions.* (London: Butterworths)
62. Yoon, Y. K. and Carpenter, G. B. (1959). *Acta Crystallogr.*, **12**, 17
63. Mullhaupt, J. T. (1959). *Diss. Abstr.*, **19**, 1581
64. Lundgren, J. O. and Olovsson, I. (1967). *Acta. Crystallogr.*, **23**, 971
65. Lundgren, J. O. and Olovsson, I. (1968). *J. Chem. Phys.*, **49**, 1068
66. Peach, M. E. and Waddington, T. C. (1965). *The Higher Hydrogen Halides as Ionizing Solvents* chapter 3. of *Non Aqueous Solvent Systems* Ed. by Waddington, T. C. (London: Academic Press)
67. Dove, M. F. A. and Hallett, J. G. (1969). *J. Chem. Soc. A*, 2781

68. Folec, B. *Nukl. Inst. Jozef Stefan, NIJS Porocilo,* P-206 C.A. 69: 54683t.
69. Lenard, J. (1969). *Chem. Revs.,* **69,** 625
70. Sakakibara, S. and Shimonishi, Y. (1965). *Bull. Chem. Soc. Jap.,* **38,** 1412
71. Sakakibara, S., Kishida, Y., Nishizawa, R. and Shimonishi, Y. (1968). *Bull. Chem. Soc. Jap.,* **41,** 1273
72. Lenard, J. and Robinson, A. B. (1967). *J. Amer. Chem. Soc.,* **89,** 181
73. Lipkin, D., Phillips, B. E. and Abrell, J. W. (1969). *J. Org. Chem.,* **34,** 1529
74. Fuoss, R. M. and Kraus, C. A. (1933). *J. Amer. Chem. Soc.,* **55,** 21, 2387
75. Salthouse, J. A. and Waddington, T. C. (1964). *J. Chem. Soc.,* 4664
76. Peach, M. E. (1971). *Inorg. Nucl. Chem. Letters,* in the press
77. Shaw, M. J., Holloway, J. H., Hyman, H. H. (1970). *Inorg. Nuclear. Chem. Letters,* **6,** 321
78. Iqbal, Z. and Waddington, T. C. (1968). *J. Chem. Soc. A,* 2958
79. Iqbal, Z. and Waddington, T. C. Unpublished results
80. Symon, D. A. and Waddington, T. C. (1971). *J. Chem. Soc. A.,* in the press
81. Salthouse, J. A. and Waddington, T. C. (1966). *J. Chem. Soc. A.,* 1188
82. Peach, M. E. (1969). *Canad. J. Chem.,* **47,** 1675
83. Frémy, E. (1856). *Ann. Chim. Phys.,* (3), **47,** 5
84. Frevel, L. K. and Rinn, H. W. (1962). *Acta. Crystallogr.,* **15,** 286
85. McGaw, B. L. and Ibers, J. A. (1963). *J. Chem. Phys.,* **39,** 2677
86. Peterson, S. W. and Levy, H. A. (1952). *J. Chem. Phys'.* **20,** 704
87. Kruh, R., Fuwa, K. and McEver, T. E. (1956). *J. Amer. Chem. Soc.,* **78,** 4256
88. Rogers, M. T. and Helmholz, L. (1940). *J. Amer. Chem. Soc.,* **62,** 1533
89. Pitzer, K. S. and Westrum, E. F. (1947). *J. Chem. Phys.,* **15,** 526
90. Ibers, J. A. (1964). *J. Chem. Phys.,* **40,** 402
91. Ketelaar, J. A. A. (1941). *Rec. Trav. Chim.,* **60,** 523
92. Ketelaar, J. A. A. and Vedder, W. (1951). *J. Chem. Phys.,* **19,** 654
93. Cote, G. L. and Thompson, H. W. (1952). *Proc. Roy. Soc. A,* **210,** 206
94. Couture, L. and Mathieu, J. P. (1949). *Compt. Rend.,* **228,** 555 idem, (1950), ibid., **230,** 1054
95. Azman, A. and Ocvirk, A. (1967). *Spectrochimica Acta.,* **23A,** 1597
96. Giguere, P. A. and Sathianandan, K. (1967). *Canad. J. Phys.,* **45,** 2439
97. Boutin, H., Safford, G. J. and Brajovic, V. (1963). *J. Chem. Phys.,* **39,** 3135
98. Waddington, T. C. (1958). *Trans. Faraday Soc.,* **54,** 25
99. Harrell, S. A. and McDaniel, D. H. (1964). *J. Amer. Chem. Soc.,* **86,** 4497
100. Bessis, G. and Bratoz, S. (1960). *J. Chim. Phys.,* **57,** 769
101. Clementi, E. and McLean, A. D. (1962). *J. Chem. Phys.,* **36,** 745
102. McLean, E. and Yoshimine (1967). *Int. J. Quantum Chem.,* **15,** 313
103. Vallee, R. E. and McDaniel, D. H. (1962). *J. Amer. Chem. Soc.,* **84,** 3412
104. Ludman, C. J. and Waddington, T. C. Unpublished results
105. West, R. (1957). *J. Amer. Chem. Soc.,* **79,** 4568
106. Schroeder, L. W. and Ibers, J. A. (1966). *J. Amer. Chem. Soc.,* **88,** 2601
107. Evans, J. C. and Lo, G. Y.-S. (1966). *J. Phys. Chem.,* **70,** 11
108. Nibler, J. W. and Pimentel, G. C. (1967). *J. Chem. Phys.,* **47,** 710
109. Ludman, C. J., Waddington, T. C. and Sterling, G. C. (1970). *J. Chem. Phys.,* **52,** 2730
110. Evans, J. C. and Lo, G. Y.-S. (1967). *J. Phys. Chem.,* **71,** 3697; idem, (1966). *J. Phys. Chem.,* **70,** 2702
111. Haas, T. E. and Welsh, S. M. (1967). *J. Phys. Chem.,* **71,** 3363
112. Ludman, C. J., Waddington, T. C., Salthouse, J. A., Lynch, R. J. and Smith, J. A. S. (1970). *Chem. Commun.,* 405
113. Swanson, J. S. and Williams, J. M. (1970). *Inorg. Nucl. Chem. Letters,* **6,** 271
114. McDaniel, D. H. and Vallee, R. E. (1963). *Inorg. Chem.,* **2,** 996
115. Salthouse, J. A. (1965). *Ph.D. Thesis,* University of Cambridge
116. Waddington, T. C. and White, J. A. (1963). *J. Chem. Soc.,* 2701
117. Harmon, K. M., Alderman, S. D., Benker, K. E., Diestler, D. J. and Gebauer, P. A. (1965). *J. Amer. Chem. Soc.,* **87,** 1700
118. Harmon, K. M. and Gebauer, P. A. (1963). *Inorg. Chem.,* **2,** 1319
119. Salthouse, J. A. and Waddington T. C. (1964). *J. Chem. Soc.,* 4664; idem (1966). *J. Chem. Soc. A.,* 28
120. Evans, J. C. and Lo, G. Y.-S. (1966). *J. Phys. Chem.,* **70,** 20
121. Schroeder, L. W. and Ibers, J. A. (1968). *Inorg. Chem.,* **7,** 594
122. Vallee, R. E. and McDaniel, D. H. (1963). *Inorg. Chem.,* **2,** 996

123. Cadý, G. H. (1934). *J. Amer. Chem. Soc.,* **56,** 1431
124. Forrester, J. D., Senko, M. E., Zalkin, A. and Templeton, D. H. (1963). *Acta Crystallogr.,* **16,** 58
125. Blinc, R., Trontelj, Z. and Volavsek, B. (1966). *J. Chem. Phys.,* **44,** 1028
126. Azman, A., Ocvirk, A., Hadzi, D., Giguere, P. A. and Schneider, M. (1967). *Canad. J. Chem.,* **45,** 1347

4
Perfluoroalkyl Derivatives of the Main Group Elements

R. J. POULET
University of Cambridge

4.1 INTRODUCTION

The discovery in 1948 that perfluoroalkyl derivatives of mercury could be prepared by the irradiation of the corresponding perfluoroalkyl iodides in the presence of mercury, indicated the possibility of a new field of chemistry[1]. Since then a vast amount of research, both academic and industrial, has been devoted to the preparation and characterisation of perfluoroalkyl derivatives of the main group elements, and its success can be measured by the wealth of new compounds that appear in the literature each year. The extent of the interest is such that it would be impossible in one short article to review all the recent advances in what has become a major field, and so I have excluded, somewhat arbitrarily, compounds having perfluoroalkyl groups bonded to hydrogen, the alkali metals (other than lithium), carbon, nitrogen (except for sulphur–nitrogen and phosphorus–nitrogen compounds), oxygen and the halogens. Several authoritative reviews have appeared in recent years, which provide an excellent background to this area of work, and some of them include comprehensive sections on derivatives of the elements listed above[3–7].

Perfluoroalkyl derivatives of some 41 elements (Figure 4.1) have so far been characterised. Those of mercury, silver and, to a lesser extent, lithium and magnesium can be considered as basic synthetic reagents and are discussed before the remainder are taken in group order.

Li												B	C	N	O	F
Ng	Mg											Al	Si	P	S	Cl
K						Mn	Fe	Co	Ni	Cu	Zn		Ge	As	Se	Br
Rb					Mo			Rh	Pd	Ag			Sn	Sb	Te	I
Cs						Re		Ir	Pt		Hg		Pb	Bi		

Figure 4.1 Elements having perfluoroalkyl derivatives

4.2 SYNTHETIC REAGENTS

4.2.1. Mercury

Perfluoroalkyl mercurials are well known for their ability to introduce the perfluoroalkyl moiety into other systems, provided that there is no Hg—C bond to be broken. The reactions of bis(trifluoromethylamide)mercury, $((CF_3)_2N)_2Hg$, and bis(trifluoromethylthio)mercury, $(CF_3S)_2Hg$, are summarised in Figures 4.3 and 4.4 in the appropriate section below.

Bis(trifluoromethyl)mercury itself has not proved to be as useful as the two mentioned, or even as the dialkyl analogue. There are few references to perfluoroalkylation reactions and even these are not conclusive[8].

The new mercurial, bis(bistrifluoromethylnitroxide)mercury, $((CF_3)_2NO)_2Hg$, is discussed in detail in Chapter 5 of this volume[9].

4.2.2 Silver

Silver salts of carboxylic acids have long been used in synthesis, the best known reaction being the pyrolysis, with iodine, to form trifluoroiodomethane,

$$AgO_2CCF_3 + I_2 \rightarrow CF_3I + CO_2 + AgI$$

(Reference 10)

A recent example is the preparation of perfluorocarboxylato disulphides, e.g.

$$AgO_2CR_f + S_2Cl_2 \rightarrow (R_fCO_2S)_2$$

(Reference 11)

$$(R_f = CF_3, C_2F_5, n\text{-}C_3F_7)$$

Silver fluoride will undergo a nucleophilic reaction across the multiple bond in perfluoro-olefins and acetylenes[12, 13],

$$AgF + \mathord{>}C{=}C\mathord{<} \rightarrow FC{=}CAg$$

e.g.

$$CF_2{=}CFCF_3 + AgF \xrightarrow{CH_3CN} (CF_3)_2CFAg \quad (1)$$

$$CF_3C{\equiv}CCF_3 + AgF \xrightarrow{CH_3CN} CF_3CF{=}C(CF_3)Ag\,(\textit{trans}) \quad (2)$$

Perfluoroisopropylsilver (1) is isolated as an adduct, $(CF_3)_2CFAg{\cdot}CH_3CN$, which can be decomposed at 100 °C to give the compound $(CF_3)_2CFCF(CF_3)_2$. *Trans*-perfluoro-1-methylpropenylsilver (2) is considerably more stable than its alkyl analogue, which decomposes at room temperature, and can be sublimed *in vacuo* at up to 175 °C. These derivatives are being used with both metallic and non-metallic halides as synthetic intermediates;

$$CF_3CF{=}C(CF_3)Ag \xrightarrow{CuBr_2/MeCN} CF_3CF{=}C(CF_3)C(CF_3){=}CFCF_3$$

Reaction with water, hydrogen chloride or bromine converts perfluoroisopropylsilver to the propanes $(CF_3)_2CFH$ or $(CF_3)_2CFBr$.

4.2.3 Magnesium

Although perfluoroalkyl Grignard reagents have been prepared and used in synthesis, they are thermally unstable, even in solution, and pose experimental problems. For this reason they have not found the popularity of their alkyl and perfluoroaryl analogues and little use has been made of them in comparison with the silver and mercury derivatives.

Bisperfluoromagnesium compounds are still unknown.

4.2.4 Lithium

Perfluoroalkyl-lithium compounds tend to suffer from the thermal instability that has precluded the widespread use of the perfluoroalkyl Grignard reagents. On warming from −78 °C, decomposition tends to occur, lithium fluoride being deposited and an olefin being formed, probably by an intermediate carbene reaction;

$$LiCF_3 \rightarrow LiF + :CF_2 \xrightarrow{:CF_2} C_2F_4$$

Both perfluorovinyl-lithium and perfluoropropynyl-lithium have recently been prepared[14–16];

$$F_2C{=}CFH + n\text{-}C_4H_9Li \rightarrow F_2C{=}CFLi + n\text{-}C_4H_{10}$$
$$F_2C{=}CFBr + MeLi \rightarrow F_2C{=}CFLi + MeBr$$
$$F_3CC{\equiv}CH + RLi \rightarrow F_3CC{\equiv}CLi + RH$$

An interesting range of perfluorocycloalkenyl-lithium compounds can be prepared by the reaction of 1-*H* perfluorocycloalkanes with methyl-lithium[17]. They may be used in ethereal solution at low temperature (up to −20 °C) to prepare stable derivatives, but if they are allowed to warm to room temperature decomposition is rapid, lithium fluoride being deposited as before. The other products tend to be polymeric, but it has recently been demonstrated that some of the transient intermediates such as a diene and a carbyne can be trapped out in furan.

4.3 COPPER AND ZINC

4.3.1 Copper

One of the few isolable copper compounds is perfluoro-t-butyl copper, which can be prepared as a 2:3 complex with dioxan from an ether–dioxan solution of perfluoro-t-butyl bromide and trifluoromethylphenyl copper at 0 °C [18].

4.3.2 Zinc

If the compound $CF_3CCl{=}CCl_2$ is reacted with zinc dust and zinc chloride in hot dimethylformamide, the product mixture is thought to contain the perfluoropropynyl derivatives of zinc, $CF_3C{\equiv}CZnCl$ and $(CF_3{\equiv}C)_2Zn$ [19]. By allowing the resulting solution to react with cupric chloride, the diyne $CF_3C{\equiv}CC{\equiv}CCF_3$, can be prepared.

4.4 BORON AND ALUMINIUM

4.4.1 Boron

Trifluoromethylboron difluoride can be prepared by the reaction of di-n-butyltrifluoromethylboron with boron trifluoride at 20 °C [20]. It may safely be stored under vacuum provided that oxygen is rigorously excluded, as even trace amounts will initiate rapid and quantitative decomposition to BF_3 and a polymer, $(—CF_2—)_n$, probably as a recombination product of difluorocarbene. CF_3BF_2 forms 1:1 adducts with triethylamine and diethyl ether, but may be recovered from the latter by the addition of trimethylamine.

Perfluoropropyl-lithium will react with boron halides to give thermally stable derivatives[21];

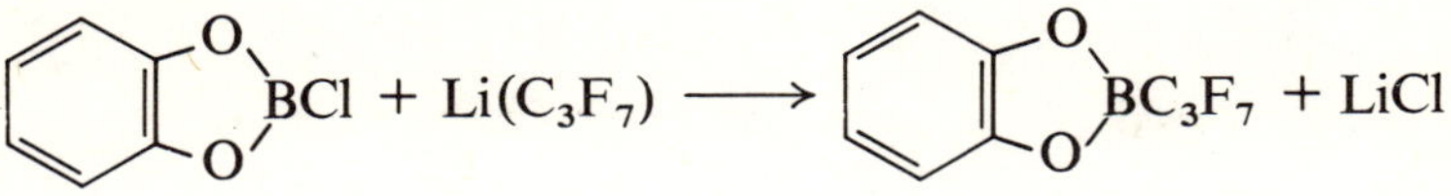

$$[(CF_3)_2N]_2BCl + Li(C_3F_7) \longrightarrow [(CF_3)_2N]_2BC_3F_7 + LiCl$$

This stability is due to a degree of π-bonding between the boron atom and the oxygen or nitrogen atoms which tends to prevent the nucleophilic attack by the α-fluorines.

The compound $B(CF{=}CF_2)_3$, prepared from perfluorovinyl-lithium and a boron trihalide, decomposes at 100 °C to form perfluorovinylboron-di fluoride and boron trifluoride. If, however, the group is incorporated into a borazine, i.e.

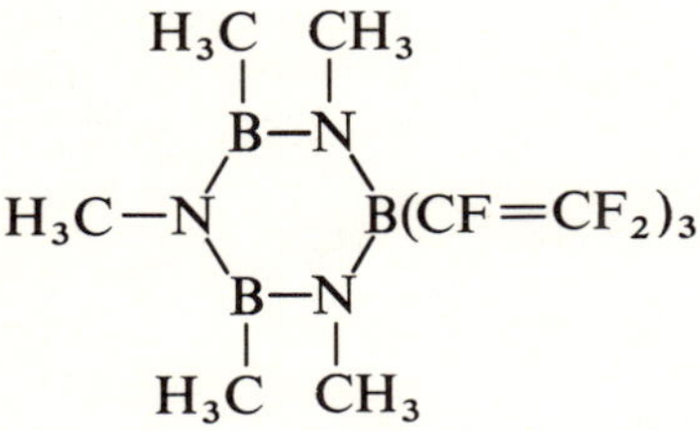

or $(CH_3NBCF{=}CF_2)_3$

the fluorine abstraction reaction is suppressed because of the low acceptor power of the boron atoms. These compounds can be readily prepared by the reaction between perfluorovinyl-lithium and the appropriate borazine[22].

4.4.2 Aluminium

The migration of fluorine atoms to boron and aluminium from the carbon atom explains the instability of these compounds, and no perfluoroalkyl derivatives of aluminium are known. The only compounds described so far have been lithium aluminium derivatives with two perfluoroalkyl groups. The compound $LiAl(C_3F_7)_2I_2$ has been reacted with triethyl-lead- and triethyltin chlorides, and perfluoropropyl derivatives of these elements are said to be produced, but the evidence for them is not conclusive as yet[23].

4.5 SILICON, GERMANIUM AND TIN

4.5.1 Silicon

One of the main interests in the derivatives of the Group IV elements has been in the preparation of compounds which, on pyrolysis, lead to the formation of carbenes. For example, difluoromethylfluorocarbene results from the heating of 1,1,2,2-tetrafluoroethyltrifluorosilane and can be trapped out with an olefin[24];

$$CHF_2CF_2SiF_3 \xrightarrow{150^\circ} SiF_4 + :CF_2CHF_2 \longrightarrow CF_2{=}CHF_2$$

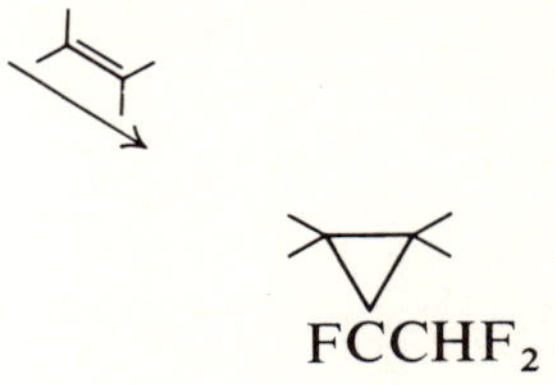

The probable mechanism for this type of reaction is an internal nucleophilic attack by an α-fluorine on the silicon atom, followed by a migration of a fluorine atom to form an olefin, provided that there is no carbene trap present. Trimethyltrifluorotin will react in a similar way on heating to 150 °C, generating difluorocarbene and trimethyltin fluoride[25]. For compounds such as $CF_3CH_2CH_2SiCCl_3$, where there is no α-fluorine, the decomposition process is somewhat different[26]. It appears to take place via a radical chain mechanism at temperatures in excess of 300 °C, compared with the much lower temperatures required for the generation of carbenes from suitable compounds.

4.5.2 Tin

On hydrolysis with aqueous alkali, the tin–carbon bond in trimethyltrifluoromethyltin is readily cleaved to give the hydroxide and fluoroform. If sodium iodide is used in place of the alkali, any olefin present in the reaction mixture will trap out the difluorocarbene that has been formed by the decomposition of trifluoromethylsodium.

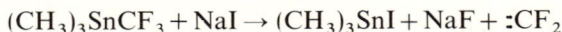

$$(CH_3)_3SnCF_3 + NaI \rightarrow (CH_3)_3SnI + NaF + :CF_2$$

This reaction has been used to prepare *gem*-difluorocyclopropanes;

$$:CF_2 + (CH_3)_2C{=}C(CH_3)_2 \longrightarrow$$

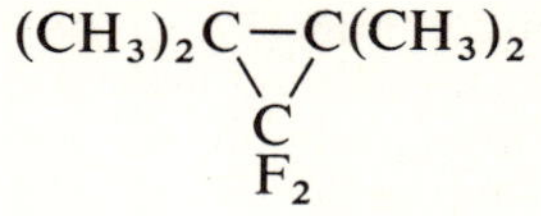

$$:CF_2 + C_6F_5CH{=}CH_2 \longrightarrow$$

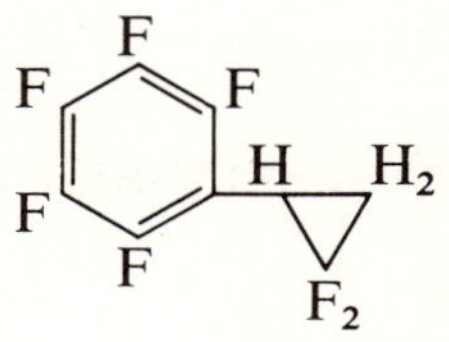

(References 27 and 28)

Trimethyltrifluoromethyltin reacts with trifluoropropynyl compounds of tin, germanium and arsenic by a carbene reaction[29].

$$(CH_3)_3SiC \equiv CCF_3 + CF_3Sn(CH_3)_3 \longrightarrow (CH_3)_3SiC = CCF_3$$
$$\overset{|}{\underset{F_2}{C}}$$

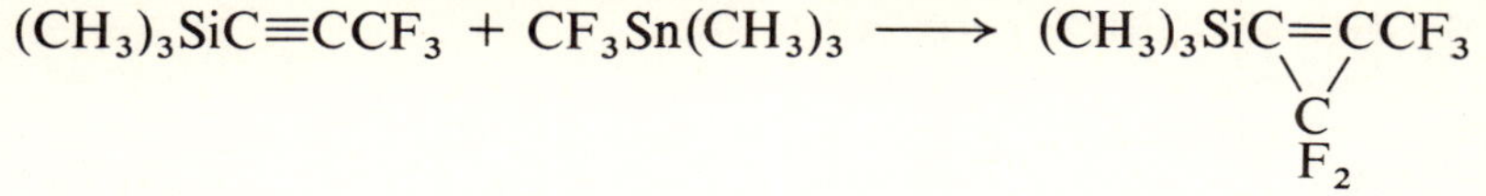

It will also react with trimethylstannane at 150 °C to form trimethyl-difluoromethyltin, presumably by a carbene insertion reaction[30];

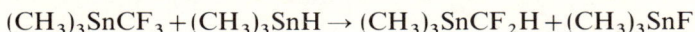

$$(CH_3)_3SnCF_3 + (CH_3)_3SnH \rightarrow (CH_3)_3SnCF_2H + (CH_3)_3SnF$$

The thermal instability of tin derivatives has not as yet lead to a detailed study of the many compounds so far prepared, only trimethyltrifluoromethyl-tin having received any serious attention. Two new derivatives are reported that were prepared by the homolytic fission of the tin–tin bond in bis(tri-methyltin) in the presence of a perfluoroalkyl iodide[30];

$$(CH_3)_3SnSn(CH_3)_3 + (CF_3)_2CFI \xrightarrow{hv} (CF_3)_2CFSn(CH_3)_3 + (CH_3)_3SnI$$
$$(CH_3)_3SnSn(CH_3)_3 + CF_2 = CFI \rightarrow CF_2 = CFSn(CH_3)_3 + (CH_3)_3SnI$$

4.5.3 Germanium

The decomposition of trifluoromethyltri-iodogermanium seems to occur in an analogous fashion to the silicon and tin compounds. At 180 °C a mixture of tetrafluoroethylene, perfluorocyclopropane and perfluoro-cyclobutane is obtained, which strongly suggests that difluorocarbene is formed by the α-elimination of fluorine.

4.5.4 Carbene generators

There are several other reactions mentioned in the literature in which perfluorocarbenes are generated. For example, the decomposition of trifluoromethylboron-difluoride in the presence of traces of oxygen to give a polymeric compound, $(-CF_2-)_n$, has been referred to earlier[21]. Interestingly, potassium trifluoromethylfluoroborate, $K^+(CF_3BF_3)^-$, decomposes at 300 °C to give tetrafluoroethylene as the major product but, if the temperature is raised to 450 °C, perfluorocyclobutane is formed[31]. No further work on the temperature dependence of this type of reaction appears to have been done.

Perfluoroalkyl derivatives of both lithium and magnesium decompose with the production of carbenes;

$$LiCF_3 \rightarrow :CF_2 + LiF$$
$$MgXCF_3 \rightarrow :CF_2 + MgXF$$

However, the best source is probably the phosphorane $(CF_3)_3PF_2$ which is readily decomposed in the gas phase as follows;

$$(CF_3)_3PF_2 \rightarrow (CF_3)_2PF_3 + :CF_2$$
$$(CF_3)_2PF_3 \rightarrow CF_3PF_4 + :CF_2$$
$$CF_3PF_4 \rightarrow PF_5 + :CF_2$$

4.6 PHOSPHORUS AND ARSENIC

4.6.1 Phosphorus

The preparation and properties of trifluoromethyl derivatives of phosphorus are set out in tabular form below;

$$CF_3I + P \overset{200°}{\rightarrow} (CF_3)_3P + (CF_3)_2PI + (CF_3)PI_2$$

(1). CF_3PI_2

$$\overset{AgCl}{\longrightarrow} CF_3PCl_2$$

$$\overset{AgBr}{\longrightarrow} CF_3PBr_2 \overset{CH_3OH}{\longrightarrow} CF_3PHO(OH)$$

$$\overset{H_2O}{\longrightarrow}$$

$$\overset{Hg}{\longrightarrow} (CF_3)_4P_4 + (CF_3)_5P_5$$

$$\overset{freeze\ dry}{\longrightarrow} CF_3PH_2$$

$$\overset{H_2O_2}{\longrightarrow} CF_3PO(OH)_2$$

(2). $(CF_3)_2PI_2$

$$\overset{HI/Hg_2}{\longrightarrow} (CF_3)_2PH$$

$$\overset{SbF_3}{\longrightarrow} (CF_3)_2PF \overset{NH_3}{\longrightarrow} (CF_3)_2PNH_2$$

$$\overset{AgCN}{\longrightarrow} (CF_3)_2PCN \overset{Cl_2}{\longrightarrow} (CF_3)_2PCl_3$$

$$\overset{AgCl}{\longrightarrow} (CF_3)_2PCl \overset{LiN_3}{\longrightarrow} (CF_3)_2PN_3$$

$$\overset{H_2O}{\longrightarrow}$$

(3). $(CF_3)_3P$

$$\overset{Cl_2}{\longrightarrow} (CF_3)_3PCl_2 \overset{H_2O}{\longrightarrow} (CF_3)_2POOH$$

$$\overset{SF_4}{\longrightarrow} (CF_3)_3PF_2$$

$$\overset{N_2O_4}{\longrightarrow} (CF_3)_3PO \overset{(CH_3)_2NH}{\longrightarrow} (CF_3)_2P(O)N(CH_3)_2 \overset{HCl}{\longrightarrow}$$

$$(CF_3)_2POCl$$

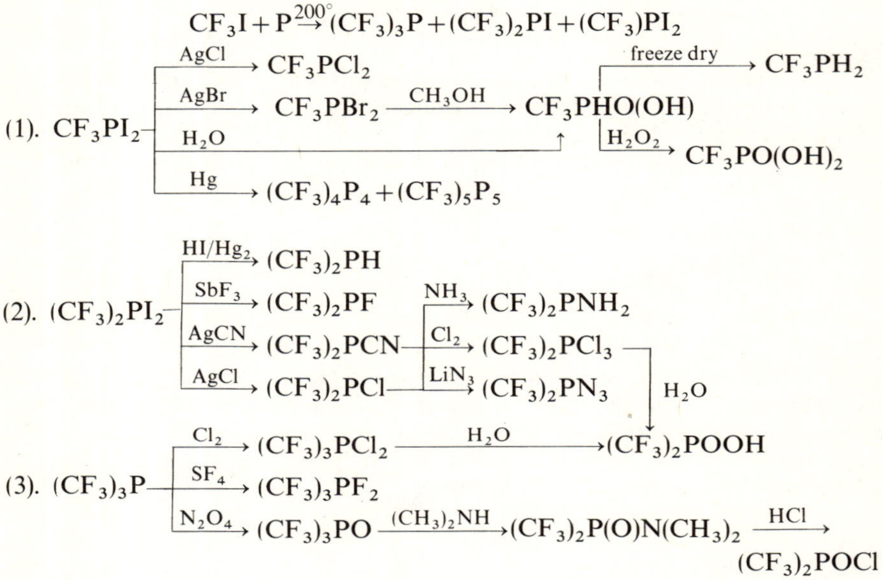

Figure 4.2   Reactions of trifluoromethyl derivatives of phosphorus[4-6]

4.6.2 Derivatives of phosphorus and arsenic

A recent method of preparing the hydrides that involves the use of the bistrifluoromethylphosphines or arsines with anhydrous hydrogen iodide in the presence of mercury results in very high yields of product (89–94%)[32];

$$2(CF_3)_2MI + 2HI + Hg \rightarrow 2(CF_3)_2MH + HgI_2 \quad (M = P\ or\ As)$$
$$(CF_3)_2MM(CF_3)_2 + 2HI + Hg \rightarrow 2(CF_3)_2MH + HgI_2$$

It is thought that the reaction of $(CF_3)_2PI$ with mercury and HI probably involves the initial production of the diphosphine which is then rapidly reduced by HI.

Derivatives of the type $((CF_3)_nMF_{6-n})^-$ (M = P, As) have been prepared as stable solids with Cs or Ag as the cation by reaction of the appropriate trifluoromethyl phosphorus or arsenic halide with a suspension of caesium fluoride in acetonitrile[36];

$$(CF_3)_2PF_4 + CsF \rightarrow Cs^+((CF_3)_2PF_4)^-$$
$$Ag[(CF_3)_2PO_2] + 2SF_4 \rightarrow Ag[(CF_3)_2PF_4] + 2SOF_2$$

The second equation shows how the general reaction can be adapted by using SF_4, known to be suitable for replacement of oxygen by fluorine, on the salts of trifluoromethyl oxyacids.

The reaction of trimethyltrifluoromethyltin with boron trifluoride is reported to give the fluoroborate $[(CH_3)_3Sn]^-(CF_3BF_3)$ by a transfer of a CF_3^- group. 1:1 adducts of this type are also suggested when PF_5, $(CF_3)_2PF_3$ and $(CF_3)_3PF_2$ are used in place of BF_3. Infrared evidence alone is cited in support of this theory because of the intractible nature of the compounds which are insoluble in organic solvents yet are very easily hydrolysed in water[36].

4.6.3 Phosphorus—sulphur compounds

Much of the recent work on perfluoroalkyl derivatives of phosphorus has been concerned with the preparation of compounds having another element, such as sulphur or nitrogen, attached to the phosphorus atom. Bis(trifluoromethyl)dithiophosphinic acid, $(CF_3)_2P(S)SH$, has been made by the reaction of H_2S with the iodide, $(CF_3)_2P(S)I$, and the diphosphine trisulphide, $((CF_3)_2PS)_2S$ [33]. The dithio acid has been shown to be monomeric, unlike the oxygen analogue.

$$2(CF_3)_2P(S)I + 2H_2S \rightarrow 2(CF_3)_2P(S)SH + 2HI$$
$$(CF_3)_2P(S)SH + 2HI \rightarrow (CF_3)_2PSH + H_2S + I_2$$

$(CF_3)_2P(S)I$ loses iodine spontaneously and, by pushing this reaction forward with mercury, a 66% yield of $((CF_3)_2P)_2S$ can be obtained. The mechanism is thought to be as follows;

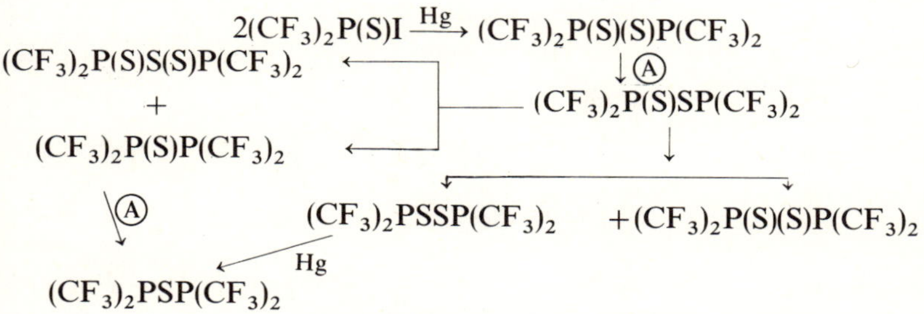

A = Anti-Arbuzov rearrangement (postulated).

The acid chloride, $(CF_3)_2P(S)Cl$, can be prepared by the action of chlorine on the diphosphine monosulphide and used to obtain a thioester by reaction with CF_3SH in the presence of trimethylamine. However, the very strong acid $(CF_3)_2PSSH$ cannot be prepared by the similar reaction with H_2S, the result being the salt $(CH_3)_3NH^+(CF_3)_2PS_2^-$.

4.6.4 Phosphorus-nitrogen compounds

On heating trifluoromethylaminophosphines with sulphur, the phosphorus atom can be oxidised to the +V state, (e.g. $(CF_3)_2P(S)NRH$, R = H, CH_3, etc.) and the action of mineral acids on these products results in the formation of compounds such as $(CF_3)_2P(S)X$ (X = F, Cl, Br, I). For example, addition of sulphur to $CF_3PFN(CH_3)_2$ gives the corresponding sulphide, $CF_3PF(S)N(CH_3)_2$ [34]. The reaction of the substituted amine $(CF_3)_2PNHR$, (R = Me, Bu^t) with boron trichloride at low temperatures gives a 1:1 adduct which breaks down on warming to give an alkylaminoboron chloride and the bistrifluoromethylphosphine chloride[35]. There appears to be a second reaction sequence in which HCl is eliminated and small quantities of the compound $(CF_3)_2PN(R)BCl_2$ are formed. The methyl compound is a dimer while the extremely unstable t-butyl compound is thought to be monomeric;

$$RNHP(CF_3)_2 + BX_3 \rightarrow \dot{R}NHP(CF_3)_2BX_3$$
$$RNHP(CF_3)_2BX_3 \rightarrow (CF_3)_2PX + RNHBX_2$$
$$RNHP(CF_3)_2 + BX_3 \rightarrow (CF_3)_2\dot{P}N(R)BCl_2 + HCl$$
$$\left(\begin{array}{l} R = Me, Bu^t \\ X = F, Cl \end{array}\right)$$

4.7 SULPHUR, SELENIUM AND TELLURIUM

4.7.1 Sulphur

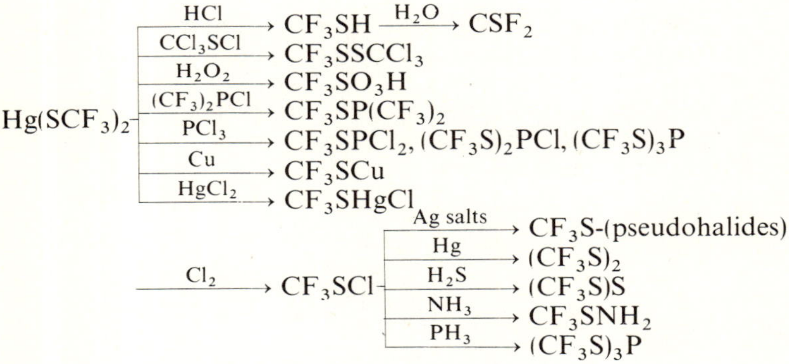

Figure 4.3 Reactions of $Hg(SCF_3)_2$ and CF_3SCl [37–40]

An additional reaction of bis(trifluoromethylthio)mercury with nitrosyl chloride has recently been reported[41]. It gives the trifluoromethylthionitrite, CF_3SNO, as a red gas at -100 °C. NOCl will also react with trifluoromethanethiol under the same conditions to give CF_3SNO, whereas with trifluoromethanesulphonic acid the reaction product is nitrosyl trifluoromethanesulphonate, $NOSO_3CF_3$ [42]. This is considered to have considerable ionic character from evidence obtained from Raman spectroscopy and from the high melting point (170 °C). It may also be prepared from bis(trifluoromethylsulphuryl)peroxide, $(CF_3SO_3)_2$, and nitric oxide at -78 °C;

$$CF_3SO_3H + NOCl \rightarrow NOSO_3CF_3 + HCl$$
$$(CF_3SO_3)_2 + 2NO \rightarrow 2NOSO_3CF_3$$

Interestingly, if trifluoromethanesulphonic acid is reacted with titanium tetrachloride, the monosubstituted compound, $TiCl_3(SO_3CF_3)$, is formed, which disproportionates on heating to ~~give~~ the disubstituted $TiCl_2$ $(SO_3CF_3)_2$ [42].

Trifluoromethylsulphur trifluoride may be prepared from liquid bis (trifluoromethyl)disulphide, $(CF_3S)_2$, and AgF_2 in rigorously anhydrous conditions to minimise the production of trifluoromethylsulphinyl fluoride CF_3SFO [43]. By storing small quantities of CF_3SF_3 in a glass bulb, CF_3SFO can be obtained quantitatively, contaminated with an equimolar amount of SiF_4 from the reaction with the glass.

The 1:1 adducts formed between CF_3SF_3 and PF_5, BF_3, AsF_5 and SbF_5 respectively are thought to have ionic character. Infrared studies[44] have shown that $CF_3SF_2^+ PF_6^-$ is very likely to be the dominant form of the adduct with PF_5.

4.7.2 Sulphur—nitrogen compounds

A considerable amount of effort has been expended in the attempts to synthesise new perfluoroalkyl compounds that contain sulphur and nitrogen.

4.7.2.1 R_fS—N compounds

Trifluoromethylsulphenylamine reacts with sulphur tetrafluoride in the presence of CsF to give trifluoromethylsulphenyliminosulphur difluoride and bis(trifluoromethylsulphenylimino)sulphur[45]:

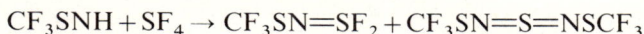

$$CF_3SNH + SF_4 \rightarrow CF_3SN{=}SF_2 + CF_3SN{=}S{=}NSCF_3$$

The alcoholysis of trifluoromethylsulphenyl isocyanates gives a range of compounds known as trifluoromethylsulphenylcarbamide acid esters[46]:

$$CF_3SNCO + ROH \rightarrow CF_3SNHCO_2R$$
$$(R = CH_3, C_3H_7, CH(CH_3)_2, C(CH_3)_3, CH_2C_6H_5, CH(C_6H_5)_2)$$

Rather more complex molecules can be prepared from bis(trifluoromethylsulphenyl)uretidinediones[46]:

$$\xrightarrow{H_2O} (CF_3SNH)_2CO$$
$$\xrightarrow{CH_3OH} CF_3SNHCONHCO_2CH_3$$
$$\xrightarrow{C_6H_5CH_2OH/H_2O} CF_3SNHCONHCO_2CH_2C_6H_5$$

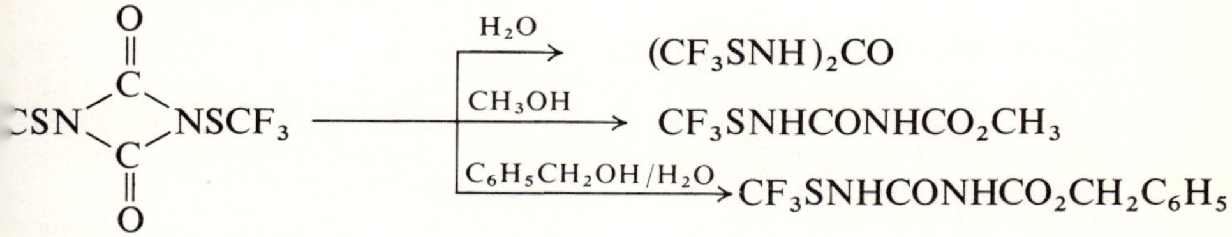

4.7.2.2 R_fN—S

The reactions of bis(trifluoromethylamide)mercury provide a simple method of preparing several of these compounds:

$$Hg[N(CF_3)_2]_2 \begin{cases} \xrightarrow{S} [(CF_3)_2N]_2S \\ \xrightarrow{SCl_2} (CF_3)_2NSCl \\ \xrightarrow{CF_3SCl} (CF_3)_2NSCF_3 \\ \xrightarrow{Cl_2} (CF_3)_2NCl \xrightarrow{SF_5Cl} (CF_3)_2NSF_5 \end{cases}$$

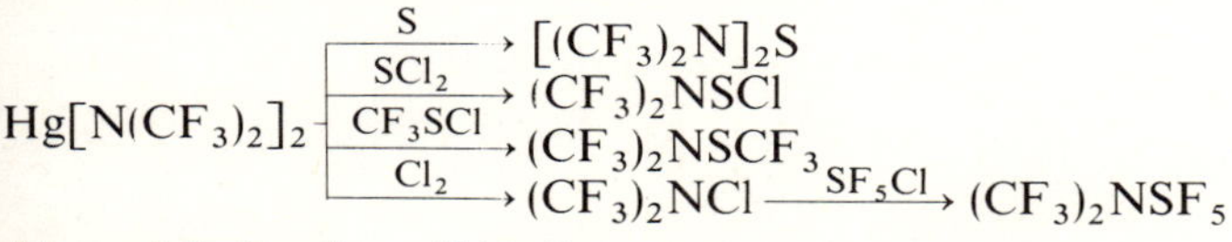

Figure 4.4 Reactions of bis(trifluoromethylamide)mercury[47-53]

$(CF_3)_2NSF_5$ has recently been prepared in 10% yield by the irradiation of *N*-chlorobistrifluoromethylamine with SF_4[51]. It is, as its structure suggests, a very stable compound, and is unaffected by acid or alkali at room temperature.

Several new *N*-perfluoroalkyliminosulphur dihalides have been found. By generating difluorocarbene from tetrafluoroethylene under a high pressure mercury lamp, insertion reactions may be observed in the N—Cl bond of chloriminosulphur difluoride, the products depending on the ratios of reactants used[54];

$$ClN{=}SF_2 + C_2F_4 \rightarrow ClF_2CCF_2N{=}SF_2$$
$$ClN{=}SF_2 + 2C_2F_4 \rightarrow ClF_2C(CF_2)_3N{=}SF_2$$

Trifluoromethyliminosulphur oxydifluoride has also been prepared. Bis(trimethylsilyl)carbodi-imide, $(CH_3)_3SiN{=}C{=}NSi(CH_3)_3$ is reacted with thionyl tetrafluoride at 60 °C, and *N*-cyanoiminosulphur oxydifluoride, $F_2S{=}NCN$, is formed as an intermediate, which reacts with more SF_4O at 250 °C to give difluorobis(iminofluorosulphur)methane $(NSF_2)_2CF_2$. If, however the reaction is continued with SF_4 in place of SF_4O, the product is the oxydifluoride, $CF_3N{=}SF_2{=}O$[54].

4.7.3 Selenium

Trifluoromethylselenyl chloride, CF_3SeCl, is proving to be as useful a reagent for transferring the CF_3Se–group as CF_3SCl is for the analogous sulphur series. The halogen atom can readily be substituted for another by reaction with the appropriate silver halide[56]. It may also be substituted for an aryl group by reaction with an aryl magnesium Grignard reagent. The resulting compound can be oxidised with chlorine and then hydrolysed to the oxide[57];

$$CF_3SeCl + ArMgBr \longrightarrow ArSeCF_3 \xrightarrow{Cl_2} ArSe(Cl_2)CF_3$$
$$\longrightarrow ArSe(O)CF_3$$

The mercurial $(CF_3Se)_2Mg$, prepared by shaking $(CF_3)_2Se_2$ with mercury, will react with PF_2I, $(CH_3)_3SiI$ and liquid HBr (at −78 °C) to give the corresponding trifluoromethylselenium compounds CF_3SePF_2, $CF_3SeSi(CH_3)_3$ and CF_3SeH respectively. The hydride, obtained quantitatively by this method, reacts with trimethyl thallium to give $CF_3SeTl(CH_3)_2$[59].

If selenium vapour is allowed to react with hexafluorobut-2-yne, the heterocyclic compound (trifluoromethyl)-1,2-selenetene, $Se_2C_2(CF_3)_2$, is formed. This will react with transition metal carbonyls to give 1,2-diselenolene complexes which tend to precipitate selenium on standing[58].

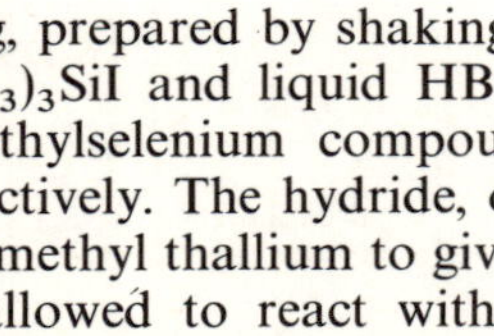

4.7.4 Tellurium

Bistrifluoromethyltellurium can be prepared as a deep-red unstable liquid by allowing $\cdot CF_3$ radicals, generated by the pyrolysis of hexafluoroacetone, to react with a tellurium mirror[60]. The product must be stored at low temperature in the dark.

References

1. Banks, R., Emeléus, H., Haszeldine, R. and Kerrigan, V. (1948). *J. Chem. Soc.*, 2188
2. Emeléus, H. and Haszeldine, R. (1949). *J. Chem. Soc.*, 2953
3. Treichel, P. and Stone, F. (1964). *Advances in Organometallic Chemistry*, edited by Stone, F. and West, R., (New York: Academic Press), **1**, 143
4. Clark, H. (1963). *Advances in Fluorine Chemistry*, edited by Stacey, M., Tatlow, J. and Sharpe, A. (London: Butterworths), **3**, 19
5. Banks, R. (1970). *Fluorocarbons and their derivatives* (London: Macdonald Technical and Scientific), 102
6. Banks, R. and Haszeldine, R. (1961). *Advances in Inorganic Chemistry and Radiochemistry*, edited by Emeléus, H. and Sharpe, A. (New York: Academic Press), **3**, 337
7. Banks, R. and Haszeldine, R. (1966). *The Chemistry of Organic Sulphur Compounds*, edited by Kharasch, N. and Meyers, C. (Oxford: Pergamon Press), **11**, 137
8. Miller, W. and Freedman, M. (1963). *J. Amer. Chem. Soc.*, **85**, 180
9. Emeléus, H., Shreeve, J. and Spaziante, P. (1969). *J. Chem. Soc.*, 431
10. Haszeldine, R. °1950). *Nature. (London)*, **166**, 192
11. Wang, C., Pullen, K. and Shreeve, J. (1970). *Inorg. Chem.*, **90**
12. Miller, W., Snider, R. and Hummel, R. (1969). *J. Amer. Chem. Soc.*, **91**, 6532
13. Miller, W. and Burnard, R. (1968). *J. Amer. Chem. Soc.*, **90**, 7367
14. Tarrant, P., Johncock, P. and Savory, J. (1963). *J. Org. Chem.*, **28**, 839
15. Drakesmith, F., Richardson, R., Stewart, O. and Tarrant, P. (1968). *J. Org. Chem.*, **33**, 286
16. Drakesmith, F., Stewart, O. and Tarrant, P. (1968). *J. Org. Chem.*, **33**, 280
17. Campbell, S., Stevens, R. and Tatlow, J. (1967). *Chem. Commun.*, 151
18. Cairncross, A. and Sheppard, W. (1968). *J. Amer. Chem. Soc.*, **90**, 2186
19. Norris, W. and Finnegan, W. (1966). *J. Org. Chem.*, **31**, 3292
20. Parsons, T., Self, J. and Schaad, L. (1967). *J. Amer. Chem. Soc.*, **89**, 3446
21. Chivers, T. (1967). *Chem. Commun.*, 157
22. Klanica, A., Faust, J. and King, C. (1967). *Inorg. Chem.*, **6**, 840
23. Dickson, R. and West, B. (1966). *Aust. J. Chem.*, **19**, 2073
24. Haszeldine, R. and Speight, J. (1967). *Chem. Commun.*, 995
25. Clark, H. and Willis, C. (1969). *J. Amer. Chem. Soc.*, **82**, 1888
26. Haszeldine, R., Robinson, P. and Simmons, R. (1967). *J. Chem. Soc. B*, 1357
27. Seyferth, D., Dertouzos, H., Suzuki, R. and Mui, J. (1967). *J. Org. Chem.*, **32**, 2980
28. Seyferth, D. and Dertouzos, H. (1968). *J. Organometal Chem.*, **11**, 263
29. Cullen, W. and Leeder, W. (1966). *Inorg. Chem.*, **5**, 1004
30. Cullen, W., Sams, J. and Waldman, M. (1970). *Inorg. Chem.*, **9**, 1682
31. Chambers, ., Clark, H. and Willis, C. (1960). *J. Amer. Chem. Soc.*, **82**, 6218
32. Cavell, R. and Dobbie, R. (1967). *J. Chem. Soc. A*, 1308
33. Gosling, K. and Burg, A. (1968). *J. Amer. Chem. Soc.*, **90**, 2011
34. Dobbie, R., Doty, L. and Cavell, R. (1968). *J. Amer. Chem. Soc.*, **90**, 2015
35. Greenwood, N. and Robinson, B. (1968). *J. Chem. Soc. A*, 226
36. Chan, S. and Willis, C. (1968). *Can. J. Chem.*, **46**, 1237
37. Jellinek, F. and Lagowski, J. (1960). *J. Chem. Soc.*, 810
38. Emeléus, H. and Pugh, H. (1960). *J. Chem. Soc.*, 1108
39. Downs, A. and Ebsworth, E. (1960). *J. Chem. Soc.*, 3516
40. Harris, J. (1967). *J. Org. Chem.*, **32**, 2063
41. Mason, J. (1969). *J. Chem. Soc. A*, 1587
42. Noftle, R. and Cady, G. (1966). *Inorg. Chem.*, **5**, 2182
43. Lawless, E. and Harman, L. (1968). *Inorg. Chem.*, **7**, 391

44. Kramar, M. and Duncan, L. (1971). *Inorg. Chem.*, **10**, 647
45. Haas, A. and Schott, P. (1967). *Angew. Chem. Int. Ed. Engl.*, **6**, 370
46. Haas, A. and Schott, P. (1966). *Chem. Ber.*, **99**, 3103
47. Emeléus, H. and Tattershall, B. (1966). *J. Inorg. Nucl. Chem.*, **28**, 1823
48. Dobbie, R. and Emeléus, H. (1966). *J. Chem. Soc. A*, 367
49. Dobbie, R. and Emeléus, H. (1966). *J. Chem. Soc. A*, 933
50. Emeléus, H. and Onak, T. (1966). *J. Chem. Soc. A*, 1291
51. Dobbie, R. (1966). *J. Chem. Soc. A*, 1555
52. Haszeldine, R. and Tipping, A. (1965). *J. Chem. Soc. C*, 6141
53. Haszeldine, R. and Tipping, A. (1968). *J. Chem. Soc. C*, 399
54. Mews, R. and Glemser, O. (1969). *Chem. Ber.*, **102**, 4185
55. Lustig, M. and Ruff, J. (1967). *Inorg. Nucl. Chem. Lett.*, **3**, 531
56. Welcman, N. and Wuif, M. (1968). *Israel J. Chem.*, **6**, 37
57. Yagupolski, L. and Voloschuke, V. (1967). *Zh. obshch. Khim.*, **37**, 1543
58. Davison, A. and Shawl, E. (1970). *Inorg. Chem.*, **9**, 1820
59. Marsden, C. (1971). Private communication
60. Bell, T., Pullman, B. and West, B. (1964). *Aust. J. Chem.*, **16**, 722

5
$(CF_3)_2NO$-Compounds

P. M. SPAZIANTE
Oronzio de Nora—Impianti Elettrochimici S.p.A.—Milano

5.1 INTRODUCTION

The chemistry of bistrifluoromethylnitroxide derivatives started in 1964 when American[1] and Russian[2] chemists first reported the preparation of the stable bistrifluoromethylnitroxide radical and this may be considered a pseudo-halogen.

In the last few years several organic and inorganic compounds containing the $(CF_3)_2NO$-group have been prepared in several ways using the *N*, *N*-bistrifluoromethylhydroxylamine-$(CF_3)_2NOH$ or the corresponding radical $(CF_3)_2NO\cdot$ generally as starting material. Therefore before discussing the $(CF_3)_2NO$-derivatives it seems useful to examine the nature of nitroxide free radicals and to describe in particular the properties of bistrifluoromethylnitroxide free radical and the different methods reported in the literature for its preparation.

5.2 NITROXIDE FREE RADICALS

Nitroxide free radicals, of the type $R_2NO\cdot$ have been known for a long time, the oldest example being the inorganic nitrosyl disulphonate radical ion $(SO_3)_2NO\cdot^{2-}$ reported for the first time by M. E. Fremy[3] in 1845 and isolated as the potassium salt, also known as Fremy's salt. This blue-violet salt, which decomposes rapidly (1/2 h) in acid and slowly (500 h) in alkaline solution, has been later characterised[4–7] and its decomposition products studied[8,9].

Another rather old example of a nitroxide radical is the diphenylnitroxide $(C_6H_5)_2NO\cdot$, first prepared by Wieland and Offenbacher[10] in 1914 by oxidation of diphenylhydroxylamine with silver (I) oxide. Its chemical and physical properties have been studied[11–14] and its half life has been reported to be 50 h in cumene at 100 °C[11]. Diphenylnitroxide has been described as an intensely red substance which decomposes on keeping, liberates iodine from an acidified solution of potassium iodide, and is reduced to the original diphenylhydroxylamine by phenylhydrazine.

Other radical derivatives of diphenylnitroxide, reported in 1920[13], are: di(*p*-toluyl)nitroxide, $(CH_3C_6H_4)_2NO\cdot$ and bis(*p*-nitrophenyl)nitroxide, $(O_2NC_6H_4)_2NO\cdot$.

In 1926 the isopropyl-phenylnitroxide radical was obtained[14] in very high yield by oxidation of the condensation products of phenylhydroxylamine and acetone, with several oxidising agents such as: potassium permanganate, potassium ferricyanide, sodium hypobromite, ammoniacal silver oxide and air; among these the most convenient is reported to be ammoniacal silver oxide. Isopropyl-phenylnitroxide has been described as a bright red, crystalline solid. Several derivatives of this radical have been reported[14]. These were obtained, either from the corresponding hydroxylamine, e.g. the potassium salt (from finely divided potassium in benzene solution), acetyl or benzoyl derivatives (from acetyl or benzoyl chloride in a pyridine solution), or from the radical with acetic or benzoic anhydride. These crystalline compounds were produced in almost quantitative yield and have been fully characterised.

In the last few years several nitroxide radicals have been isolated or observed as transient species in e.s.r. studies. Baird and Thomas[15] in 1961 made an e.s.r. study of seven di-substituted nitroxide free radicals: diphenylnitroxide, monophenylnitroxide, phenyl-ethylnitroxide, di-n-hexylnitroxide, mono-t-butylnitroxide, thiodiphenylnitroxide and dibenzylnitroxide. The authors do not report anything about the stability of such species. At the same time a study of eleven unstable free radicals of the general structure $R_2NO\cdot$, where R were different alkyl groups, was made[16].

The stable free radical di-t-butylnitroxide was synthesised in 1961 by Hoffman and Henderson[17]. This radical was described as a red volatile liquid (b.p. 74–75 °C/35 mmHg) stable to oxygen, water, aqueous alkali, and it was reported to react with hydrogen in the presence of platinum to form di-t-butylhydroxylamine (m.p. 65–66 °C) which oxidised rapidly in air to the radical. The reaction of this radical with hydrogen chloride led to the formation of di-t-butylhydroxylammonium chloride and t-nitrosobutane, while t-butylhydroxylamine reacted with hydrogen chloride giving di-t-butylhydroxylammonium chloride as the only product.

An explanation of the stability of this radical has been sought in terms of intrinsic stability of a N—O three-electron bond[18]. An additional stability is thought to arise from steric factors. The electronic structure was studied in 1962[19]. On the other hand the stability of another radical: triacetoneamine-N-oxide[20, 21] (a crystalline substance with m.p. 36 °C) reported in 1962, has been attributed to purely steric factors.

A number of different organic nitroxide radicals have been reported during the years 1964–1966[22, 37]. Twenty-two different stable nitroxide free radicals of the type RR′NO, where R was the t-butyl group and R′ a mono-substituted phenyl group, have been prepared[22] and a complete e.s.r. study has been made. Among these radicals the most stable are reported to be those in which the phenyl group is substituted in a *para* position. This stability is reported to be due to the resonance between the two structures:

$$\mathrm{N{-}O\cdot} = = = = = \mathrm{\dot{N}{-}O^-}$$

By addition of a benzene solution of lead tetra-acetate to different oximes, a series of nineteen different organic nitroxide radicals have been prepared and studied by e.s.r. spectroscopy[23].

Almost another forty different nitroxide free radicals, obtained by cathalytic oxidation of the corresponding secondary amines by means of hydrogen peroxide, were also detected by e.s.r. spectroscopy[24]. One of these radicals, the $(Me_3Si)_2NO\cdot$ is rather different from the others, because of the silicon atoms bonded to nitrogen. Another nitroxide radical: 1,1-dimethyl-3-(*N*-phenyloximino)butyl-phenylnitroxide was reported[28, 29] and used as a stabiliser for polymers.

Looking at all these nitroxide radicals containing hydrocarbon substituents, it is possible to see that their stability requires a secondary or tertiary carbon atom adjacent to the nitrogen, as in the case of di-t-butyl-nitroxide, or aromatic substituents as in diphenyl nitroxide and in its nitroxide derivatives. A nitroxide having hydrogen on the carbon adjacent to the nitrogen, e.g. $(CH_3)_2NO\cdot$, has never been isolated. The dimethylnitroxide radical, as well as its deuterated analogue, was only observed[40] as a major fragment in the mass spectroscopic studies of the reaction between methyl and deutereomethyl radicals, and nitric oxide. It was also postulated as an intermediate for the formation of the trimethylhydroxylamine.

In the bistrifluoromethylnitroxide radical, the electronegative perfluoroalkyl groups are believed to be the major factor in its stability. However, the presence of three fluorines on the α-carbon is not necessary for the stability of a nitroxide radical. This was demonstrated by Blackley in 1965[41] by the preparation of another stable nitroxide free radical containing only two fluorines attached to each α-carbon: the bis(2-chlorotetrafluoroethyl) nitroxide $(ClCF_2CF_2)_2NO\cdot$. The preparation of this radical and the evidence obtained by e.s.r. spectroscopy for other stable higher fluorocarbons (up to 7 carbon in the chain), analogues of $(CF_3)_2NO\cdot$[41], shows the existence of a general class of stable perfluoronitroxide radicals.

The bis(2-chlorotetrafluoroethyl)nitroxide was prepared by thermal decomposition (at 100 °C) of $(ClCF_2CF_2)_2NONO$, which was prepared by irradiation of a mixture of nitrosyl chloride and tetrafluoroethylene with sunlight for a few days. The reactions of this new radical with NO and NO_2, which gave the corresponding addition compounds $(ClCF_2CF_2)NONO$ and $(ClCF_2CF_2)NONO_2$, are the only ones described so far[39].

This new radical was reported as a purple liquid (b.p. 92 °C) which did not appear to dimerise on cooling, as $(CF_3)_2NO\cdot$ does, and this may be due to the steric repulsion of the larger chlorotetrafluoromethyl groups attached to the nitrogen. This steric effect is also reported to be responsible for the high stability of this radical; stability which should be decreased by the replacement of a α-fluorine with other groups.

This radical has also been reported[42] to be useful as a polymerisation inhibitor, e.g. as stabiliser and antioxidant (0.001–2 %) for nylon, rubber and methacrylate compounds. It has also been reported[42] to react with Me_2CO and C_6H_6.

5.3 BISTRIFLUOROMETHYLNITROXIDE RADICAL

The bistrifluoromethylnitroxide free radical is a purple-violet gas which, on cooling, condenses to a brown liquid b.p. −25 °C[1] (b.p. −20 °C)[2] and solidifies to a yellow solid, m.p. −70 °C[1] (m.p. −55 °C)[2]. Its ^{19}F n.m.r.[1] spectrum shows a singlet at +73.6 p.p.m.; relative to CCl_3F. Its infrared spectrum[43] shows bands at 1310, 1269, 1227, (C—F str.), 992 s (N—O str.), 728, 723, 718 cm^{-1} (triplet).

Its molecular structure has been determined in the vapour phase[44]; the principal structural parameters are in excellent agreement with the theoretical values (Table 5.1) which has been obtained using the INDO approximation[45].

Δ is the angle between CNC plane and N—O bond; the lowest energy geometry corresponds to a non-planar radical with the N—O bond being 10 degrees out of the CNC plane, but the energy required for deformations of up to 30 degrees from planarity has been found to be small (600 cal).

The chemistry of the $(CF_3)_2NO\cdot$ radical may be predicted on the basis that this radical will behave as a pseudo-halogen.

In order to bring out the halogen-like character of the bistrifluoromethylnitroxide radical, it may be of value to discuss briefly those properties which characterise this radical as a pseudo-halogen. Walden and Audrieth[46] in their review on these substances define as 'Pseudo-halogen' 'any univalent chemical aggregate which shows in the free state certain characteristics of the halogens'.

Table 5.1

Structural parameter	*Calculated values*	*Experimental values*
r(N—O)	1.26 (±0.01) Å	1.26 (±0.03) Å
r(C—N)	1.41 (±0.01) Å	1.44 (±0.008) Å
r(C—F)	1.34(±0.01) Å	1.32 (±0.004) Å
<(CNO)	117 (±1)	117.2 (±0.6)
Δ	10°	21.9 (±3)°

The similarities between the radical and the halogens are really striking and are summarised below:

(1) The radical combines with hydrogen[43] to form an acid from which can be regenerated in the free state by chemical[1,2,47] or electrochemical[48] oxidation.

Table 5.2 Physical constants of some halogens, pseudo-halogens and bistrifluoromethyl derivatives of certain non metals.

Compound	*Melting point*	*Boiling point*	*Vapour pressure* log p(mmHg)–A=B/T A	B	*Heat of vaporisation* cal/mol	*Trouton constant*
$SiCl_4$	−70	57.6				
$SiBr_4$	5.0	153.0				
$Si(OCN)_4$	34.5	247.2	9.82	3 611	16 500	31.7
$Si(NCO)_4$	26.0	185.6	9.02	2 816	12 900	26.5
$Si(NCS)_4$	143.8	313.0	8.45	3 276	13 900	23.7
$Si((CF_3)_2NO)_4$	−78	135.0	7.66	1 955	8 890	21.8
$GeCl_4$	−49.5	83.1				
$GeBr_4$	26.1	186.5				
$Ge(HCO)_4$	−8.0	204.0	8.66	2 757	12 600	26.4
$Ge((CF_3)_2NO)_4$	−22.0	138.0	7.58	1 929	8 780	21.35
$POCl_3$	2.0	105.0				
$POBr_3$	56.0	193.0				
$PO(NCO)_3$	5.0	193.1	9.17	2 931	13 410	28.8
$PO(NCS)_3$	13.8	300.0	8.53	3 240	14 820	25.8
$PO((CF_3)_2NO)_3$	30.5	113.5	7.85	2 015	9 205	22.7
$AsCl_3$	−18.0	130.0				
$AsBr_3$	32.8	221.0				
$As(NCO)_3$	97.1	224.0	8.76	2 924	14 800	26.0
$As((CF_3)_2NO)_3$	30.0	143.0	8.30	2 251	10 220	24.6

(2) The radical shows great affinity towards metals, combining in many cases directly with the formation of salts[43,49].

(3) The radical forms covalent compounds which are similar to the corresponding halide and pseudo-halide derivatives in composition and in their physical and chemical behaviour. The physical constants of a few

characteristic halogen and pseudo-halogen covalent compounds are reported in Table 5.2 together with those of the corresponding bistrifluoromethylnitroxide derivatives. Comparison of these properties shows striking similarities. They all seem to be covalent, although some may be associated in the liquid state. Chemically, the bistrifluoromethylnitroxide derivatives are similar to the anhydrous halides and pseudo-halides in the ease with which they undergo hydrolysis.

(4) The radical easily undergoes addition to an ethylenic linkage[43, 49] and is able to abstract hydrogen from several organic and inorganic substances[50–52].

(5) The radical will replace iodine, bromine, and in some cases chlorine from several metallic and metalloid halides, liberating the halogen in the free state; furthermore it is known to form at least one complex, Cs $[(CF_3)_2NO]_2I$, which is analogous to the polyhalide complexes.

The fact that the radical will oxidise iodide and bromide to iodine and bromine, and that in some cases, a derivative of the radical will react with chlorine liberating the free radical, places it in a definite position among halogens and pseudo-halogens in the sense that, representing them in order of decreasing oxidising power, the radical will be between chlorine and bromine:

$$F_2 > Cl_2 > (CF_3)_2NO\cdot > Br_2 > (CN)_2 > (SCN)_2 > I_2 > (SCSN)_3)_2 > (SeCN)_2$$

However, the fact that in some cases (e.g. HCl or BCl_3) the radical is able to oxidise a chloride to chlorine would indicate that the position of the radical and chlorine should be reversed.

The only point in which the radical differs from halogens and pseudo-halogens is in the fact that it is unable to form either inter-halogen compounds, or a stable dimer at room temperature.

The electronegativity of the bistrifluoromethylnitroxide radical may be estimated from spectroscopical data. The infrared spectra of phosphoryl and carbonyl bistrifluoromethylnitroxide compounds show strong absorp-

Table 5.3 P=O stretch frequencies

Compound	P = O *stretch* (cm^{-1})
$F_3P{=}O$	1415[55]
$[(CF_3)_2NO]_3P{=}O$	1382
$(CF_3)_3P{=}O$	1329[56]
$Cl_3P{=}O$	1309[57]
$(CH_3O)_3P{=}O$	1275[57]
$Br_3P{=}O$	1268[57]
$(CH_3)_3P{=}O$	1176[57]

tion bands which have been attributed to the P=O and C=O stretching vibration. The P=O vibration frequency is known to depend primarily upon the electronegativity of the groups bonded to phosphorus atom[53]. The high frequencies (which can be correlated with an increase of the P=O double bond character) is always associated with high electronegativity of the substituents as can be seen from Table 5.3 in which the P=O stretch vibration of several phosphoryl compounds are reported. Corresponding movements to high frequencies are observed in the carbonyl frequencies (Table 5.4); in this series COF_2 for example, absorbs at 1928 cm^{-1}, which is the highest carbonyl frequency known.

Thus, this relationship, which can be of considerable help in identifying

Table 5.4 C=O stretch frequencies

Compound	C=O *stretch*
$F_2C{=}O$	1928[57]
$(CF_3)_2NO(F)C{=}O$	1917[58]
$CF_3(F)C{=}O$	1901[53]
$[(CF_3)_2NO]_2C{=}O$	1890[58]
$CH_3(F)C{=}O$	1872[58]
$(CF_3)_2NO(CF_3)C{=}O$	1868[58]
$(CF_3)_2NO(C_3F_7)C{=}O$	1858[58]
$(CF_3)_2NO(Cl)C{=}O$	1855[58]
$(CF_3)_2NO(C_2H_5)C{=}O$	1850[53]
$(CF_3)_2NO(CH_3)C{=}O$	1845[59]
$Cl_2C{=}O$	1828[53]

the substituent groups[54], can give, to a first approximation, a measure of the electronegativity of the bistrifluoromethylnitroxide group.

From Tables 5.3 and 5.4 it is possible to see that the bistrifluoromethylnitroxide group is, after fluorine, the most electronegative group, being more electronegative than the CF_3 group. According to this rule a series of groups having decreasing electronegativity can be written in the following order:

$$F > (CF_3)_2NO\cdot > CF_3 > C_3F_7 > Cl > Br > C_2H_5 > CH_3$$

5.3.1 Preparation of bistrifluoromethylnitroxide radical

The first synthesis of bistrifluoromethylnitroxide reported by Blackley and Reinhard in 1964[1] was achieved by the following sequence of reactions:

(1) $CF_3COOAg + NOCl \rightarrow CF_3COONO + AgCl$

(2) $CF_3COONO \rightarrow CF_3NO + CO_2$

(3) $2CF_3NO \rightarrow (CF_3)_2NONO$

(4) $(CF_3)_2NONO \xrightarrow[Hg]{HCl} (CF_3)_2NOH$

(5) $2(CF_3)_2NOH \xrightarrow{Ag_2O \text{ or } F_2} 2(CF_3)_2NO$

The following reactions represent some possible alternative routes for this preparation:

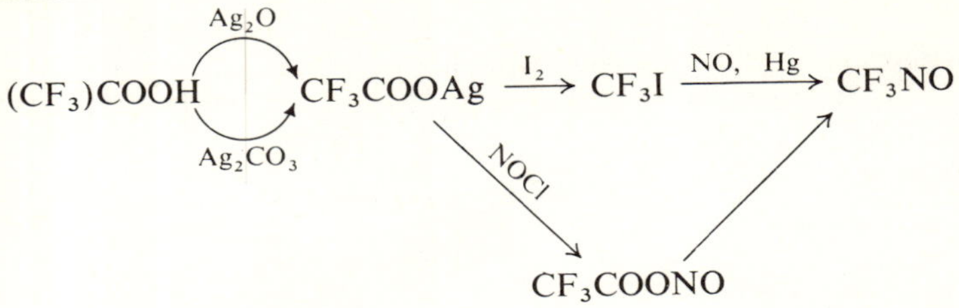
Ag2O
(CF3)COOH
CF3COOAg
Ag2CO3
I2
CF3I
NO, Hg
CF3NO
NOCl
CF3COONO

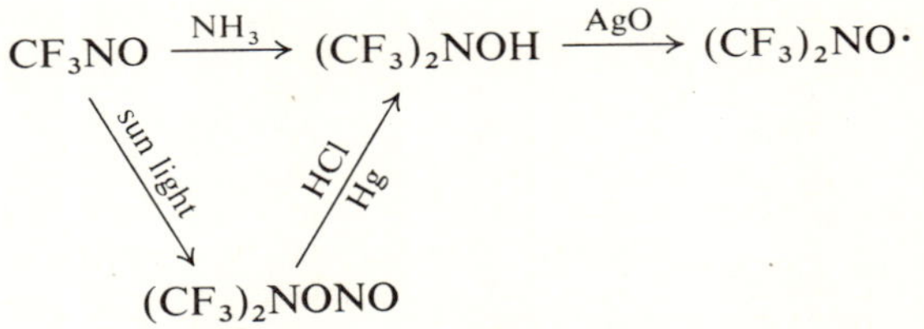
CF3NO
NH3
(CF3)2NOH
AgO
(CF3)2NO·
sun light
HCl
Hg
(CF3)2NONO

Silver trifluoroacetate can be easily prepared by dissolving either silver carbonate or silver oxide[60] in diluted trifluoroacetic acid. Both silver carbonate and silver oxide are prepared immediately before their use by mixing a solution of silver nitrate with a solution of sodium carbonate or sodium hydroxide respectively. The dry silver trifluoroacetate can be obtained by evaporating the aqueous solution on a hot plate or by extracting the silver salt by ether. The salt is then powdered and dried in a vacuum desiccator over phosphorus pentoxide.

The most important step in the synthesis of the bistrifluoromethylnitroxide is the preparation of trifluoronitrosomethane. Among the methods described in the literature for this preparation only two should be taken into consideration:

(a) Photolysis of trifluoroidomethane in the presence of nitric oxide and mercury[61, 62].

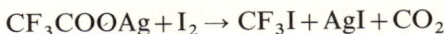

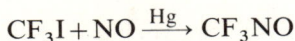

$$CF_3COOAg + I_2 \rightarrow CF_3I + AgI + CO_2$$

$$CF_3I + NO \xrightarrow{Hg} CF_3NO$$

(b) Pyrolysis of trifluoroacetilnitrite[63–65].

Although the first step of method (a), i.e. the preparation of CF_3I is much more difficult than the corresponding step of method (b), i.e. the preparation of trifluoroacetilnitrite, which is almost quantitative, difficulties are found in the pyrolysis step:

explosions, which often occur, destroy the whole apparatus with a big loss of compounds,

the trifluoronitrosomethane produced by pyrolysis contains generally a rather large amount of impurities and it is difficult and wasteful to extract them by vacuum distillation.

The photolysis carried out in the way described by Dinwoodie and Haszeldine[66] gives the best results.

For the preparation of *N,N*-bistrifluoromethylnitroxylamine, the two following methods can be used:

(1) The two steps process which consists of dimerisation of trifluoronitrosomethane, followed by reaction of the dimer with hydrochloric acid and mercury[1,67]

$$2CF_3NO \xrightarrow{\text{sunlight}} (CF_3)_2NONO$$

$$(CF_3)_2NONO \xrightarrow[\text{Hg}]{\text{HCl}} (CF_3)_2NOH$$

(2) The one-step process: the reaction of CF_3NO with ammonia

$$2CF_3NO + NH_3 \rightarrow (CF_3)_2NOH + H_2O + N_2$$

The first route, which apparently requires sunlight[1] for the dimerisation of CF_3NO and gives a very high yield of the hydroxylamine, is not very efficient when artificial light is used.

The second method can give an average yield of hydroxylamine of 70% based on the CF_3NO used. The advantages of this one-step process are in its simplicity and in its suitability for a comparatively large scale production. The two-steps method still remains useful especially when rather small amounts of compounds are involved.

The conversion of bistrifluoromethylhydroxylamine into the radical has been reported to occur with silver(I) oxide[1], fluorine[1], potassium permanganate[2], sodium fluoride[43], silver(II) oxide[47] and electrochemically[48]. The best way, which gives a quantitative conversion, is the reaction of the hydroxylamine, well dried over P_4O_{10} or H_2SO_4, with freshly prepared silver(II) oxide. This reaction is rather quick at room temperature. It is advisable not to leave the products in the reaction vessel for a long time, especially when an excess of $Ag^{II}O$ is used, since a lower yield of radical may be obtained and explosions may occur.

5.4 *N, N*-BISTRIFLUOROMETHYLHYDROXYLAMINE

N,N-bistrifluoromethylhydroxylamine is a colourless volatile liquid which boils at 32°C, it is stable at room temperature while at higher temperature it decomposes slowly.

It can be prepared by hydrogenation of the bistrifluoromethylnitroxide radical in a bed of a palladium–alumina catalyst[17]. It is also found as a product of every reaction in which abstraction of hydrogen by the $(CF_3)_2NO\cdot$ occurs.

The *N, N*-bistrifluoromethylhydroxylamine is a very weak acid in aqueous solution, its dissociation constant is 1.5×10^{-9} [18]. The reaction with NaOH in tetrahydrofurane[19] gives the sodium salt $(CF_3)_2NO^-Na^+$, a white crystalline water sensitive solid which, because of its nucleophilic reactivity, is a useful reagent for introduction of the $(CF_3)_2NO$-group into organic and inorganic molecules.

$(CF_3)_2NOH$, being a weak acid, interacts with a variety of simple organic amines[68] to form a series of weakly associated adducts. These adducts are not the high-melting solids, typical of substituted ammonium salt, but are colourless liquids or low melting, sublimable crystalline solids which exhibit varying degrees of dissociation at room temperature. These compounds are, with exception of the triethylamine adduct, stable at room temperature. They slowly decompose at 50 °C except the ammonia adduct which undergoes a reversible dissociation over a long temperature range. At room temperature its dissociation is approximately 50% while dimethylamine adduct is only about 5% dissociated under the same conditions. Pure adducts can be always recovered from the decomposition products, if some decomposition had occurred, by trap to trap distillation. Light and trace impurities of amine accelerate the decomposition process. In particular the triethylamine adduct decomposes at room temperature, to a non-volatile yellow-brown oil. The adducts are not particularly sensitive to moisture, and, initially, the solids appear to repel water. This may be due to the CF_3-groups, but after a few hours $(CF_3)_2NOH$ and free amine are formed.

The physical properties of these adducts are reported in Table 5.5.

Table 5.5

Adduct formed	*Physical state*	*m.p.* °C
$NH_3{\cdot}(CF_3)_2NOH$	Liquid	
$CH_3NH_2{\cdot}(CF_3)_2NOH$	Solid	28.0–28.5
$(CH_3)_2NH{\cdot}(CF_3)_2NOH$	Solid	35.0
$(CH_3)_3N{\cdot}[(CF_3)_2NOH]_2$	Solid	28.0–28.5
$C_2H_5NH_2{\cdot}(CF_3)_2NOH$	Liquid	
$(C_2H_5)_2NH{\cdot}[(CF_3)_2NOH]_2$	Solid	41.5–42.5
$(C_2H_5)_3N{\cdot}(CF_3)_2NOH$	Liquid	

The infrared spectra of the adducts[68] in the gas phase contain bands which can be attributed to all of the possible components, i.e. the hydroxylamine, the amine, and the adduct. From the relative strength of these bands in the gas phase the extent of dissociation of the adducts can be calculated. This is clearly shown by the spectrum of $NH_3{\cdot}(CF_3)_2NOH$ in which peaks due to NH_3, $(CF_3)_2NOH$ and $NH_3{\cdot}(CF_3)_2NOH$ exist in approximately equal proportions and indicate that the adduct is about 50% dissociated in the gas phase at room temperature.

All other adducts show weaker bands of the reagents and therefore they are dissociated in the gas phase to a smaller extent. In particular the band at 3619 cm^{-1}, which is assigned to the O–H stretching frequency of $(CF_3)_2NOH$ can be used to obtain a semi-quantitative measure of the extent of the adduct dissociation since it arise only from unassociated hydroxylamine. If association occurs through the hydroxyl proton, any associated hydroxylamine should show the O–H stretching frequency shifted from 3619 cm^{-1} to a lower frequency. In the $(CF_3)_2NO$-amine adducts the bands: at 3317 cm^{-1} for $NH_3{\cdot}(CF_3)_2NOH_{(g)}$, at 3400 cm^{-1} for $NH_3{\cdot}(CF_3)_2NOH_{(1)}$ and 3370 and 3305 cm^{-1} for $C_2H_5NH_2{\cdot}(CF_3)_2NOH_{(1)}$ are probably due to the O–H frequency shifted from 3619 cm^{-1}.

The ^{19}F n.m.r. spectra[68] of the adducts (referred to internal CCl_3F) consist of a singlet at about 69–70 p.p.m. No spin-spin coupling is observed between fluorines and protons.

The proton n.m.r. spectra[98] show a variant single resonance which can be attributed to the proton or protons on the nitrogen atom. The relative areas of the peaks of the 1:1 adducts correspond to an addition of one proton to the amine, and of the 1:2 adducts to an addition of two protons of the amine. The mass spectra of the adducts contain only peaks due to the components of the adducts.

The plots of the mole fractions (X) of amine against the pressure show for the 1:1 adduct a rise in the pressure after the $X = 0.5$ point; for 1:2 adducts there is a minimum of pressure extending from $X = 0.33$ to $X = 0.45$ and the pressure begins to rise only after $X = 0.45$.

No chemical explanation has been given to the formation of the 1:2 adducts.

The *N,N*-bistrifluoromethylhydroxylamine forms adducts either with CsF or with KF at room temperature. These adducts are white solids, $(CF_3)_2NOH{\cdot}CsF$ melts at about 70 °C with slow evolution of $(CF_3)_2NOH$. The vapour pressure of $(CF_3)_2NOH$ at room temperature is 0.5–1 mm for the CsF adduct 4 mm for that of KF.

The ^{19}F n.m.r. of the CsF adduct, like other $(CF_3)_2NOH$ adducts, shows only one signal which is shifted to downfield ($\alpha = 67.6$) compared to that of free $(CF_3)_2NOH$. This is unexpected if formation of a $(CF_3)_2NO$-ion is postulated since in this case an increase in electron density at the fluorine atoms should produce a shift towards a higher field[69].

The ^1H n.m.r. spectrum of the Cs adduct shows a single band at -12.7 p.p.m. relative to CH_3CN while $(CF_3)_2NOH$ in CH_3CN shows a band at -5.48 p.p.m. The ^1H band of the adduct should come from the proton associated with the fluoride of the CsF.

5.5 $(CF_3)_2$NO-COMPOUNDS OF GROUP I ELEMENTS

The only $(CF_3)_2$NO-compound of Group I elements known so far is the sodium bistrifluoromethylnitroxide.

The first preparation of this sodium salt[70] was achieved by treatment of a solution of *N,N*-bistrifluoromethylhydroxylamine in tetrahydrofurane with an equimolar amount of sodium hydroxide in the presence of Linde molecular sieve at 0–20 °C. The sodium salt was obtained as a white solid by evaporation of the tetrahydrofurane.

A second method, reported at the same time, was performed by passing gaseous radical over sodium wire at 20 °C/50 mm.

This method is inconvenient because it leads to a product contaminated with sodium fluoride and violent explosions may occur.

A better preparation of this sodium derivative can be achieved by reacting excess radical with powdered sodium iodide. Iodine is quickly displaced and can be easily pumped out together with the excess of the radical.

Potassium iodide does react in the same way with the radical but very slowly, while no reaction takes place with lithium iodide.

Caesium iodide, on the contrary, reacts in a different way with $(CF_3)_2NO{\cdot}$ forming an addition compound $CsI\,[ON(CF_3)_2]_2$ without liberation of iodine. However, liberation of iodine occurs in the presence of light. This compound is probably a salt of the same type as the polyhalides and this is the only evidence obtained so far for direct combination between the radical and one of the halogens.

Attempts to obtain the sodium salt by interaction between *N,N*-bistrifluoromethylhydroxylamine and sodamide according to the reaction

$$(CF_3)_2NOH + NaNH_2 \longrightarrow (CF_3)_2NONa + NH_3$$

led only to violent explosions.

Sodium bistrifluoromethylnitroxide is a white crystalline solid which reacts very rapidly with moisture and water giving $(CF_3)_2NOH$ and NaOH. Its nucleophilic reactivity makes it useful as intermediate for the introduction of the $(CF_3)_2$ NO-group into organic and inorganic molecules.

5.6 $(CF_3)_2$NO-COMPOUNDS OF GROUP II ELEMENTS

The only $(CF_3)_2$ NO-compound of Group II elements prepared so far is the mercury derivative since no synthetic route has been found for the synthesis of $(CF_3)_2$NO-derivatives of other Group II elements. The radical does not react with elemental Mg,Zn,Cd either at room temperature or at temperature as high as 200 °C and furthermore it does not react with MgI_2, BaI_2, ZnI_2, CdI_2 under the same conditions.

Therefore the investigation of the $(CF_3)_2$NO-derivatives of Group II has been concentrated on the mercury derivative: mercury(II) bistrifluoromethylnitroxide.

The perfluoro-radicals, like other free radicals are known to react readily with mercury to form stable organometallic derivatives[71,72]. Since 1931 mercury mirrors have been used to detect and identify several alkyl and aryl radicals[73,74]. Reaction occurs between bistrifluoromethylnitroxide radical and mercury either with gaseous radical or when excess of the radical in the liquid form is shaken with mercury in a sealed tube[75]. The reaction has been shown to be over 95 % quantitative if the amount of radical used is sufficient to provide an excess of liquid radical at the completion of the reaction. This preparation may be also performed in an inert solvent, containing no hydrogen, like $CFCl_3$, in which mercury and radical may be shaken in stoichiometric proportion at room temperature for 24 h.

$Hg[ON(CF_3)_2]_2$ is a white solid slightly soluble in Freons and very soluble in the liquid radical from which it separates as long colourless needles, though normally, in its preparation it is left as a snow-like solid on removal of the excess of the radical.

This mercurial is very stable at room temperature if kept in the dark. It becomes yellow and starts decomposing when exposed to daylight for a few days. It reacts readily with hydrogen-containing organic solvents, the product being mainly *N,N*-bistrifluoromethylhydroxylamine. Its solubility in those organic solvents, in which it is rather stable, e.g. carbon tetrachloride, trichlorofluoromethane, etc., is such that conventional methods of recrystallisation cannot be applied. It also becomes immediately yellow and hydrolyses in the presence of moisture and is difficult to handle in a dry box in the presence of phosphorus pentoxide. Its reaction with water readily gives yellow mercuric oxide and the hydroxylamine quantitatively. Its behaviour with moisture resembles that of di(bistrifluoromethylamino) mercury which has been reported[75] to be extremely sensitive to moisture, and to hydrolyse immediately with formation of a yellow solid.

The mercurial cannot be sublimed in vacuum since it decomposes completely at 85 °C giving mercury and radical quantitatively. Formation of the dimer or other products is not in any case observed as decomposition products. The behaviour of this mercurial, when thermally decomposed, is thus similar to that of the alkyl, thioalkyl and perfluoroalkyl mercurials, which decompose at higher temperatures yielding mercury and the corresponding radicals which, after a short life, undergo dimerisation or reaction with other molecules. Among these mercurials, the most stable has been reported[77,78] to be dimethylmercury, which decomposes only at 300 °C yielding mercury and ethane, while the corresponding perfluoroalkyl and thioalkylmercurials[79] decompose in the same way at much lower temperatures:

$$(CF_3)_2Hg \xrightarrow{170\,°C} C_2F_6 + Hg$$

$$(CH_3S)_2Hg \xrightarrow{180\,°C} CH_3SSCH_3 + Hg$$

On the other hand, mercury(II) bistrifluoromethylnitroxide differs basically in its decomposition from perfluoroalkylthio and perfluoroalkylamino mercurials, which are much more stable. Bistrifluoromethylthiomercury does not undergo appreciable decomposition at temperatures below 200 °C[80] and di(bistrifluoromethylamino) mercury does not show any permanent change after being heated at 300 °C. It dissociates at 150 °C into $CF_3N = CF_2$ and mercuric fluoride which recombine on cooling[76,81].

In some respects the decomposition of mercury(II) bistrifluoromethylnitroxide is similar to that of mercuric oxide, which has a relatively low heat of formation ($-\Delta H = 21.7$ cal mol^{-1}) compared with that of zinc oxide (83 cal mol^{-1}) and cadmium oxide (61 cal mol^{-1}). Mercuric oxide is known to be thermally unstable and its decomposition into its elements it is perceptible at 100 °C[81] and complete at 300 °C[83], its dissociation pressure being[84]: $\log p = -2513.5\,T^{-1} + 1.75 \log T - 0.001033T + 5.9461$.

The shape of the mercury(II) bistrifluoromethylnitroxide crystals and its thermal instability suggest that its structure may be built up of long chains of 4-coordinated mercury atoms linked through oxygen by double $(CF_3)_2$ NO-bridges.

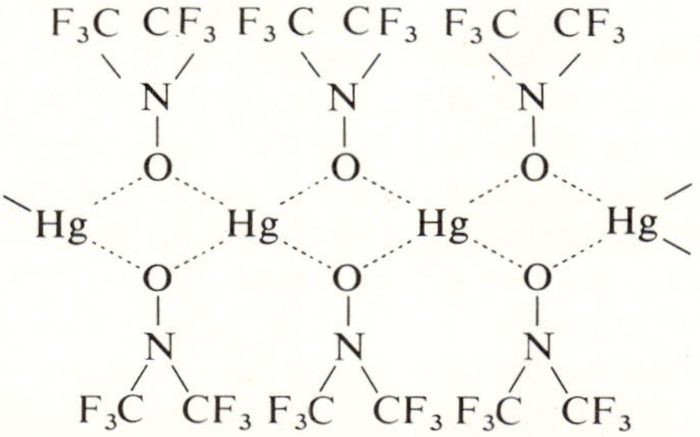

In some ways its structure will be similar to that of mercuric oxide in which the mercury atoms are bonded to oxygen in an infinite planar zig-zag chain[85].

$$\text{Hg} \overset{O}{\diagdown} \text{Hg} \overset{O}{\diagup} \text{Hg} \overset{O}{\diagup} \text{Hg} \overset{O}{\diagup} \text{Hg}$$

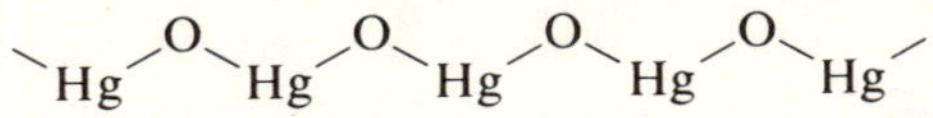

No infrared spectrum of the mercurial has been reported since when a compressed solid disc, with finely divided potassium bromide, was used, a reaction took place and the disc became brown as if bromine had separated. The spectrum, taken as a liquid film between potassium bromide plates, using Nujol or CCl_3F as solvent, was mainly that of the hydroxylamine produced either by reaction of the mercurial with Nujol or by hydrolysis.

The n.m.r. spectrum of a saturated solution of mercurial in CCl_3F showed a singlet at 71 p.p.m. relative to trichlorofluoromethane which indicates a symmetrical molecule with all fluorine atoms equivalent.

The most valuable property of mercury(II) bistrifluoromethylynitroxide is its use as an intermediate in the preparation of the bistrifluoromethyl-nitroxide derivatives. In its metathetical reactions with organic and inorganic halides, the mercury(II)bistrifluoromethylnitroxide resembles in general the alkyl, perfluoroalkyl and perfluoroalkylthio mercury, rather than bistri-fluoromethylamino mercury, because the $(CF_3)_2NO$-group is transferred readily.

Chlorine and bromine react rapidly at room temperature with mercury(II) bistrifluoromethylnitroxide, giving the mercuric halide and the free radical.

In the same way a solution of mercurial in trichlorofluoromethane reacts slowly at room temperature with iodine, liberating red mercuric iodide and radical. The slow rate of the reaction with iodine may be due to the low solubility of iodine in trichlorofluoromethane. When these reactions are carried out with an excess of the halogen, mercuric halide and a mixture of radical and excess of the halogen is recovered. No evidence is obtained in this and in other reactions that the radical would combine with chlorine, bromine or iodine.

The behaviour of mercury(II) bistrifluoromethylnitroxide towards halogens is closely similar to that of alkyl- and perfluoroalkyl-thiomercury compounds. A solution of iodine in carbon tetrachloride converts them both quantitatively to mercuric iodide and to disulphide according to the following reactions:

$$(CH_3S)_2Hg + I_2 \longrightarrow CH_3SSCH_3 + HgI_2 \text{[88]}$$
$$(CF_3S)_2Hg + I_2 \longrightarrow CF_3SSCF_3 + HgI_2 \text{[89]}$$

This mercurial reacts with a large number of organic and inorganic halides giving the corresponding bistrifluoromethylnitroxide compounds in a high yield. These reactions are rapid compared with the corresponding reactions of the other mercurials. In general with mercury(II) bistrifluoro-methylnitroxide the corresponding mercury halide is the solid product, (usually mixed with the excess of mercurial with which the reactions are generally carried out), while with the other mercurials the solid product is a mercurial of the general formula RHgX and only one of the two groups bonded to mercury is available for the reaction.

Mercury(II) bistrifluoromethylnitroxide reacts readily at room temperature with hydrogen chloride or iodide to form bistrifluoromethyl-hydroxylamine and the corresponding mercuric halide quantitatively. Its behaviour with hydrogen chloride is thus similar to that of bistrifluoro-methylthiomercury, which reacts quantitatively with dry hydrogen chloride to give trifluoromethanethiol:

$$(CF_3S)_2Hg + 2HCl \longrightarrow 2CF_3SH + HgCl_2$$

Mercury(II) bistrifluoromethylnitroxide reacts at room temperature with nitrosyl chloride giving *O*-nitrosobistrifluoromethylhydroxylamine quantitatively. This is a reaction in which the mercurial resembles di(bistrifluoromethylamino) mercury more than bistrifluoromethylthiomercury, because in the case of bistrifluoromethylthiomercury the disulphide is the major product[86, 87]:

$$(CF_3S)_2Hg + 2NOCl \longrightarrow CF_3SSCF_3 + 2NO + HgCl_2$$

$$[(CF_3)_2N]_2Hg + 2NOCl \longrightarrow 2(CF_3)_2NNO + HgCl_2$$

In the case of trifluoromethylthiomercury formation of the disulphide could be explained by the instability of *S*-nitrosotrifluoromethylthiol, which has never been prepared and may decompose according to the following reaction:

$$2CF_3SNO \longrightarrow CF_3SSCF_3 + 2NO$$

The mercury(II) bistrifluoromethylnitroxide has been found to be useful as a general reagent for the preparation of organic and inorganic derivatives only when the compound obtained is volatile enough to be pumped out from the reaction vessel, in which the other products, i.e. mercuric halide and the excess of mercurial, remain.

Another surprising property of $Hg[ON(CF_3)_2]_2$ is its unique manner of behaving towards compounds containing hydrogen bonded to the following elements: Si, Ge, N, As, Sb and in some cases C. Hydrogen atoms are abstracted to form $(CF_3)_2NOH$ and are replaced by the $(CF_3)_2NO$-group while the Hg is reduced to elemental mercury.

Mercury(II) bistrifluoromethylnitroxide dissociates into Hg and $(CF_3)_2NO\cdot$ only at elevated temperatures and it is unexpected therefore to find that it can produce abstraction reactions similar to those of the free radical at 20 °C or below, where its thermal dissociation must be negligible. Its behaviour is quite different from that of organometallic compounds in general, and must involve a mechanism which is different from that operating foı the free radical reactions.

As a matter of fact, although the radical does not react at 20 °C with ammonia, the mercurial does so at this temperature with both NH_3 and $(CH_3)NH$, and the yield of $(CF_3)_2NOH$ corresponds in both cases to the hydrogens bonded to nitrogen. The other product is an involatile yellow solid which is probably a mixture of products containing Hg–N bonds. This fluorinated mercurial seems to be unique in its behaviour of abstracting hydrogen.

5.7 $(CF_3)_2$NO-COMPOUNDS OF GROUP III ELEMENTS

The only $(CF_3)_2$NO-derivative of Group III elements so far prepared is the tris(bistrifluoromethylnitroxy)borane.

$[(CF_3)_2NO]_3B$ has been synthesised by reaction of borontrichloride with an excess of $[(CF_3)_2NO]_2Hg$ without solvents[90]. The reaction is quick at room temperature and the $[(CF_3)_2NO]_3B$ is the only volatile product which can be easily pumped out from the reaction vessel. The same product can be also obtained by reaction of $[(CF_3)_2NO]_2Hg$ with BBr_3.

Another method used[91] to prepare the boron $(CF_3)_2NO$-compound is the direct interaction of $(CF_3)_2NO\cdot$ with boron trichloride or tribromide. The reaction proceeds quickly with boron tribromide and the products are $[(CF_3)_2NO]_3B$ and elemental bromine. The separation between $[(CF_3)_2NO]_3B$ and Br_2 is rather difficult by vacuum distillation, but can be easily achieved by addition of mercury to the mixture. The mercury reacts quickly with bromine giving involatile $HgBr_2$ and the boron compound can be recovered by simple vacuum distillation. The reaction of the radical with BCl_3 is rather slow at room temperature, being completed only after several days, but in the presence of iodine the conversion of boron trichloride to tris(bistrifluoromethylnitroxy)borane occurs within a few hours. The boron compound can be separated from ICl or ICl_3 produced by the reaction, by treating the mixture with Hg. Although the presence of iodine clearly increases the reactivity of the reagents, by reacting with the chlorine produced to form ICl and ICl_3 and consequently lowering the free energy of the products, in the case in which, only a small non-stoichiometric amount of iodine is present, an increase of the reaction rate is produced and free chlorine is liberated together with ICl_3. This indicates that the iodine has a catalytic effect on the reaction but the mechanism has not yet been found. A similar reaction occurs between the radical and HCl.

No hydroxylamine is detected after 4 h at room temperature and, after 13 days, only 10% of the radical is converted to hydroxylamine. But, in the presence of 5% mole of free iodine, based on the HCl taken, this reaction has been reported to be completed in 2 h.

Iodine is converted to ICl_3, free chlorine is formed and all the hydrogen appears as hydroxylamine.

A study of the conductivity of the radical in the liquid state with and without the addition of iodine, has been made in the hope that a reaction between the radical and free iodine, involving the formation of ionic species, such as $(CF_3)_2NO)_2I^-$, I^+, could be detected. This would provide a possible mechanism for explaining the catalytic effect of iodine. No evidence for this type of interaction has been obtained, and the specific conductivity of the radical ($<10^{-7}\Omega^{-1}$ cm^{-1} at 0 °C) is not appreciably increased by the presence of iodine. The presence of an active intermediate such as $ICl[ON(CF_3)_2]_2$ could account for the catalytic effect, although no evidence has been obtained for the formation of such stable species at room temperature or at temperatures as low as −78 °C.

No reaction occurs between $LiAlH_4$ and the $(CF_3)_2NO\cdot$ radical either at room temperature or at 100 °C.

$[(CF_3)_2NO]_3B$ is a crystalline volatile solid which melts at 37 °C. The relationship between vapour pressure and temperature can be expressed by the following equation: $\log_{10}p$ (mm) = $8.42 - 2058/T$. Its boiling point at normal pressure is 99 °C. The latent heat of evaporation is 9360 cal mol^{-1} and the Trouton constant 25.2 e.u. The infrared spectrum shows the following main bands: 1443 m ($^{10}B-0$ str. as.), 1393 s ($^{11}B-0$ str. as.), 1311 vs, 1279 vs, 1260 vs and 1260 vs (C–F str.), 1049 m (N−0 str.), 976 m (C–N str.) 813 w (C–N–C bending), 716 m (CF_3 def.), 660 w ($^{11}B-0$ out of plane). The two strong bands at 1443 and 1399 cm^{-1} can be assigned to the asymmetric stretching vibration of the two isotopes of boron (i.e. ^{10}B and ^{11}B respectively) by analogy with alkoxyborane which have bands in this region undoubtedly[92] assigned to the B—O asymmetric stretch as shown in Table 5.6:

Table 5.6

	$B(OMe)_3$	$B(OEt)_3$	B(OiPr)
^{10}B—O str. as.	1361 vs	1445 vs	1433 vs
^{11}B—O str. as.	1385 m	1425 m	1422 w
^{11}B—O out of plane	664 m	673 w	664 w

In the same way the bond at 660 cm^{-1} can be assigned to ^{11}B—O out of plane vibration frequency.

The ^{19}F n.m.r. spectrum is a singlet at 70.4 p.p.m. relative to internal CCl_3F.

$[(CF_3)_2NO]_3B$ is a stable solid which hydrolyses very quickly in moist air liberating the hydroxylamine quantitatively. The reaction of $[(CF_3)_2NO]_3B$ with hydrogen chloride occurs at room temperature and gives $(CF_3)_2NOH$ and BCl_3. No reaction occurs with chlorine even at 100 °C.

On distilling the boron compound into a suspension of caesium fluoride in acetonitrile, the caesium salt dissolves and the residue, after removal of the solvent and excess of the volatile boron compound is a 1:1 adduct $CsF \cdot [(CF_3)_2NO]_3B$, an ionic species completely involatile which probably contains the $BF[ON(CF_3)_2]_3^-$ ion.

5.8 $(CF_3)_2$NO-COMPOUNDS OF GROUP IV ELEMENTS

$(CF_3)_2$NO-carbon compounds will be described in the organic compound section.

Many silane and germane containing the $(CF_3)_2$NO-group have been isolated. They can be expressed by the general formula $[(CF_3)_2NO]_4MX_{4-y}$ (where M = Si, Ge and X = Cl, Br, CH_3).

Tetrakis(bistrifluoromethylnitroxy)silane and germane may be prepared by reaction of silane and germane containing M—X and/or M—H bonds (where M = Si, Ge and X = Cl, Br, I) such as: SiX_4, SiH_4, SiH_2I_2, GeX_4, GeH_4, $GeHBr_3$, etc. with mercury(II)bistrifluoromethylnitroxide[90,91].

When no M—H bonds are present, the reaction is a simple metathetical reaction:

$$MX_4 + 2[(CF_3)_2NO]_2Hg \longrightarrow [(CF_3)_2NO]_4M + 2HgX_2$$

when M—H bonds are present, the reaction products are also $(CF_3)_2NOH$ and Hg. In this case the hydroxylamine can be easily separated from the tetrakis(bistrifluoromethylnitroxide)compound by simple vacuum distillation.

The bistrifluoromethylnitroxide radical reacts with a M—H bond at temperatures as low as −78 °C replacing the H and forming the hydroxylamine, it also reacts at room temperature with some of the silicon- and germanium-halogen bonds[91]. In this case the halogen is liberated as free halogen and is replaced by the $(CF_3)_2NO$-group. In particular Si—Br, Si—I, Ge—I bonds are attacked by the radical and the products are the tetrakis compound and the free halogen; Si—Cl, Ge—Br bonds do not react with the radical under the same conditions, so from compounds such as GeH_3Br and $SiHCl_3$, $[(CF_3)_2NO]_3GeBr$ and $(CF_3)_2NOSiCl_3$ are obtained respectively. Since the Si—H bond is much more reactive than the C—H bond, from $SiMe_2ClH$ the product is $(CF_3)_2NOSiMe_2Cl$.

Other methods are also available to produce $(CF_3)_2NO$-derivatives of Si and Ge; in particular bistrifluoromethylnitroxytrimethylsilane has been prepared in several different ways:

(a) Reaction of sodium bistrifluoromethylnitroxide with trimethylchlorosilane at room temperature in tetrahydrofurane[70].

(b) Reaction of $(CF_3)_2NOH$ with Me_3SiCl[93].

$$(CF_3)_2NOH + Me_3SiCl \longrightarrow (CF_3)_2NOSiMe_3 + HCl$$

The yield of this reaction is 50% and the other main product is hexamethyldisiloxane.

(c) Reaction between trimethylsilane and trifluoronitrosomethane under photolytic conditions[93]. The yield of $(CF_3)_2NOSiMe_3$ is 28% based on the silane used; the other main products are: the adduct $CF_3NHOSiMe_3$ 43%, trimethylfluorosilane 15%, hexafluoroazoxyethane 4%, nitrogen, hexamethyldisiloxane and various unidentified compounds.

(d) Reaction of the radical with bis(trimethylsilyl)mercury[47]

$$(Me_3Si)_2Hg + 4(CF_3)_2NO\cdot \rightarrow [(CF_3)_2NO]_2Hg + 2(CF_3)_2SiMe_3$$

Attempts to obtain $(CF_3)_2NOSiMe_3$ by reaction between $Me_3SiSiMe_3$ and radical failed as no reaction took place even at 100 °C. The germanium does not react with radical, only at 200 °C and in the presence of iodine a small yield of the tetrakis$(CF_3)_2NO$-germanium derivative can be obtained. It is evident that the reaction proceeds through the formation of Ge—I bond which reacts with the radical.

The silane and germane compounds containing the $(CF_3)_2NO$-group are colourless volatile liquids. No melting point is detectable for $[(CF_3)_2NO]_4Si$ and for $[(CF_3)_2NO]_3GeBr$, at −78 °C they are still colourless viscous oils. The physical properties of tetrakis(bistrifluoromethylnitroxy)germane (Table 5.2) are very close to those of tetramethoxy germane which has a melting point −18 °C and a boiling point +148 °C[95].

The chemical behaviour of bistrifluoromethylnitroxy silane and germane are very similar to that of tetralkoxy[95, 96] and tetraphenoxy[97] germane, which are very sensitive to moisture and hydrolyse rapidly. The occurrence of hydrolysis of the tetralkoxy germane is easily detected through observation of the 3400 cm^{-1} O—H band in their infrared spectra; in the same way the hydrolysis of bistrifluoromethylnitroxysilane and germane immediately produces absorption at 3600 cm^{-1} typical of the O—H group of the hydroxylamine. Reaction of bistrifluoromethylnitroxy silane and germane with excess of water liberates the hydroxylamine and silicon and germanium oxide, respectively. While the silicon $(CF_3)_2NO$-derivative does not react with hydrogen chloride, tetrakis(bistrifluoromethylnitroxy)germane reacts rapidly with it, liberating the hydroxylamine and germanium tetrachloride quantitatively. Bistrifluoromethylnitroxytrimethylsilane is also a colourless liquid which hydrolyses easily with water giving the hydroxylamine (90%) and hexamethyldisiloxane (85%).

The main bands of the infrared spectra are given in Table 5.7.

Table 5.7 Infrared spectra of Si- and Ge-$(CF_3)_2NO$ compounds

$[(CF_3)_2NO]_4Si$:	1311 vs, 1273 vs, 1275 sh, 1240 s, 1216 m and 1188 sh (C—F str), 1069 s (N—O str), 971 s, 926 m, (C—N str or Si—O str), 899 w and 712 ms (C—N—C bending), 617 w (CF_3 def), 480 w cm^{-1}.
$[(CF_3)_2NO]_4Ge$:	1309 vs, 1290 s, 1265 vs, 1215 s and 1185 s (C—F str), 1038 s (N—O or Ge—O str), 970 s (C—N str), 806 m and 763 m (C—N—C bending), 711 m (CF_3 def), 583 w, 473 w cm^{-1}.
$[(CF_3)_2NO]_3GeBr$:	1309 vs, 1293 s, 1231 vs and 1218 s (C—F str), 1040 s (N—O or Ge—O str), 971 (C—N str), 809 (C—N—C bending), 713 m cm^{-1} (CF_3 def).

In all the tetralkoxygermanes, such as $Ge(OC_nH_{2n+1})_4$ ($n = 1, 2, 3, 4, 5, 6$) or tetracyclohexoxy germane $Ge(OC_6H_{11})_4$, a strong absorption near 1040 cm^{-1} and a strong band near 680 cm^{-1} have been reported[95] to be characteristic of the central GeO_4 configuration. In the case of tetrakis(bistrifluoromethylnitroxy)germane no bands were observed near 680 cm^{-1} while the absorption at 1438 cm^{-1} is likely to be due to N—O stretching vibration since in all the derivatives of bistrifluoromethylnitroxide a strong band in this region appears and has been always attributed to N—O stretching vibration.

Lead and tin $(CF_3)_2NO$-compounds have been reported: direct interaction between radical and lead gives[49] the colourless crystalline compound $[(CF_3)_2NO]_2Pb$. In the same way the reaction of the radical with tin have been reported[49] to give a light grey coloured mixture of $[(CF_3)_2NO]_2Sn$ and stannous oxide in the ratio 1:3, in addition 25% of the radical introduced is converted to trifluoromethylaminoperoxide $(CF_3)_2NOON(CF_3)_2$. Reaction of $[(CF_3)_2NO]_2Hg$ with $SnCl_2$ gives a highly boiling oil as the only volatile product. This oil slowly seems to decompose at room temperature into $[(CF_3)_2NO]_2Sn$ and $(CF_3)_2NON(CF_3)_2$. The oil is probably $[(CF_3)_2NO]_4Sn$ which decomposes to a divalent state of tin, in a slow process similar to that of $SnCl_4$.

The peroxide is a colourless liquid (b.p. 48 °C) which acquires, during storage, a violet colour due to the appearance of the radical in solution. The dissociation of the peroxide into the radical is promoted by an increase in temperature.

5.9 $(CF_3)_3NO$-COMPOUNDS OF GROUP V ELEMENTS

Considerably more is known of the $(CF_3)_2NO$-compounds of Group V elements (N, P, As, Sb, Bi) than of any other groups. These compounds will be described in terms of derivatives of individual elements of this group.

5.9.1 Nitrogen$(CF_3)_2NO$-compounds

O-nitrosobistrifluoromethylhydroxylamine $(CF_3)_2NONO$ can be easily prepared by dimerisation of CF_3NO, or by reaction of the $(CF_3)_2NO\cdot$

radical with NO. The latter reaction proceeds very rapidly at room temperature being an addition of two radicals. In the same way, from $(CF_3)_2NO\cdot$ and NO_2 *O*-nitrobistrifluoromethylhydroxylamine is obtained[1]. $(CF_3)_2NONO$ is also a product of several reactions such as: reaction of $[(CF_3)_2NO]_4Ge$ with NOCl, photolysis of $(CF_3)_2NO\cdot$[49] which gives $(CF_3)_2NONO$ together with $(CF_3)_2NOON(CF_3)_2$ and $(CF_3)_2NOH$.

Di(bistrifluoromethylamino)oxide is often obtained as by-product in reactions in which reduction of $(CF_3)_2NO\cdot$ takes place, i.e. reaction of the radical with red phosphorus or with sulphur at 150 °C etc.[100]. It is also obtained when the radical reacts with trifluoronitrosomethane in the dark[43]. This reaction which gives in 3 days a 50% conversion can be explained by a free radical sequence

$$(CF_3)_2NO\cdot + CF_3NO \longrightarrow (CF_3)_2NO(O)CF_3$$

$$(CF_3)_2NO(O)CF_3 \longrightarrow CF_3NO_2 + (CF_3)_2N\cdot$$

$$(CF_3)_2N\cdot + (CF_3)_2NO\cdot \longrightarrow (CF_3)_2NON(CF_3)_2$$

The same reaction has also been reported[2] to give $(CF_3)_2NONO$, but most probably it was carried out in the presence of light which in the first place decomposes the CF_3NO. The nitroso compound was then formed according to the reactions:

$$CF_3NO \rightarrow CF_3\cdot + NO\cdot$$

$$CF_3\cdot + CF_3NO \rightarrow (CF_3)_2NO\cdot$$

$$NO + (CF_3)_2NO\cdot \rightarrow (CF_3)_2NONO$$

$(CF_3)_2NON(CF_3)_2$ is also obtained with 65% yield by reaction between the $(CF_3)_2NOH$-CsF adduct and SF_4.

The $(CF_3)_2NO\cdot$ radical does not react with ammonia even at 200 °C, but it reacts with dimethylamine[47] at room temperature giving dimethylaminobistrifluoromethylnitroxide $(CF_3)_2NONMe_2$ and $(CF_3)_2NOH$. $(CF_3)_2NONMe_2$ has been first synthesised by El-Nigumi[98] by radical exchange reaction between tri(dimethylamino)phosphine or arsine and radical:

$$(Me_2M)_3M + 2(CF_3)_2NO\cdot \rightarrow (Me_2N)_2MON(CF_3)_2 + Me_2NON(CF_3)_2$$

$$(M = P, As)$$

An interesting nitrogen $(CF_3)_2NO$-derivative is the bistrifluoromethylnitroxyfluoramide $(CF_3)_2NONF_2$. It can be prepared by reaction of the $(CF_3)_2NOH\cdot CsF$ adduct with an excess of NF_2Cl in acetonitrile at room temperature. The volatile products are: unreacted NF_2Cl, $(CF_3)_2NONF_2$, $(CF_3)_2NO\cdot$, SiF_4 and N_2F_4. $(CF_3)_2NONF$ cannot be separated from $(CF_3)_2NO\cdot$ by vacuum distillation, in addition, the $(CF_3)_2NO\cdot$ radical reacts slowly with it at room temperature giving SiF_4, $(CF_3)_2NONO$ and N_2F_4. Therefore the reported yield of 50–70% has been only estimated by indirect methods.

All the $(CF_3)_2NO$-nitrogen compounds are thermally and hydrolytically very stable liquids with the exception of $(CF_3)_2NONF_2$ which is the only unstable $(CF_3)_2NO$-derivative of nitrogen.

$(CF_3)_2NONO$ is a useful intermediate to obtain the radical, but it is not very useful to introduce the $(CF_3)_2NO$-group into molecules, this is due to the strong $(CF_3)_2NO$-N bond. This is confirmed by the high stability of the di(bistrifluoromethylamino)oxide $(CF_3)_2NON(CF_3)_2$ which does not react with water, 20% sodium hydroxide, concentrated nitric acid, concentrated hydrochloric acid at room temperature or at a temperature as high as 200 °C. $(CF_3)_2NON(CH_3)_2$ forms a I:I adduct with hydrogen chloride. At −50 °C this adduct forms needle shaped crystals which dissociate completely at 5 °C.

5.9.2 Phosphorus $(CF_3)_2NO$-compounds

A rather long series of $(CF_3)_2NO$-phosphorus compounds are known. Because of the strong oxidising properties of the $(CF_3)_2NO\cdot$ radical, the majority of the $(CF_3)_2NO$-phosphorus compounds are with P in its V oxidation state. Only by a few methods it is possible to obtain $(CF_3)_2NO$-compounds with P in the III oxidation state. These compounds have been prepared in several different ways.

The $(CF_3)_2NO\cdot$ radical reacts with elemental phosphorus in a much more complex way than the $CF_3\cdot$ radical[101–105].

The $CF_3\cdot$ radicals, produced by pyrolysis of CF_3I at 200–220 °C, react with white and red phosphorus giving trifluoromethyl derivatives of phosphorus with P in the +III oxidation state. The gaseous $(CF_3)_2NO\cdot$ radical reacts very quickly, even at −50 °C with white phosphorus, the product being tetrakistrifluoromethylhydrazine and a colourless heavy oil. The heavy oil is an intractable mixture of non-identifiable phosphorus compounds. Presumably addition of the radical to phosphorus takes place as well as oxidation of phosphorus with corresponding reduction of the radical to the tetrakistrifluoromethylhydrazine. This is proved by the presence, in the mass spectrum of the oil, of a peak at 703 m/e, which can be assigned to $[(CF_3)_2NO]_4P^+$ ion and by the presence, in the infrared spectrum of the oil, of a peak at 1382 cm^{-1} which is typical of a P=O stretch of a perfluorophosphoryl compound (cf. Table 5.3).

Furthermore, when an excess of white phosphorus is used, 40% of the radical introduced is converted to tetrakistrifluoromethylhydrazine, which represents the only component of the volatile liquid obtained from this reaction.

In contrast with the quick reaction with white phosphorus, the reaction of the radical with red phosphorus is very slow at room temperature. The main reaction which takes place is the oxidation of the phosphorus to form phosphoryl bistrifluoromethylnitroxide, with the reduction of the radical to the di(bistrifluoromethylamino)oxide.

$$P_{red} + 5(CF_3)_2NO\cdot \rightarrow [(CF_3)_2NO]_3PO + (CF_3)_2NON(CF_3)_2$$

The amount of radical which reacts with phosphorus in this way is 40% of the amount introduced, when excess of red phosphorus is used. The formation of $CF_3N{=}CF_2$, detected by its infrared spectrum, indicates that fluorination by bistrifluoromethylamino group, also takes place.

Phosphoryl, thiophosphoryl chloride and phosphorus pentachloride react readily at room temperature with mercury(II) bistrifluoromethylnitroxide[90] giving the corresponding phosphoryl, thiophosphorylbistrifluoromethylnitroxide and the pentakis(bistrifluoromethylnitroxy)phosphorane.

$$2OPCl_3 + 3[(CF_3)_2NO]_2Hg \rightarrow 2[(CF_3)_2NO]_3P{=}O + 3HgCl_2$$
$$2SPCl_3 + 3[(CF_3)_2NO]_2Hg \rightarrow 2[(CF_3)_2NO]_3P{=}S + 3HgCl_2$$
$$2PCl_5 + 5[(CF_3)_2NO]_2Hg \rightarrow 2[(CF_3)_2NO]_5P \quad + 5HgCl_2$$

These reactions are better carried out using the radical itself as solvent. Generally the phosphorus halide is added to a rather large excess of radical in which the required amount of mercury has been previously dissolved. A low yield of these compounds has been obtained when the reactants were left together at room temperature for more than 10 days; in these conditions a small amount of white volatile solid was formed. The presence in the white solid of mercury and chlorine may lead to the conclusion that a mercurial may be formed by the following reaction:

$$[(CF_3)_2NO]_2Hg + HgCl_2 \rightarrow 2(CF_3)_2NOHgCl$$

the rate of which is too small to produce an appreciable amount in a short period of time.

Abstraction of hydrogen by the $(CF_3)_2NO\cdot$ radical has been used to obtain derivatives of phosphorus. From bistrifluoromethylphosphine[47], $(CF_3)_2P[ON(CF_3)_2]_3$ is obtained:

$$(CF_3)_2PH + 3(CF_3)_2NO\cdot \rightarrow (CF_3)_2P[ON(CF_3)_2]_3 + (CF_3)_2NOH$$

In this reaction oxidation of phosphorus to P^V, by addition of the radical to the P^{III}, also occurs.

In the same way the reaction of the $(CF_3)_2NO\cdot$ radical with $(BuO)_2P(O)H$ gives $(CF_3)_2P(O)(OBu)_2$.

In contrast with germane, arsine and stibine, the reaction of mercury(II) bistrifluoromethylnitroxide with phosphine is much more complex and can be explosive. Bistrifluoromethylhydroxylamine has been found in the volatile products, but no tris(bistrifluoromethylnitroxy)phosphine has been isolated from what apparently was a mixture of volatile products. The infrared spectrum of this volatile material shows among several unidentified peaks, a band at 1395 cm^{-1} due to P=O bond, indicating that direct oxidation occurs as a side reaction.

Another example of addition of the radical to a phosphorus(III) compound is given by the reaction of $(CF_3)_2NO\cdot$ with a phosphite[98]. With trimethylphosphite, trimethoxy(bistrifluoromethylnitroxy)phosphorane $(Me_3O)_3P[ON(CF_3)_2]_2$ is obtained. An interesting analogy with this reaction is provided by the additional reaction of the peroxide MeOOMe with trimethylphosphite[106] in which the peroxide can be assumed to break down to furnish two $MeO\cdot$ radicals.

The only known example of a stable $(CF_3)_2NO$—P(III) compound is given by the reaction of $(CF_3)_2NO\cdot$ radical with tris(dimethylamino)phosphine or by the reaction of $[(CF_3)_2NO]_2Hg$ with dimethylaminochlorophosphine[98].

The former reaction[98] seems to occur through the formation of an unstable pentavalent compound $[(CF_3)_2NO]_2P(NMe_2)_3$ as an intermediate which then, for unknown reasons, splits off the $(CF_3)_2NONMe_2$ according to the reaction:

$$[(CF_3)_2NO]_2P(NMe_2)_3 \rightarrow (CF_3)_2NOP(NMe_2)_2 + (CF_3)_2NONMe_2$$

This reaction is similar to the exchange reaction of tris(trifluoromethyl) phosphine with *N*-chloro- or *N*-bromobis(trifluoromethyl)amine[107] which gives phosphine of the type $(CF_3)_{3-n}P[N(CF_3)_2]_n$ through a P^V intermediate:

$$(CF_3)_3P + n(CF_3)_2NX \rightarrow (CF_3)_{3-n}P[N(CF_3)_2]_n + nCF_3X$$

$$(X = Cl, Br; n = 1,2,3)$$

An alternative route for the preparation of a mixed dimethylamino(bistrifluoromethylnitroxy)phosphine is the reaction of mercury(II) bistrifluoromethylnitroxide with dimethylaminochlorophosphines[98]. The same reaction occurs also with arsine:

$$2(Me_2N)_{3-n}MCl_n + [(CF_3)_2NO]_2Hg \rightarrow 2(Me_2N)_{3-n}M[ON(CF_3)_2]_n + nHgCl_2 (M = P, As; n = 1,2)$$

The reaction of $(CF_3)_2NO\cdot$ radical with PCl_3 is vigorous[49] at room temperature and the main products are $(CF_3)_2NOPCl_2$ and $(CF_3)_2NOP(O)Cl_2$. Both have been converted to the corresponding esters $(CF_3)_2NOP(OC_2H_5)_4$ and $(CF_3)_2NOP(O)(OC_2H_5)_2$.

Properties—the $(CF_3)_2NO\cdot$ phosphoryl compound is the only $(CF_3)_2NO$—P compound which is solid at room temperature (m.p. 30.5 °C). Despite its high molecular weight (551), it is quite volatile and can be handled easily using standard vacuum technique.

Table 5.8 Infrared spectra of $(CF_3)_2NO$—P compounds

$[(CF_3)_2NO]_3P{=}O$:	1382 s (P=O str), 1319 vs, 1272 vs, 1235 vs, 1215 s and 1184 m (C—F str), 1029 s (N—O str), 971 (C—N str), 897 s and 800 m (C—N—C bending), 714 s (CF_3 def), 621 m, 498 m.
$[(CF_3)_2NO]_2P(OMe)_3$:	2946 m, 2906 w, 2860 w (OMe group), 1308 vs, 1267 vs and 1225 vs (C—F str), 1058 m, 1036 m (N—O str), 965 m (C—N str), 851 w (C—N—C bending), 730 m cm^{-1} (CF_3 def).

The thiophosphoryl and the phosphorane compounds do not show any melting points on cooling; at −78 °C they are still colourless viscous oils. They are thermally unstable, the phosphorane starts decomposing at 125 °C, the thiophosphoryl at lower temperature, so the boiling point of the phosphorane (188 °C) has also been calculated by extrapolation.

Very few organophosphorus compounds of the general formula $(RO)_5P$ are known. These substances are reported[108] to be rather unstable, suffering decomposition by the thermal or hydrolytic route in a manner very similar to that for penta(bistrifluoromethylnitroxy)phosphorane. A typical example of penta-aryloxyphosphorane is the pentaphenyloxyphosphorane[109] which is a very hydroscopic solid (m.p. 46 °C) which decomposes at 180 °C.

The main infrared bands of the $(CF_3)_2NO$—P compounds are reported in Table 5.8.

The strong band at 1382 cm^{-1} which appears in the infrared spectrum of $[(CF_3)_2NO]_3P{=}O$ has been attributed to the P=O stretch vibration by analogy with the spectra of other perfluorocompounds of phosphorus[110]. The assignment of the strong band at 853 cm^{-1} to the P=S stretch vibration in the infrared spectrum of $[(CF_3)_2NO]_3P{=}S$ is only tentative. This absorption may very well be due to $^{C}_{C}{>}N$ bending vibration and in this case the P=S absorption either is relatively weak or it is not detectable at all, as in several thiophosphoryl compounds reported[53].

Comparing the infrared spectra of the phosphoryl and the thiophosphoryl bistrifluoromethylnitroxide, the absence in the latter of the absorption due to P=O bond and the presence of a new band at 853 cm^{-1} may lead to

Table 5.9 N.M.R. spectra of $(CF_3)_2NO$—P compounds

	$^{\delta}(CF_3)_2NO$	$^{\delta}CF_3$	$^{\delta}CH_3$	^{J}P—H	$^{J}PONCF_3$	$^{J}PCF_3$
$[(CF_3)_2NO]_3PO$	68.7 s	—	—	—	—	—
$[(CF_3)_2NO]_5P$	67.4 s	—	—	—	—	—
$[(CF_3)_2NO]_3P(CF_3)_2$	85.5 s	61.2	—	—	—	119
$(CF_3)_2NOP(NMe_2)_2$	69.9	—	7.34 d	8.97	1	—
$(CF_3)_2P(OMe)_3$	69.9	—	6.7 d	12.1	—	—

the conclusion that this band is due to the P=S stretch: this is consistent with compounds of the type $(RO)_3PS$ in which the P=S absorption occurs between 845 cm^{-1} and 800 cm^{-1} [53].

The ^{19}F and ^{1}H n.m.r. of $(CF_3)_2NO$—P compounds shown in Table 5.9 are relative to internal CCl_3F and $SiMe_4$ respectively.
The ^{19}F n.m.r. of the $(CF_3)_2NO$ fluorine is generally a singlet although at high resolution a weak long range coupling between P and F can be seen. The structure of $[(CF_3)_2NO]_3P(CF_3)_2$, as a trigonal bipyramid in which the $(CF_3)_2NO$-groups occupies equatorial sites, is confirmed by the ^{19}F n.m.r. In the same way the ^{1}H and ^{19}F n.m.r. of $[(CF_3)_2NO]_2P(OMe)_3$ are consistent with a trigonal bipyramid structure but in which the metoxy-groups occupy equatorial sites.

All the $(CF_3)_2NO$–phosphorus derivatives are very sensitive to moisture, they hydrolyse rapidly giving bistrifluoromethylhydroxylamine quantitatively. No reaction takes place between $[(CF_3)_2NO]_3P = O$ and hydrogen chloride; this may indicate that the reverse reaction can occur.

5.9.3 Arsenic $(CF_3)_2NO$-compounds

Very few $(CF_3)_2NO$ arsenic derivatives have been obtained so far. The most important is the tri(bistrifluoromethylnitroxy)arsine which has been obtained in several different ways. The bistrifluoromethylnitroxide radical reacts slowly at room temperature with arsenic[47] giving only one volatile product, the $[(CF_3)_2NO]_3$ As in a quantitative yield. This reaction is similar to the reaction of arsenic with the $CF_3\cdot$ radicals obtained by decomposition of CF_3I [111–113] at 220 °C, in which formation of perfluoromethyliodoarsine occurs.

$[(CF_3)_2NO]_2Hg$ reacts readily at room temperature with AsX_3 (where X = Cl, Br, Ir) giving $[(CF_3)_2NO]_3As$ in a very high yield[90]. A similar reaction takes place between $[(CF_3)_2NO]_2Hg$ and AsH_3 at −78 °C[94]. The volatile products are $[(CF_3)_2NO]_3As$ (78% yield) and $(CF_3)_2NOH$. A mixture of mercury with excess of mercurial used remains in the reaction vessel.

In contrast with the reaction of mercury(II)bistrifluoromethylnitroxide and arsine, the reaction of the $(CF_3)_2NO\cdot$ radical with arsine is very quick at −78 °C and leads to 92.5% of the hydroxylamine, mixed with decomposition products (such as SiF_4, CF_4 and NO, identified by their infrared spectra) and an unidentified white solid.

The $(CF_3)_2NO\cdot$ radical reacts at room temperature with arsenic tribromide and tri-iodide liberating the free halogen and giving $[(CF_3)_2NO]_3As$ in very high yield[91].

Arsenic trichloride does not react with the radical even at temperatures up to 100 °C, this should be expected since the reverse reaction:

$$2[(CF_3)_2NO]_3As + 3Cl_2 \rightarrow 2AsCl_3 + 3(CF_3)_2NO\cdot$$

takes place.

However, in the presence of a stoichiometric amount of iodine the reaction between $(CF_3)_2NO\cdot$ and $AsCl_3$ occurs at room temperature and the main product is bis(trifluoromethylnitroxy)arsenic chloride, though a little tris (bistrifluoromethylnitroxy)arsine is also formed. Prolonged treatment replaces all the three chlorines. The role of iodine in this case is clear since chlorine is found to convert the arsenical into arsenic trichloride and its removal by iodine is essential if the reaction has to proceed. $[(CF_3)_2NO]_3As$ is also obtained by the reaction of the $(CF_3)_2NO\cdot$ radical with As_2S_3 or As_2O_3. The reaction is very slow at room temperature and gives others non-identified compounds.

Another $(CF_3)_2NO$–As derivative is bistrifluoromethylnitroxy(bistrifluoromethyl)arsine which has been obtained[47] by replacement of hydrogen from bistrifluoromethylarsine by $(CF_3)_2NO\cdot$ radical.

All the $(CF_3)_2NO$–As compounds react with hydrogen chloride giving at room temperature the hydroxylamine and arsenic trichloride quantitatively.

Tris(bistrifluoromethylnitroxy)arsine is a colourless volatile solid, which hydrolyses easily in the presence of moisture, reacts readily with water giving the hydroxylamine and arsenic oxide quantitatively.

The main bands of the infrared spectrum of $[(CF_3)_2NO]_3As$ are:
1322 str., 1304 vs, 1290 str., 1268 vs, 1259 str., 1232 vs, 1228 str. and 1210 s (C—F str.), 1034 s (N—O str.), 970 s (C—N str.), 799 m and 744 m (C—N—C bending), 720 (CF_3 def), 700 m.

The ^{19}F n.m.r. spectrum relative to CCl_3F shows two signals at $\delta = 63.0$ p.p.m. (CF_3As fluorines) and $\delta = 76.4$ p.p.m. ($(CF_3)_2NO$ fluorines). Each of these peaks are split into a: 1:6:15:20:15:6:1 septet due to a long range F–C–N–O–As–C–F (coupling constant 1.4 c.p.s.).

5.9.4 Antimony $(CF_3)_2NO$-compounds

The reaction between mercury(II) bistrifluoromethylnitroxide and antimony trichloride, carried out by reacting a solution of antimony trichloride in trichlorofluoromethane with the mercurial, gives tris(bistrifluoromethylnitroxy)stibine in about 95% yield[90].

This antimonium compound is also obtained by reaction of $[(CF_3)_2NO]_2Hg$ with stibine[94]. This reaction proceeds at low temperature according to the equation:

$$[(CF_3)_2NO]_2Hg+2SbH_3 \rightarrow 2[(CF_3)_2NO]_3Sb+6(CF_3)_2NOH+6Hg$$

In contrast with P, As and Bi, the Sb does not react with $(CF_3)_2NO\cdot$ radical at room temperature or at high temperature.

$[(CF_3)_2NO]_3Sb$ is a white crystalline solid, volatile enough to be distilled at room temperature under a good vacuum. No infrared spectrum has been recorded since it is too involatile for an infrared spectrum of the vapour and it is too hygroscopic for satisfactory mulls to be made; even in a dry box, in presence of phosphorus pentoxide, the spectrum of the hydroxylamine produced by hydrolysis, masks other features. The ^{19}F n.m.r. shows a sharp singlet at 68.9 p.p.m. relative to CCl_3F.

In contrast with the arsenic compound, which is hydrolysed completely by water to arsenic oxide and the hydroxylamine, the tris(bistrifluoromethylnitroxy)stibine is partially hydrolysable by an excess of water at room temperature. In the volatile products of the hydrolysis, two moles of the hydroxylamine per mole of compound are found, mixed with the excess of water; the weight of the involatile residue corresponds to the formation of Sb(O) $ON(CF_3)$ according to the following reaction:

$$Sb[(CF_3)_2NO]_3+H_2O \rightarrow Sb(O)ON(CF_3)_2+2(CF_3)_2NOH$$

5.9.5 Bismuth $(CF_3)_2NO$-compounds

The reaction of $(CF_3)_2NO\cdot$ radical with bismuth is rather quick at room temperature and yields tris(bistrifluoromethylnitroxy)bismuthine, a white crystalline, involatile solid which is completely soluble in the excess of the radical.

The fact that the radical reacts very quickly with bismuth, while with red phosphorus and arsenic the reaction is very slow, may be due to the metallic structure of bismuth in contrast to the tetra-atomic molecule of white phosphorus and arsenic. The same compound has also been obtained by reaction of bismuth tri-iodide with an excess of the radical; the reaction occurs at room temperature and free iodine is liberated.

The reaction of $[(CF_3)_2NO]_2Hg$ with BiX_3 (X = Cl, Br, Ir) in CCl_3F gives the same bismuth-$(CF_3)_2NO$ derivative which is very soluble in Freons and can be easily separated from the mercuric halide.

Hydrogen chloride reacts vigorously with $[(CF_3)_2NO]_3Bi$ at room temperature, liberating bismuth trichloride and the hydroxylamine. This behaviour is similar to that of tris(bistrifluoromethylnitroxy)arsine, which reacts with hydrogen chloride in the same way, but differs from that of phosphoryl bistrifluoromethylnitroxide which is stable in the presence of hydrogen chloride. The ^{19}F n.m.r. of tris(bistrifluoromethylnitroxy)bismuthine shows a singlet at 68.6 p.p.m. relative to CCl_3F.

5.10 $(CF_3)_2NO$-COMPOUNDS OF GROUP VI ELEMENTS

5.10.1 Sulphur $(CF_3)_2NO\cdot$ compounds

Several $(CF_3)_2NO$-sulphur compounds are known with sulphur in different oxidation state.

Sulphur itself does not react with the $(CF_3)_2NO\cdot$ radical at temperatures below 140 °C[100]. At 155 °C the reaction is similar to that between white phosphorus and the radical at −50 °C as the main products are $(CF_3)_2NN(CF_3)_2$ and $[(CF_3)_2NO]_2SO_2$. At 155 °C the liquid sulphur contains chains of sulphur atoms as well as rings, which are the only species present at temperatures below 155 °C. This suggests the possible explanation for the absence of reaction at room temperature. This is also proved by the fact that the radical reacts slowly at room temperature with sulphur in carbon disulphide solution, though the solvent in inert. The products of this reaction have never been identified through features in the infrared spectrum of the vapour of the volatile product, which was admixed with carbon disulphide, proved the presence of a component in which the $(CF_3)_2NO$-group was present. In this reaction it seems possible that sulphur is first oxidised directly to sulphur dioxide, which then combines with the radical. The reaction between radical and sulphur dioxide has been shown to occur at 180 °C giving $[(CF_3)_2NO]_2SO_2$, though the conversion is not quantitative under these conditions. It is also possible that the compound $[(CF_3)_2NO]_2SO_2$ may dissociate back to $(CF_3)_2NO\cdot$ and SO_2 at high temperature.

Sulphuryl bistrifluoromethylnitroxide has been prepared for the first time by the reaction of bistrifluoromethylhydroxylamine with sulphuryl fluoride in the presence of CsF[99]. This reaction takes place at 25 °C and gives bistrifluoromethylnitroxosulphonyl fluoride $(CF_3)_2NOSO_2F$ and sulphonylbistrifluoromethylnitroxide. The ratio of these two products depends on the ratio of the reagents. Using a 1:1 molar ratio of reagents, 35 % of the nitroxosulphonylfluoride and 75 % of the sulphonylbistrifluoromethylnitroxide is obtained. In the same way the reaction of the adduct $(CF_3)_2NOH{\cdot}CsF$ with SOF_2 gives bistrifluoromethylnitroxosulphinyl fluoride $(CF_3)_2NOS(O)F$ together with sulphinylbistrifluoromethylnitroxide $[(CF_3)_2NO]_2SO$. Using a molar ratio hydroxylamine/sulphinylfluoride 7:2, sulphinylbistrifluoromethylnitroxide is obtained with a yield of 95 %.

Bistrifluoromethylnitroxodisulphide $[(CF_3)_2NO]_2S_2$ can be prepared by reaction of the adduct $(CF_3)_2NOH{\cdot}CsF$ with disulphur dichloride at −20 °C[99]. The yield is not very high (25 %) the other products being $[(CF_3)_2N]_2S$ and $[(CF_3)_2NO]_2S$. Bistrifluoromethylnitroxosulphide can be also prepared by the reaction of the adduct $(CF_3)_2NOH{\cdot}CsF$ with SCl_2 at −20 °C, the yield is about 40 %. A better yield can be obtained by the reaction of the $(CF_3)_2NO\cdot$ radical with hydrogen sulphide[47]. The reaction is very quick even at very low temperature and at room temperature may lead to explosion.

The reaction of the adduct $(CF_3)_2NOH{\cdot}CsF$ with SF_4 in 2:1 ratio at −100 °C gives $(CF_3)_2NON(CF_3)$ (65 % yield), $(CF_3)_2NOS(O)F$, SOF_4 and $(CF_3)_2NO\cdot$; the biggest yield of amine oxide is obtained for a 2:1 ratio while the biggest yield of the sulphinyl compound is obtained for a 4:1 ratio. To explain the formation of $(CF_3)_2NON(CF_3)_2$ an initial formation of an unstable intermediate $[(CF_3)_2NO]_2SF_2$, which subsequently decomposes to SO_2F_2 and $(CF_3)_2NON(CF_3)_2$ has been postulated[99]. The sulphonyl fluoride should then react with any remaining hydroxylamine to give the $(CF_3)_2NOS(O)F$ as product.

Although the $=SF_2$ compounds (such as $(iC_3F_7)_2SF_2$[114], etc.) are very stable, the instability of $[(CF_3)_2NO]_2SF_2$ could come from the fact that the $(CF_3)_2NO$-group would not shield the sulphur atom from attack as in the case with the perfluoroisopropyl derivative. That the SF_4 had been hydrolysed to SOF_2 prior to reaction is precluded by the presence of the amino oxide since none is observed when the initial reactant is SOF_2.

The reaction of trifluoromethylfluorosulphate CF_3OSO_2F with the adduct $(CF_3)_2NOH{\cdot}CsF$ gives no $(CF_3)_2NOOSO_2CF_3$, but a mixture of $[(CF_3)_2NO]_2CO$, $(CF_3)_2NOSO_2F$, $(CF_3)_2NO{\cdot}SO_2$, SO_2F_2 and COF_2. This has been explained by the fact that $CF_3OSO_2F_2$ in the presence of CsF undergoes the fluorosulphurylation to give COF_2 and SO_2F_2 [115]. All the mono and bis-substituted sulphinyl and sulphonyl products have hydrolytic stability comparable to that of R_fSO_2F and $(R_f)_2SO_2$[116, 117]. In particular, $[(CF_3)_2NO]_2SO_2$ is stable to hydrolyses by water or by 0.2 M NaOH at temperature as high as 80 °C for periods up to 12 h. All the $(CF_3)_2NO$-sulphur compounds are liquid at room temperature. Their boiling points are reported in Table 5.10 together with the coefficients of the vapour pressure equation $\log_{10} P(\text{mm}) = A - B/T$.

Table 5.10

Compound	b.p. °C	A	B
$[(CF_3)_2NO]_2SO_2$	95.7	8.88	2172
$(CF_3)_2NOSO_2F$	57.0±0.2	12.51	2981
$[(CF_3)_2NO]_2SO$	79±2 (692 mm)	—	—
$(CF_3)_2NOS(O)F$	48.2	7.48	1479
$[(CF_3)_2NO]_2S_2$	102.0±0.2	9.88	2633
$[(CF_3)_2NO]_2S$	72.0±0.3	8.96	2099

The boiling points of $(CF_3)_2NO$-sulphur compounds are rather close to each other and separation can be achieved only by gas chromatography. Their infrared spectra[99] shows bands between 1190 and 1280 cm^{-1} for C—F stretch vibration, at 1023–1045 for N—O stretch frequency, at 972–980 cm^{-1} for C—N stretch vibration and at 712–729 cm^{-1} for CF_3 deformation.

5.11 $(CF_3)_2NO$-NITROGEN-SULPHUR DERIVATIVES

The reaction of the liquid $(CF_3)_2NO{\cdot}$ radical with N_4S_4 at room temperature converts it quantitatively to tetrathiazyltetra(bistrifluoromethylnitroxide) $N_4S_4[(CF_3)_2NO]_4$ [118] (the same product is also obtained when the reaction is carried out in CCl_4 at room temperature). This reaction resembles the fluorination of N_4S_4 more than the chlorination, since fluorination by a suspension of silver(II) fluoride in carbon tetrachloride gives white needles of $S_4N_4F_4$ while chlorination by gasous chlorine in CCl_4 solution, or by liquid chlorine at −78 °C, gives $N_3S_3Cl_3$ [119]. Tetrathiazyltetra(bistrifluoromethylnitroxide) can also be obtained by the reaction at room temperature between tetrasulphurtetraimide $S_4N_4H_4$ and liquid $(CF_3)_2NO\cdot$ radical. The other product is the hydroxylamine contaminated with a very small amount of water sensitive yellow solid, probably the monomer $NSON(CF_3)_2$ or the trimer $(NSONCF_3)_3$. When the same reaction is carried out with half the molecular ratio of the $(CF_3)_2NO\cdot$ radical required for the reaction to be completed from the products obtained, it appears that abstraction and addition reactions occur simultaneously

$$S_4N_4H_4 + 4(CF_3)_2NO\cdot \rightarrow S_4N_4H_2[(CF_3)_2NO]_2 + 2(CF_3)_2NOH$$

$S_4N_4[ON(CF_3)_2]_4$ is also the major product of the reaction between the $(CF_3)_2NO\cdot$ radical and trithiazyltrichloride. The white crystals obtained are also contaminated with hydrolysable yellow solid which can be washed out by water.

Reaction between $[(CF_3)_2NO]_2Hg$ and $N_3S_3Cl_3$ in CCl_3F as inert solvent, surprisingly gives again the tetramer, though fluorination of $N_3S_3Cl_3$ is known to give the trimer $N_3S_3F_3$ [120].

The trimer, trithiazyl tri(bistrifluoromethylnitroxide) can be obtained by trimerisation of the monomer $NSON(CF_3)_2$. This behaviour of the monomer to trimerise is similar to that of both thiazylchloride and fluoride which also trimerise on standing.

While the reaction of thiazylfluoride with chlorine[120] gives thiazylchloride, no reaction occurs between NSF and $(CF_3)_2NO\cdot$ radical. The only known way to obtain the monomer in a reasonable yield (over 80 %), is the reaction of the mercurial $[(CF_3)_2NO]_2Hg$ with NSF. This is an exothermic reaction which gives $NSON(CF_3)_2$ in 85 % yield in 30 min. A second product of this reaction has been tentatively identified as $NSF[ON(CF_3)_2]_2$ by its mass spectrum.

The monomer thiazyl bistrifluoromethylnitroxide polymerises to a trimeric form on standing in glass; this reaction does not appear to be quantitative as, even if the monomer is left for a month, there is always a small amount of the monomer which can be distilled off, leaving yellow, needle-shaped crystals of $N_3S_3[ON(CF_3)_2]_3$. These crystals are not very stable under vacuum and tend to decompose to the monomer and a residual green solid, but they can be stored in a desiccator over phosphorus pentoxide for a considerably long period of time.

The tetrathiazyl tetra(bistrifluoromethylnitroxide) is a white crystalline solid which is not wetted by water and is insoluble, or only slightly soluble, in acetone, ether, carbon tetrachloride, alcohols, benzene, acetonitrile and dimethylsulphoxide.

In contrast with $N_4S_4F_4$ which slowly hydrolyses in water, $N_4S_4[(CF_3)_2NO]_4$ is recovered unchanged after prolonged contact with water or concentrated hydrochloric or nitric acid at room temperature, though decomposition occurs with 10 % sodium hydroxide at 60 °C. It does not react with gaseous hydrogen chloride, but when heated in vacuum at 70 °C, 20 % of the $(CF_3)_2NO\cdot$ radical is recovered together with a mixture of non-identified compounds.

The presence of an eight rather than a six-membered ring is proved by its mass spectrum: the highest peak (at m/e 188) which corresponds to $N_4S_4((CF_3)_2NO)_3^+$ ion, is followed by the main features of the S_4N_4 mass spectrum such as $N_4S_4^+, S_3N_3^+, S_2N_2^+, SN^+$ together with peaks corresponding to $(CF_3)_2NO^+$ and fragments of it. The x-ray structural analysis[121] also shows its configuration to be tetragonal and very similar to that of $N_4S_4F_4$ with alternating N—S bonds and with $(CF_3)_2NO\cdot$ groups bonded to sulphur.·

The infrared spectrum of $S_4N_4(ON(CF_3)_2)_4$ in Nujol mull shows the following main bands: 1355 str., 1335 vs, 1267 vs, 1215 vs, 1200 vs, 1100 s, 1026 vs, 979 vs, 955 w, 810 m, 770 ms, 755 ms, 732 str., 720 s, 560 w, 535 w, 520 w, 470 s, 400 w, 350 w.

The monomer thiazyl bistrifluoromethylnitroxide is a volatile yellow liquid stable at $-78\,°C$. In contrast with the tetramer it reacts readily at room temperature with hydrogen chloride giving $S_3N_3Cl_3$ and $(CF_3)_2NOH$ and it hydrolyses readily with water giving a polymeric imide and the hydroxylamine.

$$NSON(CF_3)_2 + H_2O \rightarrow -N(H)S(O)- + (CF_3)_2NOH$$

The course of this reaction is probably through intermediates such HOSN which then undergoes a proton transfer to HNSO and polymerises. The monomer reacts with both BF_3 and BCl_3 to give unstable adducts which decompose to the boron halide and trimer $N_3S_3(ON(CF_3)_2)_3$ at or below room temperature. With BBr_3 a violent exothermic reaction occurs at $-78\,°C$ and bromine is separated. The behaviour of the trimer $S_3N_3(ON(CF_3)_2)_2$ towards water and hydrogen chloride is similar to that of the monomer as it is hydrolysed by water and by gaseous hydrogen chloride.

The chief infrared bands of the monomer are at 1400 w (N = S str), 1350 m, 1312 vs, 1269 vs and 1227 vs (C—F str), 1035 m, (N—O str), 978 s, 802 w, 712 w, 661 w, 583 m, 536 m.

Triazylchloride N_3S_4Cl does not react at room temperature with the $(CF_3)_2NO\cdot$ radical.

A very interesting reaction is that of trisulphurdinitrogendioxide $S_3N_2O_2$ with the $(CF_3)_2NO\cdot$ radical. This reaction takes place at room temperature with either the gaseous or the liquid radical. When the reaction is carried out with gaseous radical, two moles of the radical are consumed per mole of $S_3N_2O_2$ and a yellow involatile oil which crystallises to a yellow solid is formed. This is sulphanuric bistrifluoromethylnitroxide $S_3N_3O_3[ON(CF_3)_2]_3$. The following mechanism has been suggested for this reaction:

$$O=S=N-S-N=S=O \ + \ (CF_3)_2NO\cdot \ \longrightarrow \ 2(CF_3)_2NO\overset{\overset{\displaystyle O}{\|}}{S}\equiv N \ + \ S$$

$$3(CF_3)_2NO\overset{\overset{\displaystyle O}{\|}}{S}\equiv N \ \longrightarrow$$

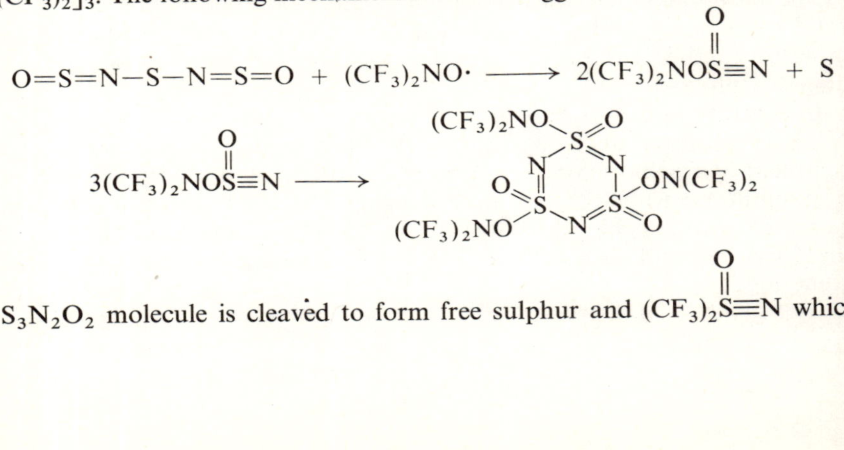

$S_3N_2O_2$ molecule is cleaved to form free sulphur and $(CF_3)_2\overset{\overset{\displaystyle O}{\|}}{S}\equiv N$ which

trimerises to a sulphanuric compound. When an excess of the liquid radical is used, a mixture of volatile products is formed. The infrared spectra of these products show that S—ON(CF$_3$)$_2$-group is present. These compounds result from further slow reaction between the radical and the free sulphur formed.

An analogous reaction occurs between S$_3$N$_2$O$_2$ and chlorine[122] where sulphanuric chloride is formed:

$$3S_3N_2O_2 + 6Cl_2 \longrightarrow 2 \quad \underset{Cl_a}{\overset{Cl_b}{}}\quad + SO_2Cl_2 + 2SOCl_2$$

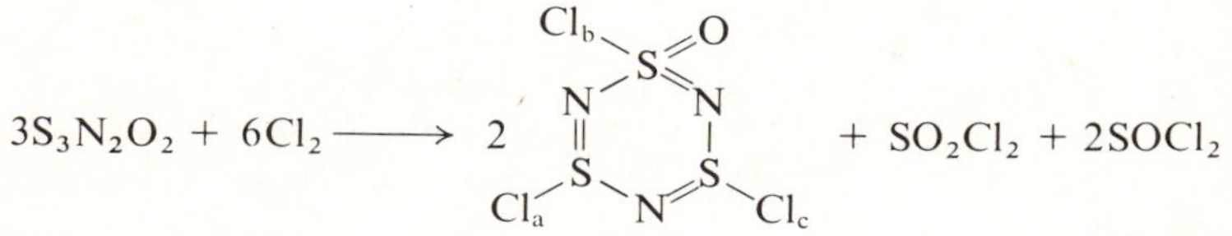

When sulphanuric chloride is treated with KF in CCl$_4$ at 145 °C, replacement of chlorine by fluorine occurs and a mixture of two stereoisometric forms of the fluoride are formed: one *cis*, which gives a singlet in the ^{19}F n.m.r. spectrum, and the other *trans*, which gives a typical AB$_2$ spectrum.

The sulphanuric bistrifluoromethylnitroxide obtained by the reaction of S$_3$N$_2$O$_3$ with the (CF$_3$)$_2$NO· radical shows a singlet at 69.3 p.p.m. relative to CCl$_3$F. After a period of 9 days at room temperature, this peak is replaced by two peaks at 69.5 and 68.1 p.p.m., the areas of which are in the ratio 2:1. This reproducible observation suggests that the sulphanuric bistrifluoromethylnitroxide exists in two forms, *cis* and *trans*, which are similar to those reported for sulphanuric fluoride. The initial reaction product which gives a singlet in the ^{19}F n.m.r. spectrum should be the *cis* form of N$_3$S$_3$O$_3$[ON

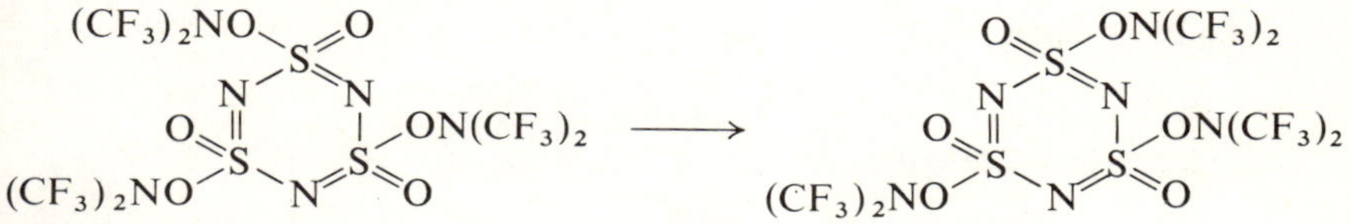

(CF$_3$)$_2$]$_3$ which slowly isomerises over a period of 9 days to the *trans* form with one of the (CF$_3$)$_2$NO-group differently orientated.

Although sulphanuric bistrifluoromethylnitroxide has a structure very similar to the sulphanuric halides, it does not possess their stability to heat and moisture.

While the sulphanuric chloride is not wetted by water, it is only very slowly hydrolysed when in contact with it, the fluoride is only effected by boiling water and both can be heated without decomposition, the fluoride up to 350 °C, the chloride up to 145 °C. The (CF$_3$)$_2$NO-derivative decomposes slowly at room temperature and is rapidly hydrolysed by moisture.

All the (CF$_3$)$_2$NO-sulphur-nitrogen compounds discussed so far can be considered as covalent compounds. Attempts to obtain a ionic compound like S$_4$N$_3$ON(CF$_3$)$_2$ by reaction of S$_4$N$_3$Cl and the (CF$_3$)$_2$NO· radical at room temperature, failed. This confirms that the (CF$_3$)$_2$NO· radical, with its odd electron in an antibonding orbital, does not have ionic properties comparable to those of the halogens.

5.12 $(CF_3)_2NO$-DERIVATIVES OF SELENIUM AND TELLURIUM

The bistrifluoromethylnitroxide radical $(CF_3)_2NO\cdot$ reacts very slowly at room temperature with Se and Te giving the tetrakis(bistrifluoromethoxy) selenide and telluride[123]. This reaction resembles more the reaction of Se and Te with chlorine than that with the $CF_3\cdot$ radical. Chlorine reacts with Se and Te giving $SeCl_4$ and $TeCl_4$ respectively while the $CF\cdot$ radicals, generated by the pyrolysis of CF_3I are known to react with selenium[124] giving two main products bistrifluoromethylselenide and bistrifluoromethyldiselenide and traces of bistrifluoromethyltriselenide. The same reaction does not occur with tellurium, only the $CF_3\cdot$ radicals obtained by pyrolysis of hexafluoroacetone, attack the tellurium minor giving bistrifluoromethylditelluride[125].

$SeCl_4$ and $TeCl_4$ do not react with the $(CF_3)_2NO\cdot$ radical, but $SeBr_4$ reacts at room temperature liberating free bromine giving $[(CF_3)_2NO]_2SeO$ and $(CF_3)_2NON(CF_3)_2$. The last two compounds are also obtained by the reaction of the $(CF_3)_2NO\cdot$ radical with Se_2Cl_2.

Selenyl chloride does not react with the $(CF_3)_2NO\cdot$ radical, but when an excess of iodine is present, reaction occurs and the products are $[(CF_3)_2NO]_2SeO$ and ICl.

Selenium and tellurium $(CF_3)_2NO$-derivatives so far prepared are colourless crystalline volatile solids, sensitive to moisture; they hydrolyse with water eliminating a quantitative amount of the hydroxylamine and the corresponding oxide. They react at room temperature with hydrogen halides liberating the hydroxylamine, with Cl_2 liberating the radical and the corresponding chloride. They react with NOCl giving $(CF_3)_2NONO$ and the chloride in very high yield. $[(CF_3)_2NO]_2SO$ reacts with BBr_3 giving free bromine and tribistrifluoromethylnitroxy borane.

5.13 $(CF_3)_2NO$-COMPOUNDS OF GROUP VII ELEMENTS

The radical failed to react with fluorine, chlorine, bromine and iodine, at temperature as high as 200 °C, under pressure, and with irradiation.

The lack of success of conventional synthetic methods in the preparation of halogen-$(CF_3)_2NO$-compounds may suggest that such compounds are inherently unstable. The only example of direct combination between the radical and one of the halogen is the addition compound $CsI[ON(CF_3)_2]_2$ which slowly decompose in the presence of light probably into $(CF_3)_2NOCs$, $(CF_3)_2NO\cdot$ and I_2.

5.14 $(CF_3)_2NO$-DERIVATIVES OF TRANSITION METALS

Only a very few $(CF_3)_2NO$-derivatives of transition metals have been prepared and described so far.

An iron salt of the radical has been reported[49] as formed when the radical was stored under pressure in a stainless steel tube at room temperature for a month, but it has not been characterised.

On the other hand, when the radical is heated at 225 °C with freshly reduced iron powder, de-oxyfluorination reaction takes place and perfluoro(methylenemethylamine) is formed in 88% yield[43] together with trifluoromethylisocyanate and silicon tetrafluoride.

The only manganese derivative $(CF_3)_2NOMn(CO)_5$ has been obtained by reaction of the $(CF_3)_2NO\cdot$ radical with $Mn(CO)_5$–H[47]. The details of this preparation and the properties of the manganese derivatives have not been published.

The cobalt derivatives: cobalt(II) bistrifluoromethylnitroxide can be obtained by reaction between anhydrous cobalt iodide and the radical in large excess. The reaction is quick at room temperature and free iodine is evolved. The cobalt salt is an involatile pink powder, rather sensitive to light, and when it is exposed to moist air it becomes immediately pale blue. It is readily decomposed by hydrogen chloride with formation of $(CF_3)_2NOH$ and $CoCl_2$ quantitatively and by chlorine with formation of the $(CF_3)_2NO\cdot$ radical.

$[(CF_3)_2NO]_2Co$ undergoes decomposition under vacuum at 80 °C. The gaseous decomposition products are $(CF_3)_2NO$, $(CF_3)_2NON(CF_3)_2$, $CF_3N = CF_2$ and small trace of $(CF_3)_2NOH$.

The low thermal stability of the $(CF_3)_2NO$-cobalt compound limits its preparation to reactions at low temperature. No reaction occurs between the radical and $Co(SCN)_2$ or TiI_4 at room temperature.

5.15 ORGANIC $(CF_3)_2$ NO-DERIVATIVES

5.15.1 Carbonyl $(CF_3)_2$NO-compounds

Several compounds having the $(CF_3)_2$NO-group attached to the carbonyl function are known. These compounds can be prepared by reaction of acetyl, perfluoroacetyl, oxalyl, carbonyl halides with either the adduct $(CF_3)_2NOH\cdot CsF$ [58], or the mercurial $[(CF_3)_2NO]_2Hg$ [90] or the sodium salt $(CF_3)_2NONa$ [70]. All these methods give the carbonyl compounds in a rather high yield.

Carbonyl(bistrifluoromethylnitroxide)[50] has been obtained either from COF_2 or from $COCl_2$. Using COF_2 as starting material, its reaction with $(CF_3)_2NOH\cdot CsF$ or with $(CF_3)_2NONa$ [70] gives a 40% yield, the other main product being $(CF_3)_2NOC(O)F$; with $COCl_2$ and the mercurial a very high yield of $[(CF_3)_2NO]_2CO$ is obtained, the only impurity being traces of $(CF_3)_2NOC(O)Cl$. $[(CF_3)_2NO]_2CO$ is a colourless liquid, very stable towards water.

Bistrifluoromethylnitroxocarbonylfluoride is the co-product in the preparation of $[(CF_3)_2NO]_2CO$ when COF_2 is used as starting material. It is a colourless liquid sensitive to moisture which hydrolyses immediately in presence of water vapour at room temperature giving $(CF_3)_2NOH$ and CO_2 as the only volatile products.

Bistrifluoromethylnitroxocarbonylchloride can be prepared by two ways:

(a) reaction of phosgene with the adduct $(CF_3)_2NOH\cdot CsF$, prepared with a large excess of the hydroxylamine to preclude the presence of CsF. This reaction proceeds only at 200 °C and a low yield of $(CF_3)_2NOC(O)Cl$ is obtained, the other product being COClF, CF_3NCO, $(CF_3)_2NOCOF$;

(b) chlorination of the fluoro-carbonyl derivative with anhydrous $AlCl_3$ and HCl. This reaction proceeds at 200 °C giving a yield of about 20%. A 10% excess of HCl improves this yield. This compound is similar to the fluoride in its behaviour towards moisture, the hydrolysis gives $(CF_3)_2NO{\cdot}$ and CO_2 as volatile products.

Trifluoromethylcarbonylbistrifluoromethylnitroxide[58] has been prepared from CF_3COCl and either the adduct $(CF_3)_2NOH{\cdot}CsF$ or the sodium salt $(CF_3)_2NONa$ in tetrahydrofurane. In both cases the yield is almost 90%. It is slowly hydrolysed by water to $(CF_3)_2NOH$ and CF_3COOH as the only volatile products. It has been shown to be useful as intermediate in the synthesis of other fluorocarbons. Pyrolysis in a platinum-lined autoclave at 220 °C gives $(CF_3)_2NOCF_3$ 50%, $(CF_3)_3N$ 33%, CO_2, CO.

Heptafluoropropylcarbonyl(bistrifluoromethylnitroxide) has been prepared by reaction of C_3F_7COCl with the adduct $(CF_3)_2NOH{\cdot}CsF$[58]. It is a very stable compound and it is not hydrolysed by water at room temperature.

Acetyl, propionyl, benzonyl and oxalyl-bistrifluoromethylnitroxide are easily obtained by reaction of CH_3COCl, C_2H_5COCl, C_6H_5COCl, $(COCl)_2$ respectively with the mercurial $[(CF_3)_2NO]_2Hg$ at room temperature[90].

The acetyl and propionyl compounds are also obtained using the corresponding chlorides and the adduct $(CF_3)_2NOH{\cdot}CsF$ [126]. All these reactions occur at room temperature with a very high yield of products.

With the only exception of the oxalyl derivative all the other carbonyl $(CF_3)_2NO{\cdot}$ compounds are colourless liquids stable at room temperature in glass. Furthermore, $[(CF_3)_2NO]_2CO$, $(CF_3)_2NOC(O)C_3F_7$, $(CF_3)_2NOC(O)C_2H_5$, $(CF_3)_2NOC(O)C_6H_5$ are stable towards water; this indicates that steric hindrance has been made for perfluoroketons[127] and for a series of compounds of the type $R_fC(NH_2) = NOC(O)R_f$ [128]; these compounds are very stable towards water when $R_f = C_2F_5$, C_3F_7 and C_7F_{15} while when $R_f = CF_3$ they are susceptible to hydrolysis.

5.15.2 $(CF_3)_2NO$-derivatives of methane

Monobistrifluoromethylnitroxymethane has been prepared for the first time by the reaction of $(CF_3)_2NONa$ with CH_3I, this reaction occurs at room temperature with 88% yield. A simple method which also gives a better yield of $(CF_3)_2NOCH_3$ is the reaction of $[(CF_3)_2NO]_2Hg$ with CH_3I.

Monobistrifluoromethylnitroxymethane is a volatile colourless liquid, stable at room temperature.

Structure I and II are possible for this compound:

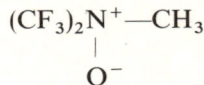

I	II
$(CF_3)_2N{\cdot}O{\cdot}CH_3$	$(CF_3)_2N^+{-}CH_3$ with O^- bonded to N (drawn below)

Evidence in favour of I is obtained from infrared spectroscopy. The infrared spectrum of the vapour of the sample are: 3020 vs, 2998 w, (C—H asym. str.), 2958 w, 2916 vw, 2815 vw (C—H sym. str.), 1304 vs, 1269 vs, 1232 vs, 1191 s, (C—F str.), 1070 s (N—O str.), 978 s, 971 vs, (C—N str.) 800 w (C—N—C bending), 701 m (CF_3 dep.) 552 w.

In the first place it is possible to observe that the stretching vibration of the methyl group, which generally occurs at 2872 cm^{-1}, is shifted to high frequency (3020–2998 cm^{-1}) as expected when the methyl group is attached to oxygen[53], furthermore, the presence of another band between 2832 and 2815 cm^{-1} is highly characteristic of a methoxy group[89]. Secondly, the spectrum shows no bands in the 1400–1700 cm^{-1} region, where a N—O bond, linked to a perfluoroamine *N*-oxide may reasonably be expected to absorb[43] (cf. $CF_3\overset{+}{N}(O^-)$—NCF_3 shows a band at 1572 cm^{-1}).

In this compound the N—O stretch at 1070 cm^{-1} is in the expected region for a structure II.

The band at 552 cm^{-1} can be interpreted as similar to the C—X vibration (ν_3) of methyl halide CH_3X

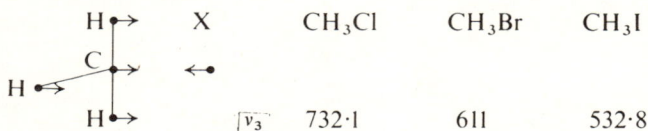

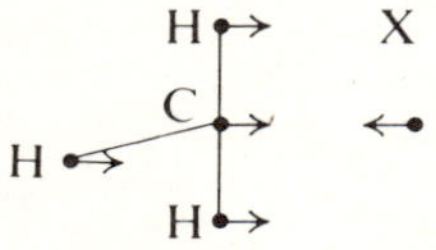

	CH_3Cl	CH_3Br	CH_3I
ν_3	732·1	611	532·8

The relative simplicity of C—F stretching vibration pattern, with only three strong bands, indicates a high degree of symmetry in the molecule[129].

The ^{19}F n.m.r. spectrum shows a single resonance at 70.5 p.p.m. to high field of CCl_3F. This is evidence for structure I, where the bistrifluoromethylamino group is bonded only to oxygen rather than to carbon, since the chemical shift of $(CF_3)_2N$—C or $(CF_3)_2\overset{+}{N}(O^-)$—C are much smaller than this[130, 131].

Other evidence in favour of I, is obtained from the mass spectrum. Parent peaks were observed at *m/e* 184 and 183 in the relative intensity 3:100. The former can be assigned to the molecule containing a ^{13}C isotope. Several other ions observed can arise from I or II; however, the complete absence of ions such as $(CF_3)_2NCH_3^+$ and its derivatives, which would only have came from a molecule of structure II, is evidence in favour of structure I.

Reaction occurs between CH_2I_2 and $[(CF_3)_2NO]_2Hg$ but the volatile liquid obtained is a mixture of compounds which are not separable by vacuum distillation, among these products, the presence of $(CF_3)_2NOH$ shows that abstraction of hydrogen takes place. A similar reaction occurs at room temperature between the $(CF_3)_2NO\cdot$ radical and di-iodomethane or iodoform. Free iodine is liberated and the presence of $(CF_3)_2NOH$ among the volatile products shows that abstraction of hydrogen takes place. Furthermore, the infrared, n.m.r. and mass spectrum of the product show the presence of $(CF_3)_2NON(CF_3)_2$ and $[(CF_3)_2NO]_2C = O$. These two compounds are also found in the products of the reaction between CI_4 and $[(CF_3)_2NO]_2Hg$ when an attempt to make the four-substituted methane-$(CF_3)_2NO$-derivative was carried out.

Most probably a compound like $[(CF_3)_2NO]_4C$ is not stable and decomposes to $(CF_3)_2NON(CF_3)$ and $[(CF_3)_2NO]_2C = O$.

Other attempts to synthesise $[(CF_3)_2NO]_4C$ have been unsuccessful since CI_4, CBr_4, CCl_4 and CH_4 failed to react with the $(CF_3)_2NO\cdot$ radical. Tristrifluoromethylhydroxylamine has been obtained by several reactions[132].

Pyrolysis of $(CF_3)_2NOCOCF_3$ also gives $(CF_3)_2NOCF_3$ with 50% yield. No reaction takes place between the $(CF_3)_2NO\cdot$ radical and CF_3I at room temperature, but an equimolar mixture of them exposed to u.v. light gives $(CF_3)_2NOCF_3$ with a 84% yield[43]. No reaction has been observed between mercury(II) bistrifluoromethylnitroxide and CF_3I, either at room temperature or at higher temperatures at which decomposition of mercurial to mercury and the radical occurs. The inability of CF_3I to react with mercurial can be attributed to the increased stability of the carbon-iodine bond in CF_3I and its failure to undergo a heterolytic fission similar to that of CH_3I (i.e. $CH_3^+I^-$). The polarisation of the carbon–iodine bond in CF_3I is in the opposite sense $CF_3^+I^-$.

The mechanism of the reaction between mercurial and CH_3I may involve an initial coordination of the more electronegative atom onto mercury, followed by a cyclic transfer of electrons.

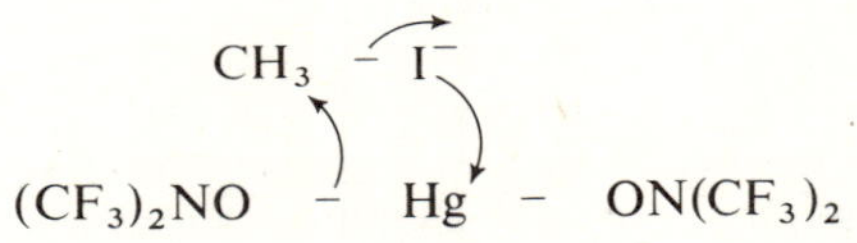

The lack of reaction between mercurial and chlorobenzene or iodobenzene may be explained in the same way. The polarisation of the carbon–halogen bond in these compounds is opposite to that of CH_3I; also the diminished chemical reactivity of the halogen atom can be attributed to a resonance hybrid of the following structures:

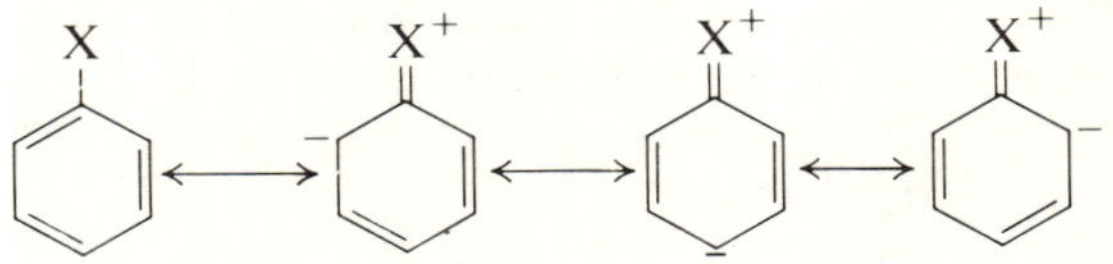

No reaction occurs between the radical and fluoroform at room temperature over a period of 3 months. Only a very small amount of $(CF_3)_2NOCF_3$ with traces of NO, LiF_4, CF_4 were collected when the mixture was left for 5 months in daylight.

On the contrary, the reaction between the $(CF_3)_2NO\cdot$ radical and chloroform or bromoform occurs at room temperature. C—Cl and C—Br bonds are not effected by the radical and only abstraction of hydrogen is observed. The products are $(CF_3)_2NOCCl_3$ and $(CF_3)_2NOCBr_3$ together with the corresponding amount of the hydroxylamine. $(CF_3)_2NOCCl_3$ is colourless stable liquid while $(CF_3)_2NOCBr_3$ is light yellow one.

Although the $[(CF_3)_2NO]_2Hg$ does not react with CCl_4, reaction occurs with CBr_4 at room temperature when excess of the radical is present. The product is probably a mixture of compounds such as $[(CF_3)_2NO]_3CBr$, $[(CF_3)_2NO]_2CBr_2$ and $(CF_3)_2NOCBr_3$.

The reactions of the $(CF_3)_2NO\cdot$ radical with olefines such as ethylene, perfluoroethylene, perfluoropropene, perfluorocyclobutene and perfluorobenzene result only in the addition of the radical across double bonds[43, 49] When the $(CF_3)_2NO\cdot$ radical and perfluoroethylene are mixed in the molar ratio 1:10 only the addition compound $(CF_3)_2NOCF_2CF_2ON(CF_3)_2$ is obtained; this fact shows that the bistrifluoromethylnitroxide is an excellent free-radical scavenger. When the $(CF_3)_2NO\cdot$ radical is mixed with an excess of an equimolar mixture of ethylene and tetrafluoroethylene, it combines almost exclusively with the fluoro-olefin. This indicates that it is a strongly nucleophilic radical.

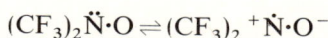

$$(CF_3)_2\ddot{N}\cdot O \rightleftharpoons (CF_3)_2\,{}^{+}\dot{N}\cdot O^{-}$$

The reaction between $(CF_3)_2NO\cdot$ and exafluorobenzene occurs only at 120 °C with formation of exafluoro(exabistrifluoromethylnitroxy)exane.

Hydrogen replacement, with formation of the hydroxylamine has been observed in several reactions of the radical with hydrogen containing molecules. Reaction with benzene[49, 51] at room temperature was completed after 3 days and yielded quantitatively the hydroxylamine and 1, 2, 4-tri bistrifluoromethylnitroxy)benzene. The reaction of the $(CF_3)_2NO\cdot$ radical with toluene is quicker, in this case only attack at the side chain occurs and the products are the hydroxylamine and *O*-benzyl-*N*,*N*-bistrifluoromethylhydroxylamine.

In the reaction of $(CF_3)_2NO\cdot$ with $C_6H_5CH_2CH_3$ only one hydrogen, bonded to the secondary carbon of the side chain, is replaced. This shows that hydrogen bonded to a carbon of an aromatic ring is more difficult to replace.

The reaction of the $(CF_3)_2NO\cdot$ radical with methanol is vigorous[51] at room temperature and results in the formation of an equivalent of the hydroxylamine and bistrifluoromethylnitroxymethanol. Derivatives of this compound have been obtained by the reactions with benzoylchloride-phenylisocyanate which gives C_6H_5—$COOCH_2ON(CF_3)_2$ and C_6H_5—$NHCOOCH_2ON(CF_3)_2$ respectively. $(CF_3)_2NOCH_2OH$ reacts violently with Na giving the sodium metoxide derivative.

In the reaction of $(CF_3)_2NO\cdot$ with ethanol a mixture of isomers are formed.

Another aromatic compound containing the $(CF_3)_2NO$ group is obtained by reaction of the salt $(CF_3)_2NONa$ with perfluoropyridine

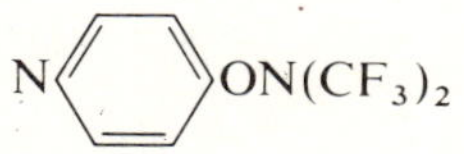

is obtained with a 28 % yield.

An unsaturated compound containing the $(CF_3)_2NO$-group $CH_2 = CH$ —$CH_2ON(CF_3)_2$ has been obtained by reaction of allylchloride with the $(CF_3)_2NOH\cdot CsF$ adduct in the presence of solvent. Allylbistrifluoromethylnitroxide is a colourless liquid at room temperature and does not hydrolyse in water even at 70 °C.

5.16 CONCLUSION

The study of $(CF_3)_2NO$-derivatives is very interesting and the results are often unexpected. This fact makes the work exciting and therefore encourages further investigation of new synthetic routes.

Much work remains to be done in the preparation and study of $(CF_3)_2NO$-compounds. Several new $(CF_3)_2NO$-derivatives may be obtained using known synthetic routes especially for the transition metal compounds which is a virtually unexplored field.

References

1. Blackley, W. D. and Reinhard, R. R. (1965). *J. Amer. Chem. Soc.*, **87,** 802
2. Makarov, S. P., Yakubovich, A. Ya., Dubov, S. S. and Medvedev, A. N. (1965). *Doklady Akad. Nauk. SSR,* **160,** 1319
3. Fremy, M. E. (1845). *Ann. Chim. Phys.*, **15,** 408
4. Rashig, F. (1887). *Ann. Chem.*, 241, 161, 223
5. Claus, A. (1871). *Ann. Chem.*, 158, 205
6. Hantzschand, A. and Semple, W. (1895). *Chem. Ber.*, 28, 2744
7. Wagner, M. (1896). *Z. Physik. Chem.*, **19,** 680
8. Haga, T. (1904). *J. Chem. Soc.*, **85,** 78
9. Murib, J. H. and Ritter, D. M. (1952). *J. Ann. Chem. Soc.*, **74,** 3394
10. Wieland, H. and Offenbacher, M. (1914). *Chem. Ber.*, **47,** 2111
11. Thomas, J. R. and Tolmann, C. A. (1962). *J. Amer. Chem. Soc.*, **84,** 2930
12. Wieland, H. and Kögl, F. (1922). *Chem. Ber.*, **55,** 1798
13. Wieland, H. and Roth, K. (1920). *Chem. Ber.*, **53,** 210
14. Banfield, F. H. and Kenyon, J. (1926). *J. Chem. Soc.*, 1612
15. Baird, J. C. and Thomas, J. R. (1961). *J. Chem. Phys.*, **35,** 1507
16. Coppinger, G. M. and Swalen, J. D. (1961). *J. Amer. Chem. Soc.*, **83,** 4900
17. Hoffman, A. K. and Henderson, A. T. (1961). *J. Amer. Chem. Soc.*, **83,** 4671
18. Linnett, J. W. (1961). *J. Amer. Chem. Soc.*, **83,** 2643
19. Lemaire, H., Rassat, A., Servoz-Gavin, P. and Bertier, G. (1962). *J. Chem. Phys.*, **59,** 1247
20. Neiman, M. B., Rozanstev, E. G. and Mamedova, Yu G. (1962). *Nature,* **196,** 472
21. (1963). *Chem. Abstr.*, **58,** 5625
22. Lemaire, H., Marechal, Y., Ramasseul, R. and Rassat, A. (1965). *Bull. Chim. France,* 372
23. Bethoux, M., Lemaire, H. and Rassat, A. (1964). *Bull. Chim. France,* 1985
24. Chapelet-Letourneux, J., Lemaire, H. and Rassat, A. (1965). *Bull. Chim. France,* 3283
25. Lemaire, H., Rassat, A. and Ravet, A. (1963). *Bull. Chim. France,* 1980
26. Brunel, Y., Lemaire, H. and Rassat, A. (1964). *Bull. Chim. France,* 1895
27. Lemaire, H., Rassat, A. (1964). *J. Chim. Phys.*, **61,** 1580
28. Briere, R., Lemaire, H. and Rassat, A. (1964). *Tetrahedron, Letters,* 1781
29. Rozantzev, E. G. and Neiman, M. B. (1964). *Tetrahedron,* **20,** 131
30. Lemaire, H., Rassat, A. and Ravet, J. P. (1964). *Tetrahedron Letters,* **47,** 3507
31. Briere, R. and Rassat, A. (1965). *Bull. Chim. France,* 378
32. Chapelet-Letourneux, G., Lemaire, H. and Rassat, A. (1965). *Bull. Chim. France,* 444
33. Chapelet-Letourneux, G., Lemaire, H., Rassat, A. and Ravet, J. P. (1965). *Bull. Chim. France,* 1975
34. Briere, R., Lemaire, H. and Rassat, A. (1965). *Bull. Chim. France,* 3273
35. Dupeyre, R. M., Rassat, A. and Rey, P. (1965). *Bull. Chim. France,* 3643
36. Chapelet-Letourneux, G., Lemaire, H. and Rassat, A. (1965). *Bull. Chim. France,* 3283
37. Lemaire, H. (1966). *J. Chim. Phys.*, **64,** 559
38. (1953). *Chem. Abstr.*, **47,** 4652 c
39. McQueen, D. M. U.S. Patent 2 619 479, Nov. 25 (1952)
40. Maschke, A., Shapiro, B. and Lampe, F. W. (1964). *J. Amer. Chem. Soc.*, **86,** 1929

41. Blackley, W. D. (1966). *J. Amer. Chem. Soc.*, 480
42. U.S. Patent 336 389, Aug. 15 (1967)
43. Banks, R. E., Haszeldine, R. N. and Stevenson, M. J. (1966). *J. Chem. Soc., (C)*, 901
44. Glidewell, C., Rankin, D. W. H., Robiette, A. G., Sheldrick, G. M. and Williamson, S. M. *J. Chem. Soc.*, (A) (in press).
45. Underwood, G. R. and Vogel, V. L. (1970). *Mol. Phys.* 19, **5,** 621
46. Walden, P. and Audrieth, L. F. (1928). *Chem. Rev.*, **5,** 339
47. Ang, H. G. (1968). *Chem. Comm.*, 1320
48. Tomilov, A. P., Smirnov Yu. D. and Videiko, A. F. (1966). *Electrokhimiya*, **5,** 603
49. Makarov, S. P., Englin, M. A., Videiko, A. F., Tobolin, V. A. and Dubov, S. S. (1966). *Doklady Akad. Nauk.*, **168,** 344
50. Jander, J. and Haszeldine, R. N. (1954). *J. Chem. Soc.*, 919
51. Makarov, S. P., Videiko, A. F., Nicolaeva, T. V. and Englin, M. A. (1967). *Zhur. Obshch. Khim.*, **37,** 9, 1975
52. Haszeldine, R. N. (1952). *J. Chem. Soc.*, 4263
53. Bellamy, L. J. (1964). The infra-red Spectra of Complex Molecules, Methuen & Co. Ltd., London
54. Daasch, L. W. and Smith, D. C. (1951). *Ind. Eng. Chem. Anal.* **23,** 853
55. Gutowsky, H. S. and Liehr, A. D. (1952). *J. Chem. Phys.*, **20,** 1652
56. Paul, R. C. (1955). *J. Chem. Soc.*
57. Soloway, A. H. and Friess, S. L. (1951). *J. Amer. Chem. Soc.*, **73,** 5000
58. Babb, D. P. and Shreeve, Jean'ne M. S. (1967). *Inorg. Chem.*, **6,** 351
59. Nielson, A. H., Burke, T. G., Woltz, P. J. and Yones, E. A. (1952). *J. Chem. Phys.*, 20, 4, 596
60. Haszeldine, R. N., (1951). *J. Chem. Soc.*, 586
61. Haszeldine, R. N. (1953). *J. Chem. Soc.*, 2075
62. Banks, R. E., Haszeldine, R. N. and McCreath, M. D. (1961). *Proc. Chem. Soc.*, 64
63. Taylor, C. W., Brice, T. T. and Wear, R. L. (1962). *J. Org. Chem.*, **27,** 1064
64. Banks, R. E., Barlow, M. G., Haszeldine, R. N. and McCreath, M. D. (1966). *J. Chem. Soc. (C)*, 1350
65. Park, J. D., Rosser, R. W. and Lacher, J. Q. (1962). *J. Org. Chem.* **27,** 1462
66. Dinwoodie, A. H. and Haszeldine, R. N. (1965). *J. Chem. Soc.*, 1675
67. Haszeldine, R. N. and Mattinson, B. J. H. (1957). *J. Chem. Soc.*, 1741
68. Flaskerud, G. G. and Shreeve, J. M. (1969). *Inorg. Chem.*, **10,** 2065
69. Gutowski, H. S. and Hoffman, C. J. (1951). *J. Chem. Phys.*, **19,** 1259
70. Banks, R. E., Haszeldine, R. N. and Hyde, D. L. (1967). *Chem. Comm.*, 413
71. Emeléus, H. J. and Haszeldine, R. N. (1949). *J. Chem. Soc.*, 2948
72. Emeléus, H. J. and Haszeldine, R. N. (1949). *J. Chem. Soc.*, 2953
73. Steacie, E. W. R., Atomic and free radical Reactions, Reinhold publ. Corp., NY., 2nd ed. 1954, vol. 1, p. 47
74. Paneth, F. A. and Lautsch, W. (1931). *Chem. Ber.*, **64,** 2702
75. Emeléus, H. J. and Spaziante, P. M. (1968). *Chem. Comm.* 770
76. Yung, J. A., Tsoukalas, S. N. and Dresdner, R. D. (1958). *J. Amer. Chem. Soc.*, **80,** 3604
77. Taylor, H. J. and Jones, W. H. (1930). *J. Amer. Chem. Soc.*, **52,** 1111n
78. Cunningham, J. P. and Taylor, H. J. (1938). *J. Chem. Phys.*, **6,** 359
79. Reid, E. E. (1958). *Organic chemistry of Bivalent Sulphur*, vol. I, Chem. Public. Co. Inc., NY
80. Haszeldine, R. N. and Kidd, J. M. (1955). *J. Chem. Soc.*, 2901
81. Emeléus, H. J. and Hurst, G. L. (1964). *J. Chem. Soc.*, 396
82. Jumbert, J. (1874). *Compt. Rend.*, **78,** 1853
83. Heslop, R. B. and Robinson, P. L. Inorganic Chemistry, Elsevier, Second edition 1963, p. 529
84. Mellor, J. W. (1935). A comprehensive treatise on Inorganic and theoretical Chemistry, Longmans, Green and Co., vol. IV
85. Aurivillius, K. (1956). *Acta Cryst.*, 9, 685
86. Man, E. H., Coffman, D. D. and Muetterties, E. L. (1959). *J. Amer. Chem. Soc.*, **81,** 3575
87. Makarov, S. P., Shpanskii, V. A., Ginsburg, V. A., Shchekotikhin, A. I., Filatov, A. S., Martynova, L. L., Pavlovskaya, I. V., Golovaneva, A. F. and Yakubovich, A. Ya. (1961). *Proc. Acad. Sci., USSR*, **62,** 142

88. Emeléus, H. J. and Pugh, H. (1960). *J. Chem. Soc.*, 1108
89. Gamboni, von G., Theus, V. and Shinz, H. (1955). *Helv. Chim. Acta*, **38,** 255
90. Emeléus, H. J., Shreeve, J. M. and Spaziante, P. M. (1969). *J. Chem. Soc. (A)*, **3,** 431
91. Emeléus, H. J., Spaziante, P. M. and Williamson, S. M. (in press). *J. Inorg. Nucl. Chem.*
92. Lehmann, W. J. Onak, T. P. and Shapiro, I. (1959). *J. Chem. Phys.*, **30,** 1219, 1222, 1226
93. Delany, A. C., Haszeldine, R. N. and Tipping, A. E. (1968). *J. Chem. Soc. (C)*, 2537
94. Emeléus, H. J., Shreeve, Jean'ne M. and Spaziante, P. M. (1969). *J. Inorg. Nucl. Chem.*, **31,** 3417
95. Johnson, O. H. and Fritz, H. E. (1953). *J. Amer. Chem. Soc.*, **75,** 718
96. Tabern, D. L., Orndorff, W. R. and Dennis, L. M. (1925). *J. Amer. Chem. Soc.*, **47,** 2039
97. Schwartz, R. and Reinhardt, W. (1932). *Chem. Ber.*, **65,** 1743
98. El Nigumi, Y. O. and Emeléus, H. J. (in press). *J. Inorg. Nucl. Chem.*
99. Lott, J. A., Babb, D. P., Pullen, K. E. and Shreeve, Jean'ne M. (1968). *J. Inorg. Chem.*, **7,** 2593
100. Coombes, J. S. and Spaziante, P. M. (1969). *J. Inorg. Nucl. Chem.* **31,** 2636
101. Bennet, F. W., Emeléus, H. J. and Haszeldine, R. N. (1953). *J. Chem. Soc.*, 1565
102. Bennet, F. W., Emeléus, H. J. and Haszeldine, R. N. (1954). *J. Chem. Soc.*, 3598
103. Bennet, F. W., Emeléus, H. J. and Haszeldine, R. N. (1954). *J. Chem. Soc.*, 3896
104. Emeléus, H. J., Haszeldine, R. N. and Paul, R. C. (1955). *J. Chem. Soc.*, 563
105. Emeléus, H. J. and Smith, J. D. (1959). *J. Chem. Soc.*, 375
106. Reid, W. and Appel, H. (1964). *Ann. Chem.*, **51,** 679
107. Ang, H. G. (1968). *Chem. Comm.*, **21,** 1320
108. Kosolapoff, G. M. (1950). Organophosphorus Compounds, J. Wiley & Sons, Inc., NY., p. 325
109. Anschütz, L. (1927). *Ann. Chem.* **454,** 71,
110. Bennett, F. W., Emeléus, H. J. and Haszeldine, R. N. (1954). *J. Chem. Soc.*, 3598
111. Emeléus, H. J., Haszeldine, R. N. and Walaschewski, E. G. (1953). *J. Chem. Soc.*, 1552
112. Emeléus, H. J., Haszeldine, R. N. and Paul, R. C. (1954). *J. Chem. Soc.*, 881
113. Brandt, G. R. A., Emeléus, H. J. and Haszeldine, R. N. (1952). *J. Chem. Soc.*, 2552
114. Muetterties, E. L. and Resemberg, R. M. (1965). *Inorg. Chem.*, **4,** 1515
115. Lustig, M. and Ruff, J. K. (1964). *Inorg. Chem.* **3,** 287
116. Banks, R. F., Haslam, G. M., Haszeldine, R. N. and Peppin, A. (1966). *J. Chem. Soc., C,* 1171
117. Temple, S. (1968). *J. Org. Chem.*, **33,** 344
118. Emeléus, H. J. Poulet, R. J. (in press). *J. Chem. Soc.*
119. Meuwsen, A. (1931). *Chem. Ber.*, **64,** 2311
120. Glemser, O. and Prichert, H. (1961). *Z. Anorg. Allgem. Chem.*, **307,** 174
121. Forder, R. A. and Sheldric, G. M. (in press). *J. Fluorine Chem.*
122. Becke-Goehring, M. and Schlater, D. (1968). *Z. Anorg. Allgem. Chem.*, **362,** 1
123. Ang, H. G., Coombes, J. S. and Sukhoverkov, V. (1969). *J. Inorg. Nucl. Chem.*, **31,** 877
124. Dale, J. W., Emeléus, H. J. and Haszeldine, R. N. (1958). *J. Chem. Soc.*, 2939
125. Bele, T. N., Pulman, B. J. and West, B. O. (1963). *Aust. J. Chem.*, **87,** 722
126. Nash, L. L., Conville, J. J. and Shreeve, J. M. (1968). *J. Inorg. Nucl. Chem.*, **30,** 3373
127. Smith, R. D., Fawcett, F. S. and Coffman, D. D. (1966). *J. Amer. Chem. Soc.*, **88,** 2459
128. Brown, H. C. and Wetzel, L. R. (1965). *J. Org. Chem.*, **30,** 3735
129. Mason, J. (1963). *J. Chem. Soc.*, 4531
130. Haszeldine, R. N. and Tipping, A. (1966). *J. Chem. Soc. (C)*, 1236
131. Haszeldine, R. N. and Tipping, A. (1968). *J. Chem. Soc. (C)*, 396
132. Dinwoodie, A. H. and Haszeldine, R. N. (1965). *J. Chem. Soc.*, 1681

6
Astatine*

E. H. APPELMAN
Argonne National Laboratory, U.S.A.

6.1 INTRODUCTION

Astatine, element 85, is the heaviest member of the halogen family of elements. Unlike other members of the family, however, it has no stable or long-lived isotope. The presently-known astatine isotopes are described in Table 6.1.

Although minute amounts of the very short-lived ^{215}At, ^{218}At and ^{219}At

*Written under the auspices of the U.S. Atomic Energy Commission.

are present in nature in equilibrium with long-lived radio elements[1–4], the only practical way of obtaining astatine is by synthesising it through nuclear reactions. The first synthesis of astatine was effected by Corson *et al.*[5], who bombarded bismuth with α-particles to obtain the isotope ^{211}At. Most subsequent chemical studies have made use of this isotope. The slightly longer-lived ^{210}At cannot be obtained isotopically pure, emits a penetrating

Table 6.1 Astatine isotopes*

Isotope	*Half-life*	*Mode of decay*	*Method of production*
^{200}At (?)	0.8 min	α	Au + ^{12}C
^{201}At	1.5 min	α	Au + ^{12}C
^{202}At	3 min	α (12%) EC (88%)	Au + ^{12}C
^{203}At	7.4 min	α (14%) EC (86%)	Au + ^{12}C
^{204}At	9.3 min	α (4.5%) EC (95.5%)	Au + ^{12}C
^{205}At	26 min	α (18%) EC (82%)	Au + ^{12}C Bi (α, 8*n*)
^{206}At	29 min	α (0.9%) EC (99.1%)	Au + ^{12}C Bi (α, 7*n*)
^{207}At	1.8 h	α (10%) EC (90%)	Au + ^{14}N Bi (α, 6*n*)
^{208}At	1.7 h	α (0.5%) EC (99.5%)	Bi (α, 5*n*)
^{209}At	5.5 h	α (5%) EC(95%)	Bi (α, 4*n*)
^{210}At	8.3 h	α (0.17% EC (99.8%)	Bi (α, 3*n*)
^{211}At	7.2 h	α (41%) EC (59%)	Bi (α, 2*n*)
^{212}At	0.3 s	α	Bi (α, *n*)
^{212m}At	0.12 s	α	Bi (α, *n*)
^{213}At	<1 s	α	descendant ^{225}Pa [^{232}Th(*p*, 8*n*)^{225}Pa]
^{214}At	2 μs	α	descendant ^{226}Pa [^{232}Th(*p*, 7*n*)^{226}Pa]
^{215}At	10^{-4} s	α	descendant ^{235}U
^{216}At	3×10^{-4} s	α	descendant ^{228}Pa [^{232}Th(*p*, 5*n*)^{228}Pa]
^{217}At	0.018 s	α	descendant ^{237}Np
^{218}At	2 s	α	descendant ^{238}U
^{219}At	0.9 min	α(97%) β^-(3%)	descendant ^{235}U

*From Hyde *et al*[56]

γ-ray, and produces the 138 day α-emitter ^{210}Po. which is often a health hazard.

The short half-lives of even the longest-lived astatine isotopes make it impractical to obtain weighable amounts of the element. With the exception of a few spectrographic and mass spectrometric studies, all investigations of astatine chemistry have utilised tracer techniques, in which the astatine has been detected only by its radioactivity. The actual concentration of astatine in these experiments has almost always been less than 10^{-10}M. At such low concentrations the effects of impurities can be very serious, especially for a halogen like astatine, which exists in several oxidation states and can form numerous organic compounds. These effects may be reduced somewhat by introducing another halogen, usually iodine, as a carrier for the astatine. However, inasmuch as the carrier is a different element, it may be fractionated from the astatine, which will then be left unprotected. Because of such effects, tracer experiments involving astatine must be interpreted very cautiously, especially when other halogens are absent, or when the astatine behaves quite differently from another halogen that is present.

Inasmuch as astatine compounds can neither be weighed nor titrated, nor, in general, can their physical properties be measured, these compounds must almost always be characterised by methods that do not take advantage of the total mass of material present. Techniques that have proved useful include co-precipitation, solvent extraction, ion exchange and other forms of chromatography, electrodeposition, electromigration and diffusion. A direct identification of some astatine compounds has been made by mass spectrometry.[6]

6.2 PREPARATION OF ASTATINE

Astatine for chemical and medical studies and for tracer use is usually prepared by bombardment of metallic bismuth or bismuth oxide with α particles of energy exceeding 20 MeV, which results in the nuclear reactions ^{209}Bi (α,xn) $^{213-x}$At. The reactions with $x = 2$, 3 and 4 have threshold energies of 20, 28 and 34 MeV, respectively[7,8]. Hence irradiation of a 0.1 mm thick bismuth target with 28 MeV α-particles will give the maximum yield of ^{211}At free of ^{210}At.

Metallic bismuth, the more common target material, is customarily fused or vaporised onto aluminium or gold backing plates. It is necessary to cool the target carefully inasmuch as astatine may be volatilised from molten bismuth. Bismuth is a poor thermal conductor, and the cooling problem increases with the thickness of the bismuth layer. The back of the target is generally water-cooled. The face is most effectively cooled by a flow of helium, although a static helium atmosphere is often used. An 0.5–1 mil stainless steel or copper cover foil pressed tightly to the surface of the bismuth helps to dissipate the heat evolved and also prevents astatine from escaping from the target.

When bismuth oxide is used, it is generally pressed into small holes drilled in the face of an aluminium plate[9]. Melting of the target material is much less likely in this case.

In all cases the beam of α-particles should be defocused as much as possible to avoid local hot spots on the target.

Various astatine isotopes are also formed by spallation reactions brought about by high energy bombardment of a variety of heavy elements.

6.3 ISOLATION AND PURIFICATION OF ASTATINE

Astatine may readily be removed from a metallic bismuth target by distillation in air from a stainless steel tube[10]. The condenser is a platinum disc clamped to a water-cooled cold finger and fitted over the mouth of the tube. Before the condenser is put in place, volatile impurities may be removed by heating to 260 °C. Astatine begins to distil at the melting point of bismuth (271 °C), and the astatine distilled at this temperature is relatively pure. However, to obtain a quantitative yield, it is necessary to heat the target to 800 °C, and at this temperature polonium and bismuth also distil. Therefore the astatine must be purified by redistillation[11]. The platinum disc is transferred to a glass vacuum system which is evacuated to $<10^{-3}$ Torr. The astatine is distilled from the disc by gradual heating to *c.* 500 °C, and the distillate is collected in a U-tube cooled with liquid nitrogen. The presence of at least 5 cm of unheated tubing between the heated zone and the U-tube prevents contamination of the distillate with polonium. Hydrocarbon greases tend to absorb astatine and should not be present between the platinum disc and the U-tube. Silicone or fluorocarbon grease is satisfactory. If an aqueous solution of astatine is desired, the element may be washed out of the U-tube with an appropriate aqueous solution. Yields around 50 % are obtained in the redistillation.

This method has been modified to obtain astatine free of organic impurities[6]. The astatine-bearing platinum disc is sealed into an oxygen-filled quartz tube. One end of the tube is connected through a breakseal to a glass vacuum line. The tube is heated to 800 °C to destroy organic matter. Then it is cooled in a liquid-solid nitrogen slush, the breakseal is broken, and the oxygen is pumped away. The astatine is then distilled by heating as before.

Distillation may similarly be used to remove astatine from gold targets.

Astatine may also be removed from targets by wet chemical methods. Typically, the bismuth target is dissolved in nitric acid, most of which is removed by boiling. The solution is then made 8 M in HCl, and the astatine is extracted into isopropyl ether. The ether is washed with an aqueous solution 8 M in HCl and 1 M in HNO_3, after which the astatine is back-extracted into aqueous 0.5 M NaOH[12]. Astatine formed in bismuth oxide targets may be isolated by dissolving the Bi_2O_3 in perchloric acid containing I_2. The bismuth may be precipitated as the phosphate, or alternatively, the astatine and iodine may be extracted into CCl_4 and back-extracted with alkali or reducing solutions[9, 13].

The nuclear chemist investigating the formation of astatine isotopes in spallation reactions is usually concerned with radiochemical rather than chemical purity. He must be able to decontaminate the astatine from a variety of radioactive nuclides. Several procedures have been designed specifically for this type of problem[14–18]; these have been reviewed in detail by Nefedov *et al.*[19]. If the astatine prepared by these procedures is to be used for chemical studies, it must be further purified from any carriers introduced during the radiochemical decontamination.

Norseev and Khalkin[20] have summarised the method used at Dubna to separate astatine from metallic thorium targets that have been irradiated with 660 MeV protons. The target is dissolved in concentrated HCl containing HF and Cl_2. Tellurium(IV) carrier is introduced, and the addition of stannous chloride precipitates metallic tellurium, which carries the astatine with it. The tellurium and astatine are re-dissolved in 8 M HCl saturated with Cl_2. The astatine is adsorbed from this solution onto a column of sulphonic acid cation exchange resin, while the tellurium is washed through with 8 M HCl. The astatine is then eluted with chlorine water or with 3 M HNO_3 containing dichromate, and an excess of Ag^+ is added to precipitate the remaining chloride. The astatine is then adsorbed on platinum at 90 °C, and the platinum is washed with 6 M HNO_3 containing dichromate. Finally the astatine is dissolved from the platinum into 1 M HNO_3 by anodic polarisation.

6.4 TECHNIQUES FOR COUNTING ASTATINE SAMPLES

Astatine-211 may be assayed by counting either its α-particles or the x-rays accompanying its electron capture. The α counting may be carried out in any conventional α-counters, such as gas-flow ionisation chambers or proportional counters. Reproducible and adherent astatine samples may be obtained by evaporating aqueous astatine solutions in about 2 M HCl to dryness on silver or platinum foils under an infrared lamp. From most other solutions such evaporations show erratic losses of astatine. Solutions of astatine in organic solvents in particular cannot usually be evaporated without substantial losses; α-counting of such solutions can best be done through the use of liquid scintillation techniques.

The requirement of virtually weightless samples to avoid self-absorption severely restricts α-counting as a means of assaying astatine. In co-precipitation experiments one may circumvent this difficulty by counting infinitely thick samples of homogeneous precipitates, i.e., samples so thick that no α-particles from the bottom of the sample are counted. However, the absorption problem can be almost completely eliminated by the use of x-ray counting methods, which permit the direct assay of solutions and of bulky and inhomogeneous precipitates. Although the x-rays may be counted with a Geiger counter, much greater efficiency is obtained with a sodium-iodide scintillator. It is advantageous to reduce the relatively high background of the scintillator by operating it as an energy analyser registering only counts of energy in the vicinity of the *c.* 90 keV K x-ray of the astatine's polonium daughter. The use of a sodium iodide crystal containing a well permits convenient counting of both solutions and solids.

^{211}At decays in part to the long-lived ^{207}Bi, which also decays by electron capture. The ratio of initial ^{211}At x-ray activity to residual ^{207}Bi activity is of the order of 6×10^4. The bismuth is usually present in colloidal form – probably adsorbed on dust particles – and will be carried along unpredictably through a surprisingly wide variety of chemical procedures. Only distillation of the astatine can be relied on to remove all of the ^{207}Bi. Samples x-ray assayed for astatine some time after purification from ^{207}Bi should be recounted after the astatine has entirely decayed away, the resulting ^{207}Bi count being subtracted from the original count of the sample. This is especially important when counting a solution from which most of the astatine has been removed, e.g., the aqueous phase from a solvent extraction.

The techniques outlined here for ^{211}At apply generally to the other astatine isotopes, with specific modifications arising from the decay scheme of the particular isotope in question. Thus, for example, ^{210}At may also be assayed by scintillation counting of its 0.25 and 1.2 MeV γ-rays.

6.5 PHYSICAL PROPERTIES OF ASTATINE AND ITS COMPOUNDS

Except for nuclear properties, the only physical property of astatine to be measured directly has been the absorption spectrum of atomic At [21]. The boiling points of organoastatine compounds have been evaluated indirectly and are discussed in Section 6.8. Other physical properties of astatine and its compounds have been predicted from theory or by extrapolation from the properties of other elements. These predictions have been reviewed in detail by Nefedov *et al.*[19] and will not be discussed further here.

6.6 AQUEOUS CHEMISTRY OF ASTATINE

6.6.1 The —1 valence state—astatide

The astatide ion is characterised by its quantitative co-precipitation with such insoluble iodides as AgI, TlI, PbI_2, or PdI_2 [22]. The activity cannot be removed from the precipitates by washing with acetone[23]. The astatide ion is not extracted into organic solvents from aqueous solution[22].

Measurements of the relative diffusion rates of I^- and At^- through a membrane indicate that the diffusion coefficient of I^- is 1.42 times that of At^- [24]. When At^- and carrier-free $^{131}I^-$ solutions are subjected to paper electrophoresis, the I^- moves toward the anode at 1.21 times the speed of the At^- [25].

Astatide ion is formed by reducing the zero state of astatine with zinc and acid, with SO_2, with stannous chloride, with arsenite at pH $\geqslant 5$ and with ferrocyanide at pH>2. On the other hand, the astatide ion may be oxidised to the zero valence state of astatine by VO_2^+, Fe^{3+}, I_2, dilute nitric acid, arsenate at pH<5, or ferricyanide at pH<3 [11, 22]. The arsenite–arsenate and ferro–ferricyanide systems are more or less reversible and may produce either At^- or At^0 depending on the pH and the ratio of arsenite to arsenate or ferrocyanide to ferricyanide. The other reactions generally go to completion, regardless of the relative concentrations of the two components of the redox couple[11]. The reduction of At^0 by SO_2, however, is sometimes slow and incomplete[17, 23].

6.6.2 The 'zero' valence state

Astatine left to its own devices in aqueous solutions is usually found in the 'zero' state. This state is characterised by volatility from solution, by extractability into organic solvents and by varying degrees of co-precipitation with metallic sulphides and hydroxides and with metallic silver or tellurium precipitated *in situ*. Partial co-precipitation with $AgIO_3$ has been reported, although co-precipitation with $Pb(IO_3)_2$ does not take place. Astatine in this state is strongly adsorbed on metallic surfaces, and somewhat less strongly on glass[11, 22, 23].

When at At^0 solution is made alkaline, the astatine ceases to be extractable into organic solvents, and it can be completely co-precipitated with AgI. Extractability is at least partially restored by re-acidification. This behaviour has been taken to indicate disproportionation of At^0 in base[22].

Zero-state astatine is reported to migrate as an anion in an electric field[22]. This observation has not been adequately explained.

If a weakly acidic aqueous solution of At^0 is extracted with successive portions of benzene or carbon tetrachloride, the distribution ratio

$$D = \frac{\text{At in organic phase}}{\text{At in aqueous phase}}$$

decreases gradually to very low values. Conversely, if the organic phase from the initial extraction is extracted with successive portions of dilute acid, *D* gradually increases to very high values[22]. This behaviour implies that the astatine is distributed among at least three different species. One of these must be highly extractable into organic solvents, one must be highly unextractable and the third must have intermediate extractability.

Zero-state astatine may be prepared from higher oxidation states by reduction with Fe^{2+}, I^- or VO^{2+} [11, 22]. It may also be obtained by oxidation of At^-, as has been described in the previous section.

The behaviour of At^0 has been characterised by severe irreproducibility. Distribution ratios and degrees of co-precipitation and adsorption vary widely and seem to depend on the history of the sample. Even qualitative agreement among independent workers is often difficult to achieve. This irreproducibility can be understood if we bear in mind that the halogens in their zero valence states tend to be fairly reactive species. In many cases the reactions involve a dissociative pre-equilibrium of the X_2 molecule, e.g.

$$X_2 \rightleftharpoons 2X$$

or

$$X_2 \rightleftharpoons X^+ + X^-$$

or

$$X_2 + H_2O \quad HOX + X^-$$

These equilibria shift to the right as the halogen concentration decreases, leading to an increase in reactivity at low concentrations. And there will certainly be no dearth of impurities with which the vanishingly small amounts of astatine can react! Even very pure water contains *c.* 10^{-6}M carbon in the form of various organic compounds. Hence, it seems likely that 'zero-state' astatine is not At_2 at all, but rather a conglomeration of organoastatine compounds, the exact nature of which varies from one solution to the next. Little wonder the astatine behaves irreproducibly!

It is interesting to note that the behaviour of iodine at very low concentrations is quite different from its normal behaviour [26–29]. Although hitherto unknown inorganic iodine species have sometimes been invoked to explain this difference, it seems more likely that the presence of organic iodine compounds is responsible.

6.6.3 Interhalogen compounds of astatine in aqueous solution

The most promising way to avoid the ambiguity of the exact nature of aqueous At^0 is to use iodine as a 'carrier' for the astatine. Molecular I_2 may be expected to react with impurities that would otherwise react with astatine, while if I^- is also present, the redox potential of the solution is controlled. Furthermore, by analogy with other halogens, the reaction of astatine with iodine to give AtI should be rapid and complete.

The distribution of astatine between CCl_4 and a slightly acidic solution containing I_2 and I^- obeys the relation

$$\frac{\text{At in } CCl_4}{\text{At in aqueous phase}} = \frac{5.5}{1+2000(I^-)}$$

where 5.5 is the value of the equilibrium distribution constant $(AtI)_{CCl_4}/(AtI)_{aq}$ and 2000 is the value of the equilibrium constant for the reaction

$$AtI + I^- \rightleftharpoons AtI_2^-$$ [30]

The relatively low distribution constant for AtI reflects its polar character. The distribution of astatine is independent of acidity between pH 1 and 5, even at an iodide concentration as low as 10^{-4} M, leading to an upper limit of 10^{-11} for the equilibrium constant of the hydrolytic reaction

$$H_2O + AtI \quad HOAt + H^+ + I^-$$ [30]

The corresponding reaction of IBr[31] has a constant of 1.5×10^{-7}.

Addition of Pb^{2+} or Tl^+ to an iodine–iodide solution containing astatine precipitates compounds of the composition $PbI_2{\cdot}xI_2$ and $TlI{\cdot}xI_2$. A large portion of the astatine co-precipitates, but it can virtually all be removed along with the I_2 by washing the precipitate with acetone. Presumably the astatine is present in the precipitate as AtI_2^- [23].

In a similar fashion, the astatine in I_2–I^- solutions co-precipitates with CsI_3. Heating of the solid to 250 °C in air volatilises both the I_2 and the astatine, leaving CsI behind[32].

Astatine extracted with iodine into chloroform has been shown to co-precipitate homogeneously with the iodine when a portion of the latter is crystallised from the chloroform solution[13]. This provides further strong evidence that the astatine is present as AtI.

Silver iodide precipitated from an I_2–I^- solution does not carry astatine, but the astatine can subsequently be completely co-precipitated with $Pb(IO_3)_2$, suggesting the possible oxidation of AtI by I_2 in the presence of Ag^+:

$$5Ag^+ + AtI + 2I_2 + 3H_2O \rightarrow AtO_3^- + 5AgI + 6H^+$$ [23]

Addition of Br^- or Cl^- to astatine in an I_2–I^- solution decreases the extractability of the astatine into CCl_4. The data have been quantitatively interpreted in terms of the reactions

$$AtI + Br^- \rightleftharpoons AtIBr^-$$

and

$$AtI + Cl^- \rightleftharpoons AtICl^-$$

with equilibrium constants of 120 and 9, respectively. If the I^- concentration is fairly low, however, it is necessary to invoke the additional reactions

$$AtIBr^- + Br^- \rightleftharpoons AtBr_2^- + I^-$$

and

$$AtICl^- + Cl^- \rightleftharpoons AtCl_2^- + I^-$$

with equilibrium constants of 1.1×10^{-3} and 2×10^{-4}, respectively[30].

When AtI in dilute acid is treated with a mixture of IBr, I_2 and Br^-, the astatine becomes much less extractable into CCl_4. The data can be interpreted in terms of the equilibria

$$AtI + IBr \rightleftharpoons AtBr + I_2$$

and

$$AtBr + Br^- \rightleftharpoons AtBr_2^-$$

The first of these reactions has an equilibrium constant of 190, the second has a constant of 320. The AtBr distributes between water and CCl_4 with an equilibrium distribution constant

$$\frac{(AtBr)_{CCl_4}}{(AtBr)_{aq}} = 0.04$$

Solvent extraction systems containing a high ratio of IBr to I_2 tend to be markedly photosensitive[30].

Table 6.2 compares the distribution of the known halogens and 1:1 interhalogens between water and CCl_4. Table 6.3 compares the formation constants of the various trihalide ions. In general, the astatine-containing species exhibit properties that are in accord with those of the more conventional molecules and ions. This bolsters our confidence in the identification of the astatine species.

Table 6.2 Distribution constants of halogens and interhalogens between water and CCl_4 at 20–25 °C

$$K_D = (XY)_{CCl_4}/(XY)_{aq}$$

XY	K_D	*Reference*
I_2	84	Seidell[57]
Br_2	27.5	Seidell[57]
Cl_2	20*	Seidell[57]
AtI	5.5	Appelman[30]
IBr	4.3	Appelman[30]
ICl	0.34	Faull[31]
AtBr	0.040	Appelman[30]

*0 °C

Neumann[12] found that both At^0 and astatine that had been oxidised with Cl_2 or with hot persulphate could be substantially extracted into isopropyl ether from 1–11 M HCl. From such HCl solutions the astatine could not be extracted into benzene or CCl_4. Neumann concluded that one or more chloroacids of astatine, such as $HAtCl_2$, were being extracted into the ether. However, he was unable to interpret his results quantitatively. Hot persulphate is known to oxidise astatine to AtO_3^- (*vide infra*)[22]. It is possible that the HCl reduces the AtO_3^- to a chloro complex of lower oxidation state.

The interpretation of Neumann's results is rendered somewhat ambiguous by the work of Wang and Khalkin[33], who showed that astatine oxidised by chlorine could be extracted into isopropyl ether from nitric acid. The

Table 6.3 Equilibrium constants for the formation of trihalide ions at 21–25 °C

Reaction	*K*(l/mol)	*Reference*
$Cl_2 + Cl^- = Cl_3^-$	0.12	Sherrill and Izard[58]
$Br_2 + Cl^- = Br_2Cl^-$	1.4	Gmelin[59]
$I_2 + Cl^- = I_2Cl^-$	3	Gmelin[60]
$AtI + Cl^- = AtICl^-$	9	Appelman[30]
$Br_2 + Br^- = Br_3^-$	17	Gmelin[59]
$IBr + Cl^- = IBrCl^-$	43	Faull[31]
$AtI + Br^- = AtIBr^-$	120	Appelman[30]
$ICl + Cl^- = ICl_2^-$	170	Faull[31]
$AtBr + Br^- = AtBr_2^-$	320	Appelman[30]
$IBr + Br^- = IBr_2^-$	440	Appelman[30]
$I_2 + I^- = I_3^-$	800	Katzin and Gebert[61]
$AtI + I^- = AtI_2^-$	2000	Appelman[30]

acidity-dependence of the extraction was very similar to that observed by Neumann for HCl solutions. The presence of some nitric acid in HCl solutions has been found to improve the extraction of the astatine into isopropyl ether[15, 34].

Wang *et al.*[18] found that chlorine-oxidised astatine could be adsorbed onto cation exchange resin from several molar HCl solutions. They noted that chloro complexes of gold, iron, gallium and thallium behave similarly.

The formation of a neutral chloride and a complex chloro anion from the recently characterised cationic astatine will be discussed in the next section.

6.6.4 Intermediate positive valence states

The evidence for intermediate positive oxidation states of astatine has traditionally been negative. The astatine has been unextractable into organic solvents, at least non-polar ones, and usually has not co-precipitated with either insoluble iodides or iodates. Astatine in such a form can be prepared by oxidation of lower states in acid with bromine or dichromate, regardless of whether or not Br^- or Cr^{3+} are present[22]. Chlorine also oxidises astatine to an intermediate state if a substantial Cl^- concentration is present. If the Cl^- concentration is very low, however, partial further oxidation to astatate takes place[11] (*vide infra*).

The intermediate states are reduced to At^0 in acid solution by I^-, Fe^{2+}, or VO^{2+} [11,22]. Arsenite–arsenate and ferro–ferricyanide mixture also reduce intermediate states to At^0 if concentrations and pH are such that further reduction to At^- does not take place (*vide supra*). The reduction by I^- takes place even if I_2 is present. In the dark, the reduction by Fe^{2+} or VO^{2+} is likewise independent of the presence of Fe^{3+} or VO_2^+, respectively. In the light, however, a mixture of VO^{2+} and VO_2^+ oxidises At^0 to intermediate positive states, and an Fe^{2+}–Fe^{3+} mixture behaves similarly if $Fe^{3+}/Fe^{2+} > 100$. These photochemical reactions are reversed when the light is removed[11].

The reactions forming intermediate positive states from At^0 and vice versa seems to be fairly independent of whether or not I_2 is present to convert the At^0 to AtI, except that the photochemical oxidation of At^0 by VO_2^+ is markedly enhanced by the presence of I_2 [11].

Appelman[23] observed that in a mixture of Br_2, IBr and Br^-, the AtBr was oxidised to an unextractable intermediate state. This implies that the bromine-oxidised product has an oxidation state greater than +1. Johnson *et al.*[22] found that bromine-oxidised astatine migrated as an anion in an electric field.

Wang *et al.*[35] investigated the behaviour of astatine that had been oxidised with dichromate in nitric acid. They found that the astatine migrated in an electric field as a cation! They further found that the astatine distributed between aqueous solutions and cation exchange resin in a manner that indicated it to be present as an ion with a +1 charge.

Norseev, *et al.* found[36] that the astatine in these solutions deposits on platinum covered with an oxide film, apparently as the result of a surface ion-exchange process. The adsorption is inhibited by the presence of macro amounts of such cations as Tl^+, Hg_2^{2+} or BiO^+. Inhibition by Cl^- has been interpreted in terms of formation of a chloride complex by the astatine species. The astatine ion cannot be completely removed from the platinum by concentrated nitric acid or by aqua regia, but it can be removed by anodic polarisation of the platinum.

Wang *et al.*[37] and Norseev *et al.*[38] also found that cationic astatine co-precipitated more or less completely with insoluble salts of singly-charged cations. The salts used were $Tl_2Cr_2O_7$, $Ag_2Cr_2O_7$, $AgIO_3$ and caesium phosphotungstate. Less efficient co-precipitation took place with rubidium and ammonium phosphotungstates. Co-precipitation of a different nature was observed with WO_3 and with $Ba(IO_3)_2$ and $Th(IO_3)_4$. The co-precipitation with iodates was interpreted to indicate that the astatine cation formed an insoluble iodate. The astatine may be leached with 1 M HCl from either caesium phosphotungstate or thallous dichromate, presumably as a chloride complex. These observations suggest adsorption by a surface ion-exchange process similar to that involved in the adsorption on platinum.

Norseev and Khalkin[20] used cation exchange to study in detail the chloride complexing suggested by the adsorption studies. They observed the step-wise addition of two chloride ions to form first a neutral molecule and then a singly-charged anion. The successive complexing constants were found to be 7×10^2 and 3.6×10^2. If the oxidation state of the astatine is +1, these constants would apply to the equilibria

$$At^+ + Cl^- = AtCl$$

and

$$AtCl + Cl^- = AtCl_2^-$$

On the other hand, the oxidation state of the astatine might be +3, and the corresponding equilibria might then be

$$AtO^+ + Cl^- = AtOCl$$

and

$$AtOCl + Cl^- = AtOCl_2^-$$

Although the anionic astatine formed by bromine oxidation may be in a different oxidation state from the cationic astatine formed by dichromate oxidation, it is also possible that they are in the same oxidation state and that the bromine solutions used by Johnson *et al.*[22] contained enough bromide to form an anionic complex. In the latter case, the oxidation state of the astatine must be greater than +1, and is probably +3, suggesting that the cationic astatine is AtO^+.

6.6.5 The +5 valence state—astatate

The astatate ion, AtO_3^-, is characterised by its quantitative co-precipitation with such insoluble iodates as $AgIO_3$, $Ba(IO_3)_2$ and $Pb(IO_3)_2$. It is necessary to use some care in interpreting these results, however, in light of the co-precipitation with insoluble iodates by cationic astatine of intermediate positive valence (*vide supra*).

Astatate may be obtained by oxidation of lower states with alkaline hypochlorite, and with periodate, persulphate or Ce^{IV} in acid solution[11, 22]. The reactions are frequently sluggish at room temperature, and heating as high as 100 °C is sometimes necessary to obtain complete oxidation. The presence of the reducing component of the redox couple used for the oxidation does not affect the results.

Astatate may be reduced in acid to an intermediate positive state by bromide or chloride, even in the presence of Br_2 or Cl_2, respectively. However, a chlorine solution to which no Cl^- has been added will only partially reduce the astatate, and such a solution appears to partially oxidise lower astatine states to astatate[11].

Nagy *et al.*[25] have carried out paper electrophoresis of AtO_3^- and carrier-free $^{131}IO_3^-$ solutions. The iodate ion migrates to the anode at 1.57 times the velocity of the astatate ion.

Wang and Khalkin[33] observed that astatate could be extracted into isopropyl ether from strong HNO_3, H_2SO_4, or $HClO_4$. This extraction differs from the extraction of chlorine-oxidised astatine from HCl or HNO_3 in that it is negligible below a characteristic acid concentration. This concentration is about 1 M for H_2SO_4, 2 M for $HClO_4$, and 3 M for HNO_3. Above this concentration the extractability rises sharply with increasing acid, reaching a maximum distribution ratio of *c.* 10 around 5 M acid. The interpretation of this behaviour is not clear. It appears that reaction with the isopropyl ether or with impurities therein may be involved.

6.6.6 The +7 valence state—perastatate

No evidence for perastatate has been found. The only reported attempt at its synthesis has involved looking for co-precipitation of vigorously oxidised astatine with KIO_4 from an iodate-periodate mixture[11]. It is possible, however, that iodate may act as a reducing agent toward perastatate. It is also possible that a 6-coordinated H_5AtO_6 would show little tendency to co-precipitate with KIO_4. Recent successes in the preparation of perbromates by oxidation of bromates with F_2 and XeF_2 [39,40] suggest that it might be worth while to try to prepare perastatate through the use of such aggressive oxidants. (See Note added in proof.)

6.6.7 Electrode potential scheme for astatine

Table 6.4 compares the electrode potentials of the astatine species in 0.1 M acid with the corresponding potentials of the other halogens. We see that astatine is the only halogen with an oxidation state between zero and +5

Table 6.4 Electrode potentials (in volts) of the halogens in 0.1 M acid at 25 °C*

$$\mathrm{At^-} \overset{0.3}{\text{——}} \mathrm{At^0} \overset{1.0\dagger}{\text{——}} \mathrm{At^{+X}} \overset{1.5}{\text{——}} \mathrm{AtO_3^-} \overset{>1.6}{\text{——}} \mathrm{H_5AtO_6}$$

$$\mathrm{I^-} \overset{0.62}{\text{——}} \mathrm{I_2(aq)} \overset{1.31}{\text{——}} \mathrm{HOI} \overset{1.07}{\text{——}} \mathrm{IO_3^-} \overset{1.58}{\text{——}} \mathrm{H_5IO_6}$$

$$\mathrm{Br^-} \overset{1.09}{\text{——}} \mathrm{Br_2(aq)} \overset{1.51}{\text{——}} \mathrm{HOBr} \overset{1.43}{\text{——}} \mathrm{BrO_3^-} \overset{1.68}{\text{——}} \mathrm{BrO_4^-}$$

$$\mathrm{Cl^-} \overset{1.40}{\text{——}} \mathrm{Cl_2(aq)} \overset{1.54}{\text{——}} \mathrm{HOCl} \overset{1.35}{\text{——}} \mathrm{ClO_3^-} \overset{1.17}{\text{——}} \mathrm{ClO_4^-}$$

$$\mathrm{F^-} \overset{2.87}{\text{——}} \mathrm{F_2(g)}$$

*Astatine potentials are taken from Appelman[11]. Halate-perhalate potentials are based on Johnson *et al.*[62] and Schreiner *et al.*[63]. Other potentials are derived from Latimer[64].
†Latimer's value of 0.8 V for the At^0–At(+X) potential was based on the oxidation of At^0 by Fe^{3+} reported by Johnson *et al.*[22]. This appears to be a photochemical reaction, however, as we have discussed in Section 6.6.4.

that is stable to disproportionation. This is consistent with the tendency among main group elements for increasing molecular weight to be accompanied by increasing stabilisation of intermediate positive states.

The fact that the potential between At^0 and At^{+X} is the same whether or not I_2 is present[11] implies that in the absence of I_2 the At^0 species is neither At_2 or At, inasmuch as at the low astatine concentrations used (*c.* 10^{-14} M) both At_2 and At should be substantially more readily oxidised than AtI.

6.7 ASTATINE COMPOUNDS IN THE VAPOUR PHASE

Astatine vapour has a strong tendency to adsorb on surfaces, a property that has severely hampered experimental studies. Thus, although astatine can be distilled in vacuum in a glass apparatus at room temperature, once it has been condensed at lower temperature, it is often necessary to heat the glass before the astatine again becomes volatile.

The most fruitful studies of vapour-phase astatine have involved mass spectrometry, using *c.* 0.05 μg quantities of astatine[6]. In the absence of macro amounts of other halogens, mass peaks were observed corresponding to the ions At^+, HAt^+ and CH_3At^+. No indication of molecular At_2 was obtained. The At^+ peak may have arisen as a fragmentation product of the other species, while the HAt may have been formed by reaction of some astatine compound with traces of water, or by substitution reactions involving organic hydrogen. A comparison experiment using a very small amount of iodine gave peaks corresponding to the ions I^+, HI^+, CH_3I^+, $C_2H_5I^+$, $C_3H_7I^+$ and I_2^+. These results go far to confirm our conviction that trace amounts of iodine and astatine are often present in the form of organic compounds.

In the presence of macro amounts of I_2, Br_2 or Cl_2, the corresponding interhalogens AtI, AtBr, and AtCl were formed and detected in the mass spectrometer. An attempt to prepare an astatine fluoride by reaction with ClF_3 led to no identifiable compound, but rendered the astatine non-volatile. It is possible that an astatine fluoride was initially formed, but that under the influence of the rather intense α-radiation it reacted with the glass apparatus to yield refractory oxy-compounds.

As of this time, astatine fluorides remain to be characterised. Studies in non-aqueous fluorinated solvents such as BrF_3 may prove rewarding, as they recently have in the case of radon fluoride[41, 42].

6.8 ORGANIC COMPOUNDS OF ASTATINE

Although organic astatine compounds have been blamed for many of the troubles encountered in astatine chemistry, only recently have such compounds been well characterised. The early and conflicting reports have been reviewed by Aten[43].

Schats and Aten[44] prepared the salts $At(pyridine)_2ClO_4$ and $At(pyridine)_2NO_3$ by reaction of AtI in chloroform with $Ag(pyridine)_2ClO_4$ and $AgNO_3$+pyridine, respectively. The compounds were identified by co-precipitation with the analogous iodine salts.

Nefedov *et al*[45] synthesised $C_6H_5AtCl_2$, $(C_6H_5)_2AtCl$, and $C_6H_5AtO_2$. The astatine was initially in the form of an aqueous At^- solution that also contained I^-. The reaction scheme used was the following:

$$(C_6H_5)_2ICl + At^- \rightarrow (C_6H_5)_2IAt + Cl^-$$

$$(C_6H_5)_2IAt \xrightarrow{170\,^\circ C} C_6H_5At + C_6H_5I$$

$$C_6H_5At + Cl_2 \rightarrow C_6H_5AtCl_2$$

$$C_6H_5AtCl_2 + (C_6H_5)_2Hg \rightarrow (C_6H_5)_2AtCl + C_6H_5HgCl$$

$$C_6H_5AtCl_2 + OCl^- + 2OH^- \rightarrow C_6H_5AtO_2 + 3Cl^- + H_2O$$

Macro amounts of the homologous iodine compound were formed along with the astatine compound at each stage of the reaction. These iodine compounds were then used as carriers for the astatine compounds. The astatine compounds were identified by paper chromatography, their chromatographic behaviour being very similar to that of the corresponding iodine compounds. Norseev and Khalkin[46] report the preparation of the analogous *p*-tolyl compounds of astatine in the same manner.

Samson and Aten[47,48] used a variety of methods to synthesise astatobenzene:

(a) Preparation of $(C_6H_5)_2IAt$ by reaction of At^- with $(C_6H_5)_2I{\cdot}I$ in hot 50% aqueous ethanol, followed by decomposition of the $(C_6H_5)_2IAt$ at 175 °C, as in the scheme of Nefedov *et al.*[45]:

$$(C_6H_5)_2I{\cdot}I + At^- \rightarrow (C_6H_5)_2IAt + I^-$$
$$(C_6H_5)_2IAt \xrightarrow{175°C} C_6H_5I + C_6H_5At$$

(b) Reaction between iodobenzene and AtI dissolved in it, with and without the presence of intense γ radiation:

$$C_6H_5I + AtI = C_6H_5At + I_2$$

(c) Reaction of AtI with phenylhydrazine in an aqueous solution containing I_2 and I^-:

$$C_6H_5NHNH_2 + AtI_2^- + I_3^- \rightarrow C_6H_5At + N_2 + 5I^- + 3H^+$$

(d) Reaction of At^- with phenyldiazonium chloride in aqueous solution:

$$C_6H_5N_2Cl + At^- \rightarrow C_6H_5At + N_2 + Cl^-$$

(e) Reaction at 130–200 °C between iodobenzene vapour and At^- adsorbed on Kieselguhr along with KI:

$$C_6H_5I + At^- = C_6H_5At + I^-$$

(f) A hot-atom synthesis in which triphenylbismuth was irradiated with α-particles.

In all but the last case macro amounts of iodobenzene were either present initially or were formed along with the astatobenzene. Samson and Aten characterised the astatobenzene by gas chromatography, and from its chromatographic behaviour they estimated its normal boiling point to be 212 ± 2 °C.

Nefedov *et al.*[49] prepared astatobenzene by the electron-capture decay of ^{211}Rn to ^{211}At in benzene. They characterised their product by both thin-layer and gas chromatography, and also by treating it with Cl_2 to form $C_6H_5AtCl_2$. Kuzin *et al.*[50] used gas chromatography to estimate the boiling point of astatobenzene formed both in this manner and by heating $(C_6H_5)_2IAt$. They arrived at a value of 222 ± 3 °C, somewhat higher than the estimate of Samson[47].

Samson and Aten[51] synthesised each of the normal alkyl astatides from CH_3At to n-$C_6H_{13}At$ by reaction at 80–180 °C between the vapour of the analogous alkyl iodide and At^- adsorbed on Kieselguhr along with KI. These compounds, too, were characterised and separated from the corresponding iodides by gas chromatography. Chromatographic estimates of the normal boiling points of all of these compounds except CH_3At appear in Table 6.5. (See also Samson[47].) (See Note added in proof.)

Table 6.5 Normal boiling points of aliphatic astatides (Samson and Aten[51])

Compound	B.P.(°C)
C_2H_5At	98 ± 2
n-C_3H_7At	123 ± 2
n-C_4H_9At	152 ± 3
n-$C_5H_{11}At$	176 ± 3
n-$C_6H_{13}At$	201 ± 2

Samson and Aten[52] prepared $AtCH_2COOH$ by the reaction between At^- and aqueous ICH_2COOH. They measured the acidity dependence of the extraction of the compound into isopropyl ether and thereby determined its acid dissociation constant. Values for pK_A of 3.78, 3.70 and 3.73 were obtained at 0°, 22° and 27 °C, respectively. Iodoacetic acid has pK_A's of 3.14 at 0 °C and 3.12 at 22 °C. The equilibrium distribution constant $(AtCH_2COOH)_{ether}/(AtCH_2COOH)_{aq}$ has the values 1.72, 1.41, and 1.23 at 0°, 22° and 27 °C, respectively. For ICH_2COOH the corresponding values are 3.40 at 0 °C and 2.50 at 22 °C.

6.9 BIOLOGICAL BEHAVIOUR OF ASTATINE

Astatine resembles iodine in that it concentrates in the thyroid gland. A substantial portion, however, distributes throughout the body, where it acts as an efficient internal irradiation source[53–55]. The biological studies of astatine have been reviewed elsewhere[19, 43], and inasmuch as no significant contributions have been made in recent years, this work will not be discussed in detail in the present article.

6.10 ACKNOWLEDGEMENTS

The author wishes to thank Mrs. Alberta Martin for carrying out the literature search required for the preparation of this review, and Mrs. Gerda Siegel for assistance with translations.

Note added in proof

Khalkin *et al.*[65] have recently used xenon difluoride in alkaline solution to oxidize astatine to the +7 valence state. (See Section 6.6.6). They characterised their product by its behaviour during paper electrophoresis and by its co-precipitation with potassium and caesium metaperiodates. The co-precipitation obeyed the Berthelot–Nernst homogeneous distribution law, implying isomorphous replacement of the periodate ions by the perastatate ions.

Gesheva *et al.*[66] have recently extended the work of Samson and Aten[51] on the alkyl astatides to include the isopropyl, isobutyl, isoamyl, and primary active amyl astatides. They have estimated the respective boiling points of these compounds to be 112 ± 2, 142 ± 3, 163 ± 3, and 165 ± 3 °C.

References

1. Karlik, B. and Bernert, T. (1943). *Naturwiss.*, **32**, 44
2. Karlik, B. and Bernet, T. (1944). *Z. Physik*, **123**, 51
3. Walen, R. J. (1949). *J. Phys. Radium*, **10**, 95
4. Hyde, E. K. and Ghiorso, A. (1953). *Phys. Rev.*, **90**, 267
5. Corson, D. R., MacKenzie, K. R. and Segré, E. (1940). *Phys. Rev.*, **58**, 672
6. Appelman, E. H., Sloth, E. N. and Studier, M. H. (1966). *Inorg. Chem.*, **5**, 766
7. Kelly, E. and Segré, E. (1949). *Phys. Rev.*, **75**, 999
8. John, W., Jr. (1956). *Phys. Rev.*, **103**, 704
9. Aten, A. H. W., Jr., Doorgeest, T., Hollstein, U. and Moeken, H. P. (1952). *Analyst*, **77**, 774
10. Barton, G., Ghiorso, A. and Perlman, I. (1951). *Phys. Rev.*, **82**, 13
11. Appelman, E. H. (1961). *J. Amer. Chem. Soc.*, **83**, 805
12. Neumann, H. M. (1957). *J. Inorg. Nucl. Chem.*, **4**, 349
13. Aten, A. H. W., Jr., van Raaphorst, J. G., Nooteboom, G. and Blasse, G. (1960). *J. Inorg. Nucl. Chem.*, **15**, 198
14. Kurchatov, B. V., Mekhedov, V. N., Chistyakov, L. V., Kuznetsova, M. Ya., Borisova, N. I. and Solovyev, V. G. (1958). *Zh. Exp. Teor. Fiz.*, **35**, 56
15. Belyaev, B. N., Wang, Yun-Yui, Sinotova, E. N., Nemet, L. and Khalkin, V. A. (1960). *Radiokhimiya*, **2**, 603
16. Lefort, M., Simonoff, G. and Tarrago, X. (1959). *Compt. Rend.*, **248**, 216
17. Lefort, M., Simonoff, G. and Tarrago, X. (1960). *Bull. Soc. Chim. France, 1960*, 1726
18. Wang, Fu-Chiung, Kang, Meng-Hua and Khalkin, V. A. (1962). *Radiokhimiya*, **4**, 94
19. Nefedov, V. D., Norseev, Yu. V., Toropova, M. A. and Khalkin, V. A. (1968). *Usp. Khim.*, **37**, 87
20. Norseev, Yu. V. and Khalkin, V. A. (1968). *J. Inorg. Nucl. Chem.*, **30**, 3239
21. McLaughlin, R. (1964). *J. Opt. Soc. Am.*, **54**, 965
22. Johnson, G. L., Leininger, R. F. and Segré, E. (1949). *J. Chem. Phys.*, **17**, 1
23. Appelman, E. H. (1960). *U.S. At. Energy Comm. UCRL*-9025
24. Durbin, P. W. (1955). *University of California Radiation Laboratory Medical and Health Physics Quarterly Report, Jan.-Mar., UCRL*-3013, 34. *(U.S. At. Energy Comm.)*
25. Nagy, G. A., Khalkin, V. A. and Norszejev, J. V. (1967). *Magy. Kem. Foly*, **73**, 191
26. Kahn, M. and Wahl, A. C. (1953). *J. Chem. Phys.*, **21**, 1185
27. Good, M. L. and Edwards, R. R. (1956). *J. Inorg. Nucl. Chem.*, **2**, 196
28. Wille, R. G. and Good, M. L. (1957). *J. Amer. Chem. Soc.*, **79**, 1040
29. Eiland, H. M. and Kahn, M. (1961). *J. Phys. Chem.*, **65**, 1317
30. Appelman, E. H. (1961). *J. Phys. Chem.*, **65**, 325
31. Faull, J. H. (1934). *J. Amer. Chem. Soc.*, **56**, 522
32. Brinkman, G. A., Veenboer, J. Th. and Aten, A. H. W., Jr. (1963). *Radiochim. Acta*, **2**, 48
33. Wang, Yung-Yu and Khalkin, V. A. (1961). *Radiokhimiya*, **3**, 662
34. Appelman, E. H. (1960). *U.S. At. Energy Comm. NAS-NS* 3012
35. Wang, Fu-Chun, Norseev, Yu. V., Khalkin, V. A. and Ch'ao, Tao-Nan (1963). *Radiokhimiya*, **5**, 351
36. Norseev, Yu. V., Ch'ao, T'ao-Nan and Khalkin, V. A. (1966). *Radiokhimiya*, **8**, 497
37. Wang, Fu-Chun, Krylov, N. G., Norseev, Yu. V., Ch'ao, Tao-Nan and Khalkin, V. A. (1965). *Soosazhdenie i Adsorbtsiya Radioaktivn. Elementov, Akad. Nauk SSSR, Otd. Obshch. i Tekhn. Khim.*, 80-8 (Russ.)
38. Norseev, Yu. V., Khalkin, V. A. and Ch'ao, T'ao-Nan. (1965). *Izv. Sibirsk. Otd. Akad. Nauk SSSR, Ser. Khim. Nauk*, 21
39. Appelman, E. H. (1968). *J. Amer. Chem. Soc.*, **90**, 1900
40. Appelman, E. H. (1969). *Inorg. Chem.*, **8**, 223
41. Stein, L. (1969). *J. Amer. Chem. Soc.*, **91**, 5396
42. Stein, L. (1970). *Science*, **168**, 362
43. Aten, A. H. W., Jr. (1964). *Advanc. Inorg. and Radiochemistry*, **6**, 207

44. Schats, J. J. C. and Aten, A. H. W., Jr. (1960). *J. Inorg. Nucl. Chem.*, **15**, 197
45. Nefedov, V. D., Norseev, Yu. V., Savlevich, Kh., Sinotova, E. N., Toropova, M. A. and Khalkin, V. A. (1962). *Dokl. Akad, Nauk SSSR*, **144**, 806
46. Norseev, Yu. V. and Khalkin, V. A. (1967). *Chem. Zvesti*, **21**, 602
47. Samson, G. (1969). *Chem. Weekbl.*, **65**, 27
48. Samson, G. and Aten, A. H. W., Jr. (1970). *Radiochim. Acta*, **13**, 220
49. Nefedov, V. D., Toropova, M. A., Khalkin, V. A., Norseev, Yu. V. and Kuzin, V. I. (1970). *Radiokhimiya*, **12**, 194
50. Kuzin, V. I., Nefedov, V. D., Norseev, Yu. V., Toropova, M. A. and Khalkin, V. A. (1970). *Radiokhimiya*, **12**, 414
51. Samson, G. and Aten, A. H. W., Jr. (1969). *Radiochim. Acta*, **12**, 55
52. Samson, G. and Aten, A. H. W., Jr. (1968). *Radiochim. Acta*, **9**, 53
53. Hamilton, J. G., Asling, C. W., Garrison, W. M. and Scott, K. (1953). *The Accumulation, Metabolism and Biological Effects of Astatine in Rats and Monkeys* (Berkeley and Los Angeles, California: University of California Press)
54. Hamilton, J. G., Durbin, P. W., Asling, C. W. and Johnston, M. E. (1955). *Proc. Intern. Conf. Peaceful Uses Atomic Energy, Geneva*, **10**, 175 (published 1956)
55. Durbin, P. W., Asling, C. W., Johnston, M. E., Parrott, M. W., Jeung, N., Williams, M. H., Hamilton, J. G. (1958). *Radiation Res.*, **9**, 378
56. Hyde, E. K., Perlman, I. and Seaborg, G. T. (1964). *The Nuclear Properties of the Heavy Elements*, Vol. II, 1081–1082 (Englewood Cliffs, N.J. Prentice-Hall)
57. Seidell, A. (1958). *Solubilities of Inorganic and Metal Organic Compounds*, Vol. I (New York: D. Van Nostrand Co.)
58. Sherill, M. S. and Izard, E. F. (1928). *J. Amer. Chem. Soc.*, **50**, 1665
59. *Gmelins Handbuch der Anorganischen Chemie* (1931) System No. 7, 283 (Berlin: Verlag Chemie)
60. *Gmelins Handbuch der Anorganischen Chemie* (1933) System No. 8, 427 (Berlin: Verlag Chemie)
61. Katzin, L. I. and Gebert, E. (1955). *J. Amer. Chem. Soc.*, **77**, 5814
62. Johnson, G. K., Smith, P. N., Appelman, E. H. and Hubbard, W. N. (1970). *Inorg. Chem.*, **9**, 119
63. Schreiner, F., Osborne, D. W., Pocius, A. V. and Appelman, E. H. (1970). *Inorg. Chem.*, **9**, 2320
64. Latimer, W. M. (1952). *The Oxidation States of the Elements* (New York: Prentice Hall)
65. Khalkin, V. A., Norseev, Yu. V., Nefedov, V. D., Toropova, M. A. and Kuzin, V. I. (1970). *Dokl. Akad. Nauk SSSR*, **195**, 623
66. Gesheva, M., Kolachkovsky, A. and Norseev, Yu. (1971). *Preprint, Joint Institute for Nuclear Research*, Dubna, P6–5683

7
Halogen Cations

R. J. GILLESPIE and M. J. MORTON
McMaster University, Hamilton, Ontario

7.1 INTRODUCTION

Since the previous review of this area 5 years ago in *Halogen Chemistry*[1], the topic of halogen cations has continued to be highly productive as well as provoking lively controversy. At that time, the species responsible for the intense blue colour obtained when iodine is dissolved in strong acids had just been re-identified as the I_2^+ cation[2], rather than the I^+ cation which in the previous few years had been the subject of a series of papers including a Quarterly Review[3].

In the last 5 years, the identity of the I_2^+ cation has been confirmed by a wide variety of techniques and by several authors, and the previous evidence in favour of the I^+ cation has been re-interpreted[4–8]. Indeed, it is now clear that there is no evidence for the existence of the I^+ cation, or for the Br^+ or Cl^+ cations, as stable species in condensed phases even in the strongest acids known, and hence there is little justification for these species to be postulated as reaction intermediates in solution or as products in self-ionisation reactions such as

$$2ICl \rightleftarrows I^+ + ICl_2^-$$

The identification of I_2^+, and other iodine cations, was the result of a detailed study of the physical and chemical properties of the fluorosulphuric acid solvent system, which enabled an accurate interpretation to be made of the results of measurements of conductivities, freezing points, magnetic properties and spectra of the solutions that were obtained on the oxidation of iodine in the acid. Similarly, the recent development of the chemistry of related super-acid media has led to the identification not only of bromine and chlorine cations, but also of polyatomic cations of other non-metals such as sulphur, selenium and tellurium[10].

The highly electrophilic halogen cations can only exist in media of extremely low basicity such as fluorosulphuric acid, antimony pentafluoride, iodine pentafluoride and sulphur dioxide, and in the presence of anions of extremely low basicity. Thus, in the search for the more highly electrophilic cations of bromine and chlorine attention has been concentrated on the solvent systems which give the most weakly basic anions such as $Sb_3F_{16}^-$ and $SbF_2(SO_3F)_4^-$. As a result, evidence has been obtained for the existence of the Br_2^+[11] and Br_3^+[11] and Cl_3^+[12] cations, and the controversy over the iodine cation has been replaced by one on the evidence for the Cl_2^+ and ClF^+ cations, whose existence in super-acid systems has been proposed on the basis of e.s.r. spectra[13–17].

The development of laser Raman spectroscopy has made a particularly significant contribution in the structural identification of halogen cation salts, especially with simple anions such as AsF_6^-, BF_4^-, and SbF_6^-. Taken in conjunction with infrared spectra, it has often been possible to assign vibrational frequencies with confidence, and to make some estimates of the bond angles, whereas such analysis based on infrared alone has often been misleading. Thus, both stretching frequencies first reported for the ClF_2^+ cation on the basis of the infrared spectrum have now been re-assigned to the anion, and the bending frequency has also been observed as a result of laser Raman spectroscopy[18,19]. Similarly, the Cl_2F^+ cation, formed in the reaction between chlorine monofluoride and arsenic pentafluoride at low temperatures, was given the symmetrical bent structure Cl—F$^+$—Cl rather than the unsymmetrical structure Cl—Cl$^+$—F as no typical Cl—F stretching frequency at approximately 800 cm^{-1} was observed in the infrared[20]. A later laser Raman study however gave a strong band in this region, and the combined spectra could be reasonably assigned to the asymmetric Cl—Cl$^+$—F cation[12]. A further feature of laser Raman spectra is that it is often possible to obtain well-defined spectra with very little experimental difficulty, provided the samples are not highly coloured: for example, the three vibrational frequencies of the Cl_3^+ cation[12] in $Cl_3^+AsF_6^-$ were readily identified in a spectrum obtained from a solid sample which had been prepared *in situ* in a cooled chloro-fluoro carbon (Kel-F) tube. Even when the samples are highly coloured, it is sometimes possible to obtain Raman spectra by choosing an exciting wavelength which is not absorbed and does not cause fluorescence. In certain cases, moreover, excitation with the strongly absorbed wavelength gives rise to intense resonance Raman spectra which enable the absorbing species, e.g. I_2^+ or Br_2^+, to be identified at extremely low concentrations[7,11].

The number of x-ray structural determinations has increased in recent years from one (ICl_2^+)[21] to four (ClF_2^+[22], BrF_2^+[23], Br_2^+[24]), but even so this is a small fraction of the 15 or so halogen cations which have been identified. This at least partly reflects the difficulty of obtaining and manipulating single crystals of salts of these highly reactive and sensitive species.

The total number of possible halogen cations assuming that they have structures analogous to the known cations is probably around 30, of which only about half have been prepared to date.

7.2 DIATOMIC CATIONS

The di-iodine and dibromine cations I_2^+ and Br_2^+ are now well-established, but there is no good evidence for the existence of Cl_2^+ as a stable species in solution. There is also no evidence for the possible heteroatomic species IBr^+, ICl^+ and $BrCl^+$.

7.2.1 The di-iodine cation I_2^+

In 1966, Gillespie and Milne[2] showed by conductimetric, spectrophotometric and magnetic susceptibility measurements in fluorosulphuric acid that the blue iodine species is I_2^+ and not I^+. When a solution of iodine in fluorosulphuric acid was oxidised with peroxydisulphuryl difluoride, the concentration of the blue iodine cation was found to reach a maximum at the 2:1 mole ratio

$$2I_2 + S_2O_6F_2 \rightarrow 2I_2^+ + 2SO_3F^-$$

rather than at the 1:1 mole ratio expected for the formation of I^+ according to the equation

$$I_2 + S_2O_6F_2 \rightarrow 2I^+ + 2SO_3F^-$$

Supporting evidence for the formation of I_2^+ was obtained from conductivity measurements, which showed that the conductivity of a 1:1 $S_2O_6F_2/I_2$ solution was much less than that expected for the formation of I^+, and that 2:1 solutions had conductivities at low concentrations close to the conductivities of solutions of the same concentration of KSO_3F, indicating the formation of one mole of SO_3F^- per mole of I_2. Magnetic measurements showed that the blue solutions of iodine in fluorosulphuric acid, or in oleum, contain a paramagnetic species which has a magnetic moment of 2.0 ± 0.1 BM as expected for the $^2\Pi_{\frac{3}{2}}$ ground state of the I_2^+ cation. The I_2^+ cation has characteristic peaks in its visible absorption spectrum at 640, 490 and 410 nm, and has a molar extinction coefficient at 640 nm of 2560.

The I_2^+ cation is not completely stable in fluorosulphuric acid, and undergoes some disproportionation to the more stable I_3^+ species and $I(SO_3F)_3$ according to the equation

$$8I_2^+ + 3SO_3F^- \rightleftarrows I(SO_3F)_3 + 5I_3^+$$

as is shown for example by the appearance of the characteristic peak of the I_3^+ cation at 305 nm in solutions of $I_2^+SO_3F^-$ in fluorosulphuric acid, and by conductivity measurements which show that there is less than one mole of SO_3F^- per mole of iodine as the concentration of iodine is increased. This disproportionation is largely prevented in a 1:1 $I_2/S_2O_6F_2$ solution in which $I(SO_3F)_3$ is formed

$$5I_2 + 5S_2O_6F_2 \rightarrow 4I_2^+ + 4SO_3F^- + 2I(SO_3F)_3$$

and does not occur if the fluorosulphate ion concentration in the fluorosulphuric acid is lowered by the addition of antimony pentafluoride

$$SbF_5 + SO_3F^- \rightarrow SbF_5(SO_3F)^-$$

or in the less basic solvent 65% oleum.

In 100% H_2SO_4, the disproportionation of I_2^+ to I_3^+ and an iodine(III) species, probably $I(SO_4H)_3$, is essentially complete, and only traces of I_2^+ can be detected by means of its resonance Raman spectrum.

Blue solutions of iodine in oleum, which were originally supposed to contain I^+, have been re-investigated by conductimetric, spectrophotometric and cryoscopic methods, and the results confirm that the I_2^+ cation is formed[25]. In 65% oleum, the oxidation of iodine to I_2^+ is rapid and complete, according to the equation

$$2I_2 + 5SO_3 + H_2S_4O_{13} \rightarrow 2I_2^+ + 2HS_4O_{13}^- + SO_2$$

Iodine and iodine monochloride dissolve in antimony pentafluoride to give blue solutions that contain the I_2^+ cation[4]. Blue solutions of I_2^+ can also be obtained by dissolving iodine in iodine pentafluoride in the presence of a trace of moisture[26] and iodine can be oxidised to I_2^+ in solution in IF_5 by any of the pentafluorides of P, As, Sb, Nb and Ta[4]. None of the blue solutions of iodine in any of these solvents give e.s.r. spectra, and this is in accordance with the expectation that strong spin–orbit coupling in the I_2^+ cation would broaden the spectrum beyond detection.

A blue solid of apparent composition $(SbF_5)_2I$ was first isolated by Ruff and his co-workers as long ago as 1906[27], and the same material has been obtained more recently by Kemmitt *et al.*[4]. Although this material clearly contains I_2^+, it appears to be a mixture which also contains Sb^{III}. However, from the reaction of iodine with antimony pentafluoride and with tantalum pentafluoride in solution in iodine pentafluoride, these workers obtained blue crystalline solids which had compositions close to IMF_6 or $I_2M_2F_{11}$. Since it appeared that the iodine was in the $+\frac{1}{2}$ oxidation state, these compounds may presumably be formulated as $I_2^+Sb_2F_{11}^-$ and $I_2^+Ta_2F_{11}^-$. Unfortunately, it was apparently not possible to obtain suitable crystals for an x-ray crystallographic study, so the bond length in I_2^+ is not known.

On cooling the bright blue solutions of the iodine cation I_2^+ in fluorosulphuric acid, there is a rapid change of colour to an intense red near the freezing point. It has been shown by means of spectroscopic, cryoscopic, conductimetric and magnetic susceptibility measurements carried out at low temperatures that the iodine cation dimerises to the I_4^{2+} cation, with $\Delta H_d = -10 \pm 2$ kcal mol^{-1} [6].

$$2I_2^+ \rightleftarrows I_4^{2+}$$

The I_4^{2+} cation is diamagnetic and has intense absorption maxima at 470, 357 and 290 nm, with molal extinction coefficients at -86 °C of 11 000, 46 000 and 25 000 respectively. Nothing is known of the structure of the I_4^{2+} cation.

Initial attempts to observe the vibrational frequency of the I_2^+ cation by Raman spectroscopy were unsuccessful due to the absorption of the exciting radiation by the highly-coloured solutions[4]. Recently, it has been shown[7] that by using the strongly absorbed 6328 Å He/Ne laser excitation resonance Raman spectra can be obtained from very dilute solutions of I_2^+. Such resonance Raman spectra of dilute solutions of the I_2^+ cation in fluorosulphuric acid show, in addition to the fundamental at 238 cm^{-1}, intense overtones which become progressively broader and weaker. These spectra of the I_2^+ cation can be used to detect the cation at a mole ratio as low as 10^{-6} in fluorosulphuric acid. The increase in the vibrational frequency from 215 cm^{-1} for the iodine molecule to 238 cm^{-1} for I_2^+ is consistent with the loss of an antibonding electron from I_2.

7.2.2 The dibromine cation Br_2^+

A compound formulated as SbF_5Br was prepared by Ruff in 1906 [28] by the reaction of Br_2 with SbF_5, but the nature of this compound remained a mystery. No further studies of the bromine–antimony pentafluoride system were made until 1966, when McRae[28] reported evidence that Br_3^+ was formed in this system, but no evidence for Br_2^+ was obtained. However, in 1968 Edwards *et al.*[25] prepared $Br_2^+Sb_3F_{16}^-$ and determined its crystal structure. At the same time, Gillespie and Morton reported[11] the characterisation of the Br_2^+ and Br_3^+ cations in the fluorosulphuric acid solvent system.

The $Sb_3F_{16}^-$ salt of Br_2^+ was prepared by Edwards *et al.*[24] by the reaction of bromine with bromine pentafluoride and antimony pentafluoride

$$9Br_2 + 2BrF_5 + 30SbF_5 \rightarrow 10(Br_2^+Sb_3F_{16}^-)$$

It is a scarlet-red paramagnetic solid that is immediately decomposed by moisture. The x-ray crystallographic study[24] showed the solid to consist of the Br_2^+ cation and the *trans*-$Sb_3F_{16}^-$ anion. The bond length in Br_2^+ was found to be 2.13 Å, compared with 2.27 Å in the neutral molecule Br_2, and a stretching frequency of 368 cm^{-1} was observed in the Raman spectrum, compared with the frequency of 317 cm^{-1} for Br_2 (Table 7.1). The decrease in the bond length, and the increase in the stretching frequency, are consistent with an increase in bond order resulting from the loss of an antibonding electron from Br_2.

Because they are more electrophilic than the corresponding iodine cations, a more weakly basic solvent than is necessary for the iodine cations is required in order to obtain the bromine cations in solution. Thus, it has been found[11] that when bromine is oxidised with $S_2O_6F_2$ in solution in the super acid HSO_3F—SbF_5—$3SO_3$, Br_3^+, Br_2^+, $BrOSO_2F$ and $Br(OSO_2F)_3$ are all formed, the relative amounts depending on the amount of added $S_2O_6F_2$. The equilibria in this system are somewhat complex, and even in this extremely

Table 7.1 Stretching frequencies, absorption maxima and bond lengths of the halogens and the diatomic halogen cations

	Stretching frequency cm^{-1}	*Principle absorption* *Max.* nm	*Bond length* Å	*Ionisation energy* (eV)†
Cl_2	564.9*	330	1.98	11.50
Cl_2^+	645.3*	—	—	—
Br_2	320	410	2.27	10.51
Br_2^+	360	510	2.13	—
I_2^+	215	500	2.66	9.33
I^+	238	640	—	—

*G. Herzberg. (1960). *Molecular Spectra and Molecular Structure*, Vol. 1, Princeton, N.J.: Van Nostrand
†D. C. Frost, C. A. McDowell and D. A. Vroom. (1967). *J. Chem. Phys.*, 4255

weakly basic medium the Br_2^+ ion is not completely stable, as it undergoes appreciable disproportionation to the more stable Br_3^+ and $BrOSO_2F$

$$2Br_2^+ + 2HSO_3F \rightleftarrows Br_3^+ + BrOSO_2F + H_2SO_3F^+$$

Moreover, $BrOSO_2F$ itself undergoes some disproportionation to Br_2^+, Br_3^+ and $Br(OSO_2F)_3$

$$5BrOSO_2F + 2H_2SO_3F^+ \rightleftarrows 2Br_2^+ + Br(OSO_2F)_3 + 4HSO_3F$$
$$4BrOSO_2F + H_2SO_3F^+ \rightleftarrows Br_3^+ + Br(OSO_2F)_3 + 2HSO_3F$$

Solutions in super acid containing the Br_2^+ ion have a characteristic cherry-red colour, with a maximum absorption at 510 nm and a characteristic band in the Raman spectrum at 360 cm^{-1}. Using various excitation frequencies the intensity of this band was found to be a maximum with 5145 Å radiation, indicating that the Br_2^+ ion like the I_2^+ ion exhibits a resonance Raman effect.

7.2.3 The dichlorine cation Cl_2^+

The ionisation energy of Cl_2 is considerably greater than that of Br_2 or I_2 (Table 7.1), but it is nevertheless 17 kcal mol^{-1} less than that of O_2^+ which is known as a stable species in the form of several salts such as $O_2^+SbF_6^-$. However, the lattice energies of Cl_2^+ salts, estimated from Kapustinskii's second equation, are approximately 20 kcal mol^{-1} less than those of the corresponding O_2^+ salts, giving overall heats of formation some 3 kcal mol^{-1} less favourable than for the O_2^+ salts. This indicates that it may be extremely difficult to obtain stable salts of Cl_2^+. No evidence for the existence of Cl_2^+, even in the super acid $HSO_3F—SbF_5—3SO_3$, was obtained from a study of the conductivities and Raman spectra of solutions of ClF and Cl_2 in this solvent[12]. The only stable species formed appeared to be chlorine(I) fluorosulphate. Thus, although the Cl_2^+ cation has been identified in the gas phase at very low pressures, there is no evidence for it as a stable species in solution or in the solid state. Recent claims by Olah and Comisarow[13,14] to have identified the Cl_2^+ and ClF^+ cations on the basis of the e.s.r. spectra of solutions of ClF_3 and ClF_5 in SbF_5, HSO_3F/SbF_5 and HF/SbF_5 are apparently not correct, and have been criticised by several workers. Symons *et al.*[30] claim that the e.s.r. spectrum assigned to the ClF^+ cation arises from the $ClOF^+$ cation, as the sum of the spin densities on chlorine and fluorine calculated from the observed spectrum is only 0.62, and the e.s.r. parameters fit in well with those of the isostructural FOO, ClOO, ClO_2 and NF_2 radicals. By analogy, they assign the e.s.r. spectrum assigned to Cl_2^+ by Olah and Comisarow to the $ClOCl^+$ cation. Christe and Muirhead[31] claim that radicals are not produced in the reaction of highly-purified SbF_5 and ClF_3 or ClF_5, and that the radicals observed by Olah and Comisarow must have been due to impurities. Gillespie and Morton[30] observed an increase in intensity by a factor of several hundred in the e.s.r. signal previously assigned to the ClF^+ cation on adding a small amount of water to a sample of $ClF_2^+SbF_6^-$ in SbF_5, which supports the assignment to an oxy-radical, which they argue is more likely to be the $FClO^+$ radical, isoelectronic with ClO_2, or the ClO_2F^+ radical, isoelectronic with ClO_3, rather than the $ClOF^+$ radical proposed by Symons and co-workers[31].

7.3 TRIATOMIC CATIONS

7.3.1 The tri-iodine cation I_3^+

The first evidence for the existence of a stable iodine cation was obtained by Masson in 1938 [32]. He postulated the presence of the I_3^+ and I_5^+ cations in solutions of iodine and iodosyl sulphate in concentrated sulphuric acid in order to explain the stoichiometry of the reaction of such solutions with chlorobenzene to form both iodo and iodoso derivatives. In 1962, Arotsky, Mishra and Symons[34] gave conductimetric evidence for I_3^+ formed from iodic acid and iodine in 100% sulphuric acid. Gillespie and co-workers confirmed the formation of I_3^+ from HIO_3 and I_2 in 100% sulphuric acid by detailed conductimetric and cryoscopic measurements[34].

$$HIO_3 + 7I_2 + 8H_2SO_4 \rightarrow 5I_3^+ + 3H_3O + 8HSO_4^-$$

It has also been shown that the I_3^+ cation may be prepared in fluorosulphuric acid solution[2] by the reaction:

$$3I_2 + S_2O_6F_2 \rightarrow 2I_3^+ + 2SO_3F^-$$

Solutions of the red-brown I_3^+ cation have characteristic absorption maxima at 305 and 470 nm, with a molar extinction coefficient of 5200 at 305 nm. No data on the vibrational frequencies or the structure of this cation have been published.

Ruff[27] first reported the formation of a brown solid with the apparent composition SbF_5I when excess iodine was reacted with SbF_5. This reaction was confirmed by Kemmitt *et al.*[4], who also showed from the absorption spectrum of a solution of the product in liquid AsF_3 that it contains the I_3^+ cation. However, like the supposed blue $(SbF_5)_2I$ compound, this material is probably also complex and almost certainly contains Sb^{III}.

7.3.2 The tribromine cation Br_3^+

In 1968, Gillespie and Morton[11] showed conclusively by conductimetric and absorption spectra measurements that Br_3^+ is formed quantitatively in $HSO_3F—SbF_5—3SO_3$ super-acid medium by the reaction $3Br_2 + S_2O_6F_2 \rightarrow 2Br_3^+ + 2SO_3F^-$. The solutions are brown in colour, and have a strong absorption at 300 nm with a shoulder at 375 nm. Solutions of Br_3^+ can also be obtained by oxidising bromine with $S_2O_6F_2$ in solution in HSO_3F, but they are not completely stable in this more basic solvent, and undergo some disproportionation according to the equation

$$Br_3^+ + SO_3F^- \rightleftharpoons Br_2 + BrOSO_2F$$

Solutions of Br_3^+ in HSO_3F have a characteristic band at 290 cm^{-1}, which in view of the fact that Br_2 has a stretching frequency at 320 cm^{-1} may reasonably be assigned to the Br_3^+ cation. As Br_3^+ is presumably a bent molecule like Cl_3^+, it is reasonable to suppose that the 290 cm^{-1} band is due to both the symmetric and antisymmetric stretching frequencies ν_1 and ν_3, while the bending frequency ν_2, which would be expected to have a frequency of ~ 100 cm^{-1}, was not observed.

In 1969, Glemser and Smalc[35] obtained the compound $Br_3^+AsF_6^-$ as a product of the reaction of dioxygenyl hexafluorarsenate(V) with bromine

$$2O_2^+ \cdot AsF_6^- + 3Br_2 \rightarrow 2Br_3^+ \cdot AsF_6^- + 2O_2$$

and by the reaction of bromine pentafluoride, bromine and arsenic pentafluoride. The compound has a chocolate-brown colour, and in solution in HSO_3F has the same two bands in the absorption spectrum at 300 and 375 nm as were obtained by oxidising bromine to Br_3^+ with $S_2O_6F_2$.

7.3.3 The trichlorine cation Cl_3^+

Although conductimetric and absorption spectra measurements have given no evidence for the formation of the Cl_3^+ ion in solution, even in the super acid solvent $HSO_3F—SbF_5—3SO_3$ [29], it has been possible to prepare a solid compound of Cl_3^+, although it is only stable at low temperatures.

The yellow solid $Cl_3^+AsF_6^-$ has been prepared from chlorine, chlorine monofluoride and arsenic pentafluoride at -76 °C [12].

$$Cl_2 + ClF + AsF_5 \rightarrow Cl_3^+AsF_6^-$$

At room temperature, the salt is completely decomposed to give Cl_2, ClF and AsF_5. The Raman spectrum shows the active bands of the AsF_6^- ion, together with three relatively intense bands, at 490 (split to 485 and 492), 225 and 508 cm^{-1}, which are assigned to ν_1, ν_2 and ν_3 respectively of the bent Cl_3^+ cation. The assigned frequencies are very close to the vibrational frequencies of the isoelectronic SCl_2 molecule (514, 208 and 535 cm^{-1})[36] which has a bond angle of 103 degrees, and it is concluded that the Cl_3^+ cation has a similar structure. Using a simple valence force field gives good agreement for the observed frequencies of the Cl_3^+ cation with a bond angle of 100 degrees, a stretching force constant b = 2.5 mdyn $Å^{-1}$ and a bending force constant d = 0.36 mdyn $Å^{-1}$.

The Cl_3^+ cation has also been identified, by means of its Raman spectrum, in the yellow solid which precipitates out of a solution of Cl_2 and ClF in HF/SbF_5 at −76 °C. At room temperature, the Cl_3^+ cation is completely disproportionated in this solvent to give chlorine and ClF_2^+ salts. The salt $Cl_3^+BF_4^-$ is not formed from mixtures of chlorine, chlorine monofluoride and boron trifluoride at temperatures down to −130 °C[29].

7.3.4 Triatomic interhalogen cations

The following are the possible triatomic interhalogen cations: ClF_2^+, BrF_2^+, IF_2^+, Cl_2F^+, Br_2F^+, I_2F^+, $ClBrF^+$, $ClIF^+$, $BrIF^+$, $BrCl_2^+$, ICl_2^+, Br_2Cl^+, I_2Cl^+, $BrICl^+$, IBr_2^+ and I_2Br^+, assuming, as seems reasonable, that the least electronegative halogen occupies the central position, where it carries a formal positive charge. Of these 16 cations, only ClF_2^+, BrF_2^+, IF_2^+, Cl_2F^+ and ICl_2^+ have so far been definitely established, and crystal structure determinations have been published only for ICl_2^+, BrF_2^+ and ClF_2^+. All three ions have been found to have angular structures in accordance with the fact that the valence shell of the central atom has four electron-pairs which would be expected to have a tetrahedral arrangement.

7.3.4.1 ClF_2^+

Adducts of ClF_3 with AsF_5, SbF_5 and BF_3 have been known for some time[1], and more recently it was established by infrared and Raman spectral studies that these compounds are best formulated as salts of the ClF_2^+ cation, e.g. $ClF_2^+AsF_6^-$ [18,19]. The spectroscopic data indicate a bent structure. Vibrational frequencies and force constants are listed in Table 7.2. The strong splitting of ν_2 of the AsF_6^- anion, and its appearance as a strong band in the infrared as well as in the Raman spectrum contrary to the mutual exclusion rule, has been attributed to a lowering of symmetry of the octahedral AsF_6^- as a consequence of fluorine bridging between AsF_6^- and the ClF_2^+ cation[19]. Very recently, the crystal structures of $ClF_2^+SbF_6^-$ and of $ClF_2^+AsF_6^-$ have been determined[22,37]. In the former, the ClF_2^+ ion has a bond angle of 95.9 degrees and a bond length of 1.58 Å, while for the latter compound the bond angle and the bond length, respectively, have the rather different values of 103.2 degrees and 1.54 Å. We note that the shorter bond length is found with the larger bond angle, in agreement with the predictions of the valence-shell electron-pair repulsion theory[38], according to which electron-pairs of

Table 7.2 Vibrational frequencies (cm^{-1}) and force constants of the ClF_2^+ and BrF_2^+ cations

	$ClF_2^+SbF_6^-$ *Raman*	$ClF_2^+AsF_6^-$ *Raman*	$ClF_2^+AsF_6^-$ *Infrared*	$ClF_2^+BF_4^-$ *Raman*	$ClF_2^+BF_4^-$ *Infrared*		BrF_2^+ *Raman*	SbF_6^- *Infrared*	BrF_2^+ *Raman*	AsF_6^- *Infrared*
ν_1	{805 809	{806 809	810	{788 798	798	ν_1	705	705	706	713
ν_2	387	384		{373 394		ν_2	362	—	360	—
ν_3	830	821	818	808	813	ν_3	702	692	703	698
Simple valence force field force constants (mdyn $Å^{-1}$) f	4.8	4.7		4.6		Modified valence force field constants $f_{r\alpha} = 0$ and $\alpha = 95$ degrees f_r	4.60			
d	0.63	0.62		0.61		f_{rr}	0.21			
						f_α	0.47			

shorter stronger bonds are closer together and repel each other more than those of weaker longer bonds. There is strong fluorine bridging between the anion and the cation in both cases, and the two fluorine bridges formed by each ClF_2^+ give rise to a very approximately square coordination of fluorine around chlorine, which is indeed the geometry predicted by the valence-shell electron-pair repulsion theory for AX_4E_2 coordination, where X is a ligand and E a lone pair[38] (Figure 7.1).

7.3.4.2 BrF_2^+

The 1:1 adduct of BrF_3 with SbF_5 has been shown by x-ray crystallography[39] to contain BrF_2^+ and SbF_6^- ions held together by fluorine bridging in such a way that bromine acquires a very approximately square-planar configuration. Each bromine atom has two fluorine atoms at 1.69 Å and making an angle of 93.5 degrees at bromine, and two other neighbouring fluorine atoms at 2.29 Å which complete a very distorted square coordination around bromine (Figure 7.1).

The infrared and Raman spectra of $BrF_3{\cdot}SbF_5$, $BrF_3{\cdot}AsF_5$ and $(BrF_3)_2GeF_4$ have been reported very recently[40,41]. The vibrational frequencies given by Christe and Schack[40] for the BrF_2^+ ion are given in Table 7.2. There is, however, some disagreement between this work and that of Surles *et al.*[41], and the latter authors tentatively assign a frequency of 308 cm^{-1} to the bending mode ν_2. The bands in the spectrum of $(BrF_3)_2GeF_4$, that are

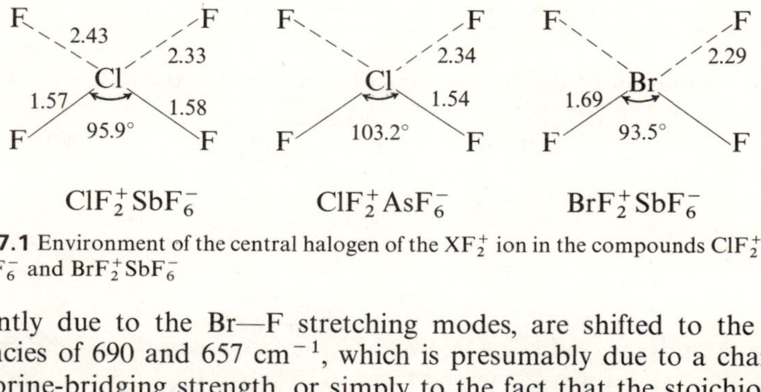

$ClF_2^+SbF_6^-$ $ClF_2^+AsF_6^-$ $BrF_2^+SbF_6^-$

Figure 7.1 Environment of the central halogen of the XF_2^+ ion in the compounds $ClF_2^+SbF_6^-$, $ClF_2^+AsF_6^-$ and $BrF_2^+SbF_6^-$

apparently due to the Br—F stretching modes, are shifted to the lower frequencies of 690 and 657 cm^{-1}, which is presumably due to a change in the fluorine-bridging strength, or simply to the fact that the stoichiometry necessitates a different structure for the crystalline solid. It is interesting to note that Surles *et al.*[41] found an even larger shift to 635 and 625 cm^{-1} for the stretching frequencies of the BrF_2^+ ion in solution in BrF_3. They attributed this shift to strong solvation of the BrF_2^+ ion.

The electrical conductivity of liquid bromine trifluoride[42] (specific conductance = 8×10^{-3} ohm^{-1} cm^{-1}) may at least partly be attributed to the self-ionisation $2BrF_3 \leftrightarrows BrF_2^+ + BrF_4^-$.

7.3.4.3 IF_2^+

The salts IF_2^+ AsF_6^- and $IF_2^+SbF_6^-$ have been prepared from IF_3 and AsF_5, and from IF_3 and SbF_5 in AsF_5 as solvent at -70 °C[43]. The compound

$IF_2^+SbF_6^-$ is stable to 45 °C, and the solid gives two broad overlapping ^{19}F n.m.r. signals whose relative intensities were estimated to be 1 : 2.6, and which were assumed to arise, therefore, from fluorine on iodine and fluorine on antimony respectively. $IF_2^+AsF_6^-$ was found to be stable only to −20 °C.

7.3.4.4 ICL_2^+

Raman spectra of the adducts $AsF_5 \cdot 2ClF$ and $BF_3 \cdot 2ClF$ have established that these compounds contain the unsymmetrical $ClClF^+$ cation[12], and not the symmetrical $ClFCl^+$ cation previously reported on the basis of the infrared spectrum alone[20]. The vibrational frequencies and assignments are listed in Table 7.3, and as with the salt $ClF_2^+AsF_6^-$, the splitting of ν_2 of AsF_6^- and its appearance in the infrared spectrum can reasonably be attributed

Table 7.3 Vibrational frequencies and assignments of the Cl_2F^+ cation

Assignment (cm^{-1})	$Cl_2F^+AsF_6^-$ *Raman*	$Cl_2F^+AsF_6^-$ *Infrared*	$Cl_2F^+BF_4^-$ *Raman*	$Cl_2F^+BF_4^-$ *Infrared*
ν_1(Cl—F str)	744		743	
ν_2(Cl—Cl str)	528 535	527 535	516 540	511* 519 528 532
ν_3(bend)	293 299	293	296	

*These bands contain ν_4 of BF_4^-, which appears at 519, 529 cm^{-1} in the ClF_2^+ salt, as well as ν_2 of the Cl_2F^+ cation

Table 7.4 Variation of the frequencies of ν_1 and ν_2 for the AsF_6^- ion with the nature of the cation

	$ClF_2^+AsF_6^-$	$Cl_2F^+AsF_6^-$	$Cl_3^+AsF_6^-$
ν_1	693	685	674
ν_2	544 602	563 581	571

to fluorine bridging. Table 7.4 shows how the frequency ν_1 and the splitting of ν_2 of the AsF_6^- ion vary in the series $Cl_3^+AsF_6^-$, $Cl_2F^+AsF_6^-$ and $ClF_2^+AsF_6^-$, in which an increasingly strong fluorine bridging is expected.

The Cl_2F^+ ion appears to be unstable in solution, and was found to be completely disproportionated in SbF_5/HF even at −76 °C.

$$2Cl_2F^+ = ClF_2^+ + Cl_3^+$$

The Cl_3^+ cation disproportionates further at room temperature to give chlorine and ClF_2^+ [17].

7.3.4.5 ICl_2^+

X-ray crystallographic investigations[21] of the adducts of ICl_3 with $SbCl_5$ and $AlCl_3$ have shown that they may be regarded as ionic compounds, i.e.,

$ICl_2^+SbCl_6^-$ and $ICl_2^+AlCl_4^-$, although there is considerable interaction between the two ions via two bridging chlorines, which gives an approximately square-planar arrangement of four chlorines around the iodine atom similar to the arrangement of fluorines around bromine and chlorine in $BrF_2^+{\cdot}SbF_6^-$ and $ClF_2^+{\cdot}SbF_6^-$. The bond angle and bond length for ICl_2^+ were found to be 92.5 degrees and 2.31 Å in $ICl_2^+{\cdot}SbCl_6^-$, and 96.7 degrees and 2.28 Å in $ICl_2^+AlCl_4^-$.

The electrical conductivity of liquid ICl_3 (specific conductance = 9.85 $\times 10^{-3}$ ohm^{-1} cm^{-1})[44] can be attributed to the self-ionisation

$$2ICl \rightleftarrows ICl_2^+ + ICl_4^-$$

7.3.4.6 I_2Cl^+

There is no certain evidence for the I_2Cl^+ cation, but presumably the electrical conductivity of liquid ICl (specific conductance = 4.52×10^{-3} ohm^{-1} cm^{-1} at 31 °C)[44], which has previously been ascribed to the self-ionisation

$$2ICl \rightleftarrows I^+ + ICl_2^-$$

is in fact due to a self-ionisation which produces the I_2Cl^+ ion according to the equation

$$3ICl \rightleftarrows I_2Cl^+ + ICl_2^-$$

The I_2Cl^+ cation, however, is possibly extensively disproportionated to give the known I_3^+ and ICl_2^+ cations

$$2I_2Cl^+ \rightleftarrows I_3^+ + ICl_2^+$$

7.4 PENTA- AND HEPTA-ATOMIC CATIONS

Solutions of I_3^+ in 100% H_2SO_4, or in fluorosulphuric acid, dissolve at least one mole of iodine per mole of I_3^+, and a new absorption spectrum is obtained with bands at 470, 340 and 270 nm[34,2]. At the same time, there is no change in either the conductivity or the freezing point of the solutions. Consequently, it has been concluded that I_3^+ reacts with iodine to form the I_5^+ cation

$$I_3^+ + I_2 \rightleftharpoons I_5^+$$

There is no structural information on the I_5^+ cation, but a plausible structure would be

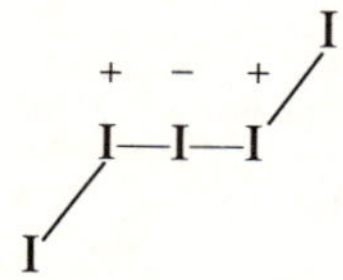

Some further iodine will dissolve in a solution of I_5^+, and there may possibly be some formation of I_7^+.

There is no evidence for the formation of the Br_5^+ or Cl_5^+ cations, and they are presumably unstable with respect to the triatomic cation and the diatomic molecule.

Chlorine pentafluoride forms 1:1 adducts with AsF_5 and SbF_5, and a preliminary interpretation of the Raman spectra of these compounds indicates that they can probably be formulated as $ClF_4^+AsF_6^-$ and $ClF_4^+SbF_6^-$ [45]. Bromine pentafluoride forms the adducts $BrF_5 \cdot 2SbF_5$ and $BrF_5 \cdot SO_3$ [46]. These may presumably be formulated as $BrF_4^+Sb_2F_{11}^-$ and $BrF_4^+SO_3F^-$, although the latter compound might be the covalent BrF_4SO_3F. Iodine pentafluoride forms 1:1 adducts with SbF_5 [47] and PtF_5 [48]. There has been a

Table 7.5 Vibrational frequencies and assignments of the IF_6^+ cation in $IF_6^+AsF_6^-$

Assignment	Observed frequency cm^{-1} Raman	Infrared
$\nu_1(A_{1g})$	708	
$\nu_2(E_g)$	732	
		797
$\nu_3(F_{1u})$		790
$\nu_4(F_{1u})$		404
$\nu_5(F_{2g})$	340	

Valence force constants (mdyn $Å^{-1}$) f_r, 5.60; f_{rr}, −0.06; $f_{rr'}$, 0.27

preliminary report[49] of the crystal structure of $IF_5 \cdot SbF_5$ which is shown to be $IF_4^+ \cdot SbF_6^-$. The IF_4^+ ion has a structure like SF_4, with two fluorines occupying the axial positions of a trigonal-bipyramid, and two fluorines and a lone pair occupying the equatorial positions. The electrical conductivity of liquid IF_5 (specific conductance $= 2.30 \times 10^{-5}$ ohm^{-1} cm^{-1}) [44] can presumably be attributed to the self-ionisation

$$2IF_5 = IF_4^+ + IF_6^-$$

In the Raman spectrum of $IF_4^+SbF_6^-$, nine lines have been assigned to the IF_4^+ cation[50], which is consistent with the C_{2v} structure found by x-ray crystallography.

Iodine heptafluoride has been shown to form the complexes $IF_7 \cdot AsF_5$ and $IF_7 \cdot 3SbF_5$ [51]. The Raman and infrared spectra of $IF_7 \cdot AsF_5$ in the solid state, and in solution in HF, show that it should be formulated as $IF_6^+AsF_6^-$ [52, 53]. Presumably, $IF_7 \cdot 3SbF_5$ is the ionic compound $IF_6^+Sb_3F_{16}^-$. Vibrational assignments and force constants are listed in Table 7.5.

References

1. Gutman, V. (Ed.). (1967). *Halogen Chemistry*, Vol. 1, pp.28 and 259 (London and New York: Academic Press)
2. Gillespie, R. J. and Milne, J. B. (1966). *Inorg. Chem.*, **5**, 1577
3. Arotsky, J. and Symons, M. C. R. (1962). *Quart. Rev. Chem. Soc.*, **16**, 282
4. Kemmitt, R. D. W., Murray, M., McRae, V. M., Peacock, R. D. and Symons, M. C. R. (1968). *J. Chem. Soc.*, 862
5. Gillespie, R. J. and Malhotra, K. C. (1969). *Inorg. Chem.*, **8**, 1751

6. Gillespie, R., J., Milne, J. B. and Morton, M. J. (1968). *Inorg. Chem.*, **7**, 2221
7. Gillespie, R. J. and Morton, M. J. (1969). *J. Mol. Spec.*, **30**, 178
8. Adhami, G. and Haslem, H. (1970). *J. Electroanal. Chem.*, **26**, 363
9. Gillespie, R. J. (1968). *Accounts Chem. Res.*, **1**, 202
10. Gillespie, R. J. and Passmore, J. (1971). *Accounts Chem. Res.*, in press
11. Gillespie, R. J. and Morton, M. J. (1968). *Chem. Commun.*, 1565
12. Gillespie, R. J. and Morton, M. J. (1970). *Inorg. Chem.*, **9**, 811
13. Olah, G. A. and Comisarow, M. B. (1968). *J. Amer. Chem. Soc.*, **90**, 5033
14. Olah, G. A. and Comisarow, M. B. (1969). *J. Amer. Chem. Soc.*, **91**, 2172
15. Eachus, R. S., Sleight, T. P. and Symons, M. C. R. (1969). *Nature (London)*, **222**, 769
16. Christe, K. O. and Muirhead, J. S. (1969). *J. Amer. Chem. Soc.*, **91**, 7777
17. Gillespie, R. J. and Morton, M. J. (1971). *Inorg. Chem.*, in press
18. Christe, K. O. and Sawodny, W. (1967). *Inorg. Chem.*, **6**, 313
19. Gillespie, R. J. and Morton, M. J. (1970). *Inorg. Chem.*, **9**, 616
20. Christe, K. O. and Sawodny, W. (1969). *Inorg. Chem.*, **8**, 212
21. Vonk, C. G. and Wiebenga, E. H. (1959). *Acta Crystallogr.*, **12**, 859
22. Edwards, A. J. and Sills, R. J. C. (1970). *J. Chem. Soc.*, 2697
23. Edwards, A. J. and Jones, G. R. (1969). *J. Chem. Soc.*, 1467
24. Edwards, A. J., Jones, G. R. and Sills, R. J. C. (1968). *Chem. Commun.*, 1527
25. Gillespie, R. J. and Malhotra, K. C. (1969). *Inorg. Chem.*, **8**, 1751
26. Aynsley, E. E., Greenwood, N. N. and Wharmby, D. H. W. (1963). *J. Chem. Soc.*, 3369
27. Ruff, O., Graf, H., Heller, W. and Knock. (1906). *Chem. Ber.*, **39**, 4310
28. McRae, V. M. (1966). *Ph.D. Thesis*, University of Melbourne
29. Gillespie, R. J. and Morton, M. J. (1971). *Inorg. Chem.*, in press
30. Eachus, R. S., Sleight, T. P. and Symons, M. C. R. (1969). *Nature, (London)*, **222**, 769
31. Christe, K. O. and Muirhead, J. S. (1969). *J. Amer. Chem. Soc.*, **91**, 7777
32. Masson, I. (1938). *J. Chem. Soc.*, 1708
33. Arotsky, J., Mishra, H. C. and Symons, M. C. R. (1962). *J. Chem. Soc.*, 2582
34. Garrett, R. A., Gillespie, R. J. and Senior, J. B. (1965). *Inorg. Chem.*, **4**, 563
35. Glemser, O. and Smalc, A. (1969). *Angew. Chem. Int. Ed. Engl.*, **8**, 517
36. Siebert, H. (1966). *Anwendungen der Swingungsspektroscopie in der Amorganischen Chemie*, 48, (Berlin: Springer-Verlag)
37. Gillespie, R. J. and Morton, M. J. (1970). *Inorg. Chem.*, **9**, 616
38. Gillespie, R. J. (1970). *J. Chem. Educ.*, **47**, 18
39. Edwards, A. J. and Jones, G. R. (1969). *J. Chem. Soc.*, 1467
40. Christe, K. O. and Schack, C. J. (1970). *Inorg. Chem.*, **9**, 2296
41. Surles, T., Hyman, H. H., Quarterman, L. A. and Popov, A. I. (1970). *Inorg. Chem.*, **9**, 2726
42. Banks, A. A., Emeléus, H. J. and Woolf, A. A. (1949). *J. Chem. Soc.*, 2861
43. Schmeisser, M., Ludovici, W., Naumann, D., Sartori, P. and Scharf, E. (1968). *Chem. Ber.*, **101**, 4214
44. Greenwood, N. N. and Emeléus, H. J. (1950). *J. Chem. Soc.*, 987
45. Christe, K. O. and Pilipovich, D. (1969). *Inorg. Chem.*, **8**, 391
46. Schmeisser, M. and Pammer, E. (1957). *Angew. Chem.*, **69**, 281
47. Woolf, A. A. (1950). *J. Chem. Soc.*, 3678
48. Bartlett, N. and Lohman, D. H. (1962). *J. Chem. Soc.*, 8253; (1964). *J. Chem. Soc.*, 619
49. Baird, H. W. and Giles, H. F. (1969). *Acta Crystallogr.*, **A25**, S115
50. Shamir, J. and Yaroslavsky, I. (1969). *Israel J. Chem.*, **7**, 495
51. Seel, F. and Detmer, O. (1958). *Angew. Chem.*, **70**, 163; (1959). *Z. Anorg. Allg. Chem.*, **301**, 113
52. Christe, K. O. and Sawodny, W. (1967). *Inorg. Chem.*, **6**, 1783
53. Christe, K. O. (1970). *Inorg. Chem.*, **9**, 2801

8
Oxides and Oxyacids of the Halogens

B. J. BRISDON
University of Bath

8.1 INTRODUCTION

Considerable advances have been made in this area over the last few years. Most notable of these is the pioneering work by Appelman on perbromates which had been rationalised out of existence for so long.

The impact of physical methods has been considerable over the whole of this area and these techniques have proved invaluable in the study of unstable molecules. Results of such studies will be given prominence in this review.

8.2 OXYGEN FLUORIDES

Eight binary compounds of oxygen and fluorine have been reported, namely O_2F, OF, OF_2, O_2F_2, O_3F_2, O_4F_2, O_5F_2 and O_6F_2. Of these, the radicals O_2F and OF, as well as the more stable compounds OF_2 and O_2F_2, have been well characterised. The composition and existence (as molecular entities) of the higher oxygen fluorides is still uncertain.

These compounds are of theoretical interest because of their structural and spectral properties which indicate unusual bonding in some instances, and practically, as potential rocket fuels.

Recent reviews of this subject are given in references 1–3.

8.2.1 Dioxygen monofluoride O_2F

This free radical was first characterised by the matrix isolation technique[4–6]. O_2F radicals were produced and trapped during the photolysis of OF_2–O_2 or F_2–O_2 mixtures in O_2, N_2 or Ar matrices at 4 K. Infrared studies on isotopic mixtures indicated that O_2F was a bent triatomic molecule with a strong O—O bond and a weak O—F linkage. For normal coordinate analysis purposes the molecular parameters of O_2F_2 were assumed to apply to O_2F. Force constants and infrared data are tabulated in Table 8.1.

At low temperatures on diffusion, the O_2F radicals form loosely bound aggregates $(O_2F)_n$. In the infrared spectrum of the new species only the O—O stretching frequency at 1510 cm^{-1} is noticeably different from the bands observed in the isolated O_2F radical, which suggests that the intermolecular forces must be very weak. The apparent lack of reactivity of O_2F

Table 8.1 Force constants and infrared fundamentals of $^{16}O_2F$[6]

Force constants					*Infrared frequencies*
k_{OO} (mdyn Å^{-1})	k_{OF} (mdyn Å^{-1})	$k_{OO—OF}$ (mdyn Å^{-1})	k_{OOF} (mdyn Å rad^{-2})	$k_{OF—OOF}$ (mdyn rad^{-1})	(cm^{-1})
10.50	1.32	0.300	1.008	0.027	1495.0 ν_1 O—O stretch 584.5 ν_2 F—O stretch 376.0 ν_3 F—O—O bend

is consistent with the existence of nearly molecular oxygen in this radical as the intermolecular bond strength of O_4 is only 535 kJ mol^{-1}.

In view of the relative stability of O_2F, it is not surprising to find e.s.r. evidence of its formation under a variety of conditions. For example, the electron irradiation of liquid CF_4 containing traces of oxygen[7] the photolysis of O_2–F_2 mixtures[8], of liquid OF_2 or of OF_2–O_2 mixtures[9].

The last of these methods leads to a high concentration of O_2F radicals, probably due to the principal reactions

$$OF_2 \xrightarrow{h\nu} F + OF$$
$$F + O_2 \longrightarrow O_2F$$

while the slow decay of the O_2F e.s.r. signal may involve

$$2O_2F \rightarrow (O_4F_2) \rightarrow O_2F_2 + O_2$$

Malone and McGee[10] were able to measure some thermochemical properties of the dioxygen fluoride radical and hence verify the bonding description suggested by infrared results. Direct measurement of the appearance potentials of OF^+ and O_2F^+ obtained by pyrolysis of O_2F_2 were used to calculate bond dissociation energies D_{O_2-F} and D_{O-OF} of 77 and 463 kJ mol^{-1} respectively.

8.2.2 The oxygen monofluoride radical OF

OF was postulated as an intermediate in the thermal and photolytic decomposition of OF_2; however, its transient existence necessitated use of the matrix isolation technique to confirm its formation[11]. Photolysis of a dilute mixture of OF_2 in solid argon at 4 K resulted in the development of a new infrared band at 1028.5 cm^{-1}, which showed an isotopic shift of 31.1 cm^{-1} when ^{18}O-enriched OF_2 was used. This compares well with the calculated value of 31.5 cm^{-1} expected for ^{18}OF.

Further work by Arkell[12] has shown that photolysis of OF_2—N_2O (or OF_2—CO_2) combinations in a nitrogen matrix at 4 K gives increased yields of OF radicals compared to undiluted OF_2. Diffusion at 29 K caused complete loss of OF radicals with the formation of O_2F_2 and some OF_2. In the absence of N_2O, diffusion of OF radicals produced mainly OF_2 by recombination of OF with F atoms. Thus it seems likely that a large accumulation of OF radicals is made possible by the presence of dinitrogen oxide which reacts with F atoms.

$$F + N_2O \rightarrow OF + N_2$$

In the absence of F atoms dimerisation of FO on diffusion would be expected.

The observation of OF radicals as precursors of OF_2 during photolysis of F_2—N_2O mixtures supports this mechanism and sets a lower limit of about 167 kJ mol^{-1} for the bond energy of the OF radical[9]. This is compatible with the thermochemical estimates of almost 210 kJ mol^{-1} but much higher than the value of 121 kJ mol^{-1} obtained from electron impact studies[13]. Since F_2 contains one more antibonding electron than OF, the O—F bond would be expected to be stronger than the F—F bond, and this is borne out by the force constants and bond lengths of the two molecules.

Table 8.2 Summary of OF molecular parameters

	Calculated[15]	*Experimental*[2, 11]
r_e (Å)	1.321	
ω_e (cm^{-1})	1211	>1028.5*
$\omega_e x_e$ (cm^{-1})	5.15	
B_e (cm^{-1})	1.104	
α_e (cm^{-1})	0.0097	
k_e (mdyn $Å^{-1}$)	7.5	5.4
I.P. (kJ mol^{-1})	1270±50	1270±30
μ (D)	−0.361	

*The experimental infrared absorption frequency refers to OF in an argon matrix[11], ω_e should be at a higher frequency. O'Hare and Wahl[15] suggest a correction of about 22 cm^{-1} by analogy with NF in two phases.

Consequently an electron impact value of less than 164 kJ mol^{-1} (D_{F_2}) does not seem acceptable.

In view of the importance of OF in many of the reactions of oxygen fluorides, and because of its intractable behaviour, several papers have dealt with calculations of its thermochemical properties. Molecular orbital calculations by Mortimer[14] yielded a value of 209 kJ mol^{-1} for D_{OF}, while a very recent paper by O'Hare and Wahl[15] gives computed values for the spectroscopic constants, bonding energy, ionisation potential, electron affinity, dipole and quadrupole moments of OF. Theoretical values are summarised in Table 8.2 and compared with experimental values where available.

8.2.3 Oxygen difluoride OF_2

A simple preparation of this the most stable oxygen fluoride has recently been described[16], in which fluorine is reacted with water in the presence of an alkali fluoride. Yields of over 70% have been reported in some instances, and the product was free from HF and SiF_4.

A reinvestigation of the infrared spectrum of gaseous OF_2 has been carried out by Nebgen and co-workers[17], who considered Fermi resonance in the interpretation of the infrared spectrum (ν_1 and $2\nu_2$, which both belong to the A_1 species of the C_{2v} point group, are within 6 cm^{-6} of each other). This effect was responsible for the complicated spectral patterns exhibited in the overtones and combinations observed, and also accounted for the splitting of ν_1 in the low temperature matrix spectrum of OF_2 [18]. A summary of the spectral parameters of OF_2 is given in Table 8.3.

The kinetics and mechanism of the thermal decomposition of F_2O have been thoroughly investigated[20–24]. The products of decomposition are mainly fluorine and oxygen in the stoichiometric ratio. The activation

Table 8.3 Spectral parameters of $^{16}OF_2$

	Fundamental infrared frequencies ν_1	ν_2	ν_3	*Reference*	*Force constants** k_1	k_2	k_3	k_4	*Reference*
Solid	925 915	461	821	18	4.1 ± 0.3	1.0 ± 0.2	0.25 ± 0.1	0.8 ± 0.1	18
Gas	928	461	831	17	3.95	0.81	0.14	0.72	19

k_1, k_2, k_3 and k_4 in mdyn Å^{-1} are used as abbreviations for f_d, the valence force constant, f_{dd} and $f_{\alpha d_0}$, the interaction force constants, and $f_{\alpha d_0^2}$, the deformation force constant respectively.

energy for the overall process is within the range 146 ± 21 kJ mol^{-1}, and OF radicals have been postulated as the intermediates which lead to the formation of oxygen and fluorine. Data obtained from shock tube studies[24] from 770–1390 K were accounted for by the scheme.

$$F_2O + Ar \overset{k_1}{\rightleftharpoons} F + OF$$
$$2OF \xrightarrow{k_2} 2F + O_2$$
$$2F + Ar \rightleftharpoons F_2 + Ar$$

with

$$k_1 = 10^{17.3}\,e^{-42.5/RT}\ cm^3\ mol^{-1}\ s^{-1} \text{ and } k_2 = 10^{21.1}\ cm^3\ mol^{-1}\ s^{-1}.$$

These results yield a value of 178.5 ± 17.1 kJ mol^{-1} for the bond dissociation energy (D_{FO-F}) of oxygen difluoride. A re-determination of the heat of formation (ΔH^0_{f298}) of OF_2 by constant pressure flame calorimetry gave a value of 24.52 ± 1.59 kJ mol^{-1} [25] and $\Delta H^0_{f0(OF_2)}$ is calculated to be 26.75 kJ mol^{-1} by means of the enthalpy functions for OF_2 [26]. In combination with D_{FO-F} this allows the calculation of D_{OF}.

$$D_{OF} = \Delta H^0_f(O, g) + 2\Delta H^0_f\,(F, g)^* - \Delta H^0_f\,(OF_2) - D_{FO-F}$$
$$= 195.2 \pm 17.1 \text{ kJ mol}^{-1}.$$

*The recently reported value of D°_{F2} of 129.2 ± 2.9 kJ mol^{-1} [27] has not been used because it is inconsistent with previous independent determinations. Confirmation is needed before such a value can be fully accepted.

Several of the chemical reactions of OF_2 have also received attention. In many of its chemical reactions, oxygen difluoride acts as a powerful fluorinating agent but with a higher activation energy than fluorine. At elevated temperatures it reacts simply as a 2:1 mixture of fluorine and oxygen. In a few instances photochemical reactions involving the addition of OF_2 are known. Solomon *et al.*[28] were able to demonstrate conclusively that the OF radical was an important intermediate in one such reaction. Labelled $^{17}OF_2$ and ordinary SO_3 produced peroxysulphuryl difluoride ($FSO_2O^{17}OF$) in which ^{17}O appeared exclusively in the –OF position. Alternatively, labelled $S^{17}O_3$ produced $FS^{17}O_3OF$ in which no ^{17}O appeared in the –OF position. Thus the hypofluorite oxygen must come only from OF_2. The possibility of an oxygen atom transfer mechanism can be rejected; in this case oxygen atoms would be expected to exchange with the sulphate oxygens. Consequently the mechanism suggested by previous workers[29] is confirmed

$$OF_2 \xrightarrow{h\nu} F + OF$$
$$F + SO_3 \rightarrow FSO_3$$
$$FSO_3 + OF \rightarrow FSO_2OOF$$

Another photochemical reaction involving FO radicals was studied by Guisberg *et al.*[30] in which fluorochlorinated ketones and OF_2 were irradiated with u.v. light to produce mixtures of fluoroxy compounds.

Anderson and Fox[31] reported an unusual addition reaction in which COF_2 added OF_2 at room temperature in the presence of a CsF catalyst to yield bis(trifluoromethyl)trioxide (CF_3OOOCF_3). The dependence of the reaction on the presence of an alkali metal fluoride and the readiness with which reaction occurred in the dark suggested an ionic mechanism

$$COF_2 + MF \rightarrow M^+OCF_3^-$$
$$OF_2 + CF_3O^- \rightarrow CF_3OOF + F^-$$
$$CF_3OOF + CF_3O^- \rightarrow CF_3OOOCF_3 + F^-$$
$$CF_3OOF + COF_2 \rightarrow CF_3OOOCF_3$$

Spectroscopic measurements were consistent with a structure containing a chain of three oxygen atoms, whereas the corresponding fluorine compound is unknown.

Attempts to prepare salts of the OF^+ cation by the reaction of OF_2 with Lewis acids such as AsF_5 and SbF_5 resulted in the formation of dioxygenyl salts[32].

$$4OF_2 + 2AsF_5 \xrightarrow{200\,^\circ C} 2O_2AsF_6 + 3F_2$$

As the temperature at which the reaction was carried out was above the decomposition temperature of OF_2, it seems likely that the reaction involves molecular oxygen, fluorine and the Lewis acid. This is supported by the production of dioxygenyl salts using a mixture of fluorine, oxygen and AsF_5 under similar conditions to those employed above.

Henrici, Lin and Bauer[33] extended their studies on the thermal decomposition of F_2O to include the kinetics and mechanism of the F_2O–CO reaction in shock tubes between 800 and 1400 K. Besides the formation of CO_2 a significant part of the carbon monoxide reacted to form F_2CO.

A preliminary report on the thermal and photochemical reactions of

OF_2 with COS in the temperature range 50–90 °C has also appeared[34]. The main course of the thermal reaction is indicated by the equation,

$$F_2O + COS \rightarrow F_2SO + CO$$

with small quantities of SOF_4, SO_2F_2, COF_2 and SiF_4 formed at high temperatures. The photochemical reaction produced CO and SOF_2 initially, and in the presence of excess F_2O further reaction of SOF_2 occurred with the formation of SOF_4 and SO_2F_2.

8.2.4 Dioxygen difluoride O_2F_2

A new method of preparing O_2F_2 has been reported[35] in which radiolysis of liquid mixtures of O_2 and F_2 at 77 K with 3 MeV 'bremsstrahlung' produces a mixture of O_2F_2 and $(O_2F)_n$ which can be separated to yield pure O_2F_2 as a yellow diamagnetic solid melting at 119 K (9 K higher than previously reported).

The infrared spectrum of the solid at 77 K[36] indicates that F_2O_2 has the same C_{2v} symmetry as H_2O_2, but with a much shorter O—O distance. It is initially surprising to find that the O—O stretching vibration does not appear around 1500 cm^{-1}, as in O_2F. However, normal coordinate analysis explains the low O—O stretching frequency (1306 cm^{-1}) as a consequence of the large values found for the off-diagonal force constants. The low value found for the O—F stretching constant is in keeping with the considerably longer O—F bonds in O_2F_2 compared with those found for OF_2 [19]. A comparison of structural parameters of O_2F_2 and related molecules is given in Table 8.4.

Table 8.4 Structural parameters of O_2F_2 and related molecules[36]

Molecule	$r_{(O—O)}$Å	$r_{(O—X)}$Å	k_{OO} mdyn Å^{-1}	k_{OF} mdyn Å^{-1}
OF_2	—	1.41	—	3.95
O_2F_2	1.22	1.58	10.25	1.50
O_2F	—	—	10.50	1.32
H_2O_2	1.48	0.95	4.6	—
O_2	1.21	—	11.4	—

The results of semi-empirical calculations on O_2F and O_2F_2 confirm earlier descriptions of their bonding. The strength of the O—O bond is explained by the interaction of fluorine with the unpaired π^* orbitals on the oxygens to form a three centre molecular orbital which is still antibonding with respect to O—O but bonding with respect to O—F. Thus the overall O—O bond order remains virtually the same as in O_2; the O—F bonds are weak compared with a normal electron pair bond.

Further experimental evidence in favour of such a bonding scheme is provided by the results of mass spectrometric studies[37]. Thus the bond dissociation energy of the O—O bond in O_2F_2 was found to be 432.6 kJ mol^{-1} whereas the energy required to break an O—F bond in the compound was estimated to be only about 75 kJ mol^{-1}. ^{19}F n.m.r. measurements clearly indicate that the F nuclei of O_2F_2 are in a very different electronic environment from those of F_2 and OF_2. The chemical shift of liquid O_2F_2 [38,39] is further downfield than any other simple fluoride. Turner *et al.*[9] made e.s.r. and n.m.r. measurements on the same samples of O_2F_2 in CF_3Cl and found that the paramagnetic species O_2F, (invariably present because of the low O_2—F bond dissociation energy), did not significantly affect the ^{19}F chemical shift.

O_2F_2 reacts readily at −150 °C or below, and even at these temperatures very violent reactions may occur with hydrogen containing materials. The interaction of O_2F_2 with certain Lewis acids, with oxides of sulphur and nitrogen and with perfluorocompounds are some of the more tractable of its reactions.

Solomon[40] showed that O_2F_2 interacted with BF_3 or PF_5 to produce compounds containing the dioxygenyl cation (O_2^+). Subsequent work by the same author[41] using isotopic tracer methods indicated that the O_2F radical is an intermediate in both the preparation and decomposition of O_2BF_4.

The same final product was obtained by Groetschel *et al.*[35], but by using a mixture of oxygen fluorides obtained from irradiated mixtures of F_2 and O_2, they reported evidence for the formation of other less stable compounds, tentatively suggested to be O_4BF_4 (formed from O_4F and BF_3) and possibly even O_6BF_4.

Bantov *et al.*[42] extended this reaction to other Lewis acids.

Solomon *et al.*[43,44] have also investigated the reaction of O_2F_2 with sulphur oxides. At −160 °C or below the main reaction of O_2F_2 with SO_2 is represented by the equation

$$SO_2 + O_2F_2 \rightarrow F_2SO_2 + O_2$$

Isotopic studies showed that this is a simple fluorination reaction. Besides sulphuryl fluoride, smaller quantities of fluorosulphuryl hypofluorite (peroxysulphuryl difluoride) are formed via an OOF intermediate together with lesser quantities of disulphuryl fluoride and FSO_2OF. The corresponding reaction with SO_3 yields equimolar amounts of FSO_3F and $(SO_3F)_2$.

Dinitrogen tetroxide reacts with O_2F_2 at low temperatures to form a mixture of nitrosyl fluoride, nitryl fluoride and nitroxyl fluoride[45]. With hexafluoropropylene O_2F_2 is reported to yield a mixture of $CF_3CF_2CF_2OOF$ and $CF_3CF(OOF)CF_3$ [45].

8.2.5 Higher oxygen fluorides

8.2.5.1 O_3F_2

Recent work on O_3F_2 has not confirmed its existence as a distinct molecular entity. Cryogenic mass spectrometric results were interpreted[37] in favour of a model consisting of loosely bound O_2F and OF radicals. Metz and her co-workers[38] suggested that 'O_3F_2' was O_2F_2 containing 'interstitial' oxygen, whereas Solomon *et al.*[39] favoured a system described by

$$(O_2F_2, O_4F_2) \rightleftharpoons 2OOF \rightarrow O_2F_2 + O_2$$

A ^{17}O n.m.r. spectrum of three lines was obtained for O_3F_2 [39], the largest of which was attributable to O_2F_2, and the other two of equal intensity to $(OOF)_n$. None of the models possible for molecular O_3F_2 are consistent with this result. Metz *et al.*[38] reported a single line in the ^{19}F n.m.r. spectrum of O_3F_2 in accordance with their model, but Solomon *et al.*[46] managed to resolve their signal into two separate lines. Further, reaction of O_3F_2 with BF_3 produced O_2BF_4 without significant evolution of oxygen as expected from the 'interstitial' oxygen model. Thus Metz's model appears untenable at this time.

8.2.5.2 O_4F_2

Arkell observed bands that he attributed to O_4F_2 in his matrix studies on the O_2F radical[4], and Spratley *et al.*[5] assigned bands at 376, 586 and 1510 cm^{-1} to $(O_2F)_n$, formed by polymerisation of O_2F radicals. A similar product is obtained during the radiolysis of a 3:1 F_2—O_2 mixture, and from a consideration of its physical and spectral properties, Goetschel *et al.*[35] suggested that $(O_2F)_n$ is best described as an equilibrium mixture of O_2F and O_4F_2 with only a low equilibrium concentration of O_2F in O_4F_2 near the melting point.

8.2.5.3 O_5F_2 *and* O_6F_2

Sufficient data is still not available to reach any definite conclusions about the nature of O_5F_2 or O_6F_2.

8.3 CHLORINE OXIDES

The properties of eight binary compounds of chlorine and oxygen are discussed of which ClO, ClOO and Cl_2O_3 are short lived under normal experimental conditions. ClO and ClOO have been examined at low temperatures in an inert gas matrix and by rapid spectroscopic techniques suited to their transient existence.

All of the chlorine oxides are reactive and powerful oxidising agents, generally more stable than their fluorine analogues.

Reference 1 reviews the chemistry and general properties of these oxides.

8.3.1 Chlorine monoxide radical ClO

The ClO radical can be produced in a variety of chemical reactions including
(a) thermal decomposition of ClO_2 or ClO_3,
(b) decomposition of $FClO_3$ in an electrical discharge,
(c) microwave or radio-frequency discharge through Cl_2–O_2 mixtures,
(d) reaction of Cl atoms with ClO_2 or O_3 at 300 K,
(e) gas phase photolysis of Cl_2O, ClO_2 or Cl_2–O_2 mixtures.

The mechanism suggested for the flash photolysis of Cl_2–O_2 mixtures[47] involved the initial formation of the ClOO radical.

$$Cl_2 \xrightarrow{h\nu} Cl + Cl \tag{8.1}$$
$$Cl + O_2 + M \rightleftharpoons ClOO + M \tag{8.2}$$
$$Cl + ClOO \rightarrow 2ClO \tag{8.3}$$
$$Cl + ClOO \rightarrow Cl_2 + O_2 \tag{8.4}$$

Reaction (8.4) is faster than (8.3) and consequently only a small fraction of ClOO radicals reacts to give ClO.

Kinetic studies[48] revealed that the decay reaction of ClO radicals is second order in (ClO) over the temperature range 294–495 K, and the probable decay mechanism again involves short lived peroxy radicals.

$$ClO + ClO \rightleftharpoons Cl + ClOO$$
$$ClOO + Cl \rightarrow Cl_2 + O$$
$$ClOO + M \rightarrow Cl + O_2 + M$$

Direct spectral evidence for the peroxy radical was obtained by Morris and Johnston[49] in their absorption spectroscopy studies, and a further very comprehensive report by the same authors[50] on the kinetics of the Cl_2–O_2

Table 8.5 Molecular constants of ^{35}ClO

	Microwave[53]	E.S.R.[52]	*Other references*
r_e (Å)	1.569 ± 0.0010	1.571	
ω_e (cm^{-1})	616 ± 61		~970 [54]
B_O (cm^{-1})	0.6206	0.622	
μ (D)	1.239 ± 0.010	1.26	
eqQ(MHz)	-87.02 ± 1.53	-88 ± 6	
I.P. (kJ mol^{-1})			46.64 [55]
D_O (ClO) (kJ mol^{-1})			265 ± 10 [56]

reaction was interpreted in terms of a complex mechanism involving ten elementary reactions and four intermediates, Cl, ClOO, ClO and Cl_2O_2, two of which (ClOO and ClO) were observed directly. Seven of the ten elementary rate constants were evaluated.

$$Cl_2 + h\nu \rightarrow Cl_2 + Cl$$
$$Cl + O_2 + M \rightleftharpoons ClOO + M$$
$$Cl + ClOO \rightleftharpoons ClO + ClO$$
$$Cl + ClOO \rightarrow Cl_2 + O_2$$
$$ClO + ClO + M \rightleftharpoons Cl_2O_2 + M$$
$$Cl_2O_2 + M \rightarrow Cl_2 + O_2 + M$$

Clyne and Cruse[51] suggested that the mechanism involving Cl_2O_2 is important in the high pressure region in which the rate of decay of ClO is first order in third body concentration, whereas at low pressures the decay mechanism involving the peroxy radical ClOO occurs.

ClO is a relatively simple molecule with a longer lifetime than OF and, as such, is amenable to many physical techniques which can provide a great deal of physical and chemical information.

The ground state of ClO is expected to be

$$(z\sigma)^2(y\sigma)^2(x\sigma)^2(w\pi)^4(v\pi)^3 \quad 2\pi_i$$

and an examination of the gas phase electron resonance spectrum by Carrington *et al.*[52] established that the ground state is $^2\pi_{\frac{3}{2}}$ (inverted doublet) and led to values for several of the molecular constants of ClO. Confirmation of many of these values was obtained by a detailed examination of the microwave spectra of this radical[53]. A summary of these values is given in Table 8.5.

Low temperature photolysis of matrix isolated Cl_2O produced new infrared spectral features near 960 and 375 cm^{-1} [54, 57]. A combination of growth plots during photolysis, concentration dependence and chlorine isotopic shifts were used to show that three new types of molecule were produced.

A group of bands centred around 945 cm^{-1} were assigned to $(ClO)_2$, which indicates that pairs of ClO radicals are only weakly linked, unlike the corresponding fluorine compound. A second molecule, ClClO, formed by the weak interaction of a Cl atom with a ClO radical, was also recognised and assigned an angular structure.

8.3.2 The chlorine peroxide radical ClOO

The existence of the ClOO radical as an intermediate in the formation and decay of ClO has already been discussed. Arkell and Schwager[58] were able to isolate and identify this peroxy radical, which they produced by the photolysis of ClO_2 in an argon matrix or chlorine in an argon matrix at 4 K.

Table 8.6 I.R. fundamentals and force constants for ClOO

Fundamentals for	$^{35}Cl^{16}O^{16}O$	*Assignments*	*Force constants** (*units as in Table* 8.1)	
ν_1	1440.8	O—O stretch	k_{OO}	9.654
ν_2	407	Cl—O stretch	k_{OCl}	1.290
ν_3	373	Cl—O—O bend	k_{OOCl}	1.038
			$k_{OCl—OOCl}$	0.545
			$k_{OO—OCl}$	0.070

*These values were calculated using r_{OO} = 1.23 Å, r_{OCl} = 1.83 Å and α_{OOCl} = 110 degrees.

Three fundamental infrared active absorptions were found for ClOO and force constant calculations indicated that the bonding in this molecule is quite similar to that found in O_2F. (cf. Table 8.1). An e.s.r. study by Symons *et al.*[59] showed that the g-tensor values for ClOO were very similar to those for the isostructural fluorine compound.

8.3.3 Chlorine monoxide Cl_2O

The structural parameters of Cl_2O have been redetermined recently. Microwave studies were made on the three chlorine isotopic species of Cl_2O and from the rotational and centrifugal force constants obtained, values were calculated for the structural parameters and harmonic force constants of the molecule[60]. The rotational data was also used to calculate vibrational frequencies and hence provide an independent check on the vibrational assignments of Rochkind and Pimentel[61].

An electron diffraction study of gaseous Cl_2O [62] confirmed the results of one of the previous studies but gave more accurate interatomic distances and angles. A summary of these results is given in Table 8.7.

Mass spectral studies by Fisher[55], on the pyrolysis products of Cl_2O and $HClO_4$ gave thermodynamic data for several chlorine–oxygen species. Pyrolysis of Cl_2O (decomposition temperature about 350 °C) produced mainly chlorine and oxygen together with a small yield of ClO radicals. The ionisation potentials of Cl_2O and ClO were found to be 46.65 and 46.40 kJ mol^{-1} respectively, which gave a value of 135 ± 8 kJ mol^{-1} for D_{Cl-OCl}, in fair agreement with the thermochemical value of 143 kJ mol^{-1}.

A series of papers by Freeman and Phillips[64–66] have dealt with the reactions of oxygen, nitrogen and hydrogen atoms with chlorine monoxide in a fast flow system monitored by mass spectrometry. For oxygen and nitrogen, primary reactions were

$$O + Cl_2O \rightarrow 2ClO$$

and

$$N + Cl_2O \rightarrow NCl + ClO$$

but for hydrogen an additional reaction was considered besides the Cl abstraction.

$$H + Cl_2O \rightarrow HOCl + Cl$$

Subsequent work by Perona *et al.*[67] confirmed that both types of reaction were important for the H/Cl_2O reaction.

Synthetic reactions involving Cl_2O include reports on the preparation of chloryl fluoride from Cl_2O and AgF_2 [68] and a reinvestigation of the reaction of Cl_2O with AsF_5 [69]. Previous reports of $ClOAsF_5$ appear to be in error and the product $ClO_2^+AsF_6^-$ is formed by the reaction

$$5Cl_2O + 3AsF_5 \rightarrow 2ClO_2^+AsF_6^- + AsOF_3 + 4Cl_2$$

8.3.4 Chlorine(III) oxide Cl_2O_3

Photolysis of gaseous ClO_2 in a large vessel, part of which was cooled to −45 °C, resulted in the condensation of Cl_2O_6 and a new chlorine oxide[70]. Decomposition was very slow at −45 °C but on warming the mixture to 0 °C, the new oxide decomposed to chlorine and oxygen in the ratio 1:1.5 and left Cl_2O_6 (m.p. +3 °C). Because of the low stability of the compound and contamination with Cl_2O_6, only the empirical formula $ClO_{1.5}$ was obtained, but by a consideration of the chemical reactivity and physical characteristics of the product a structure containing a weak Cl—Cl linkage rather than a Cl—O—Cl linkage was suggested[70].

Further studies on the factors affecting the explosive decomposition of ClO_2 over ranges of temperature and pressure[71], indicated that the new oxide, Cl_2O_3, was formed in the induction period prior to the explosion.

Table 8.7 Force constants and structural parameters of Cl_2O

Cl—O *distance* (Å)	Cl—O—Cl *bond angle (degrees)*	*Force field* k_1	k_2	k_3	k_4	*Investigational method*	*Reference*	I.R. *fundamentals* ($^{35}Cl_2O$)	*Observed (solid)*	*Calculated (ref. 59)*
$r_o = 1.700\,38 \pm 0.000\,69$	110.96 ± 0.08	2.88	0.31	0.17	0.42	Microwave	60	ν_1	630.7	619.2 ± 5
$r_s = 1.700\,38 \pm 0.000\,43$	110.86 ± 0.04							ν_2	296.4	292.5 ± 1
		2.75	0.40	0.15	0.46	Infrared	61			
		2.85	0.51	0.18	0.43		63			
$r_g(1) = 1.693 \pm 0.003$	111.2 ± 0.3					Electron diffraction	62	ν_3	670.8	705 ± 12

Table 8.8 Structural parameters of ClO_2 .

Cl—O *distance* (Å)	O—Cl—O *bond angle (degrees)*	*Force field* k_1	k_2	k_3	k_4	*Investigational method*	*Reference*	*Fundamental i.r. absorptions* ($^{35}ClO_2$)	*i.r.*	*u.v.*
$r_s = 1.473 \pm 0.01$	117.6 ± 1					Microwave	74	ν_1	945.2	945.3
$r_o = 1.472 \pm 0.005$	117.4 ± 0.2					U.V. and i.r. spectroscopy	75	ν_2	447.3	447.9
		7.01	−0.164	−0.002	0.649		76			
$r_g(1) = 1.475 \pm 0.003$	117.7 ± 1.7					Electron diffraction	72	ν_3	1110.8	—
		7.23	−0.018	0.248	0.626		77			

The experimental results were consistent with the following scheme:

Initiation reactions:

$2ClO_2 \xrightarrow{wall} ClO + ClO_3$	$\Delta H \approx 253$ kJ mol^{-1}
$ClO_2 + ClO \rightarrow Cl_2O_3$	$\Delta H \approx -92$ kJ mol^{-1}
$ClO + Cl_2O_3 \rightarrow ClO_2 + ClOOCl$	$\Delta H \approx -26$ kJ mol^{-1}
$ClOOCl \rightarrow Cl + OOCl$	$\Delta H \approx 184$ kJ mol^{-1}
$Cl + OOCl \rightarrow 2Cl + O_2$	$\Delta H \approx 184$ kJ mol^{-1}
$Cl + ClO_2 \rightarrow 2ClO$	$\Delta H \approx 138$ kJ mol^{-1}

Termination reactions:

$Cl + ClO_2 \rightarrow Cl_2 + O_2$	$\Delta H = -1242$ kJ mol^{-1}
$2Cl \rightarrow Cl_2$	

8.3.5 Chlorine dioxide ClO_2

Chlorine dioxide is by far the most documented of the chlorine oxides and has many industrial applications as a bleaching agent and bacteriocide. From the structural and chemical point of view it is an interesting example of the small class of inorganic radicals which are reluctant to dimerise.

Most of the recent work on this oxide has been concerned with structural and mechanistic aspects. Clark and Beagley[72] have reinvestigated the electron diffraction pattern from gaseous chlorine dioxide and found internuclear distances (Cl—O) of 1.475±0.003 Å and an OClO bond angle of 117.7±1.7 degrees. These results, together with data provided by microwave, u.v. and infrared spectroscopy are reported in Table 8.8. The comparative stereochemistries of ClO_2 (nineteen valence electrons) and OCl_2 (twenty valence electrons) have been successfully explained by Walsh's correlation diagram[73].

Spectroscopic investigations have also been carried out on the excited states of ClO_2. The visible bands of this oxide are attributable to a $^2A_2 \leftarrow ^2B_1$ transition, representing a π–π excitation in a molecular orbital description. Previous investigators, (Coon *et al.*[79]), had suggested the possibility of an unsymmetrical excited state structure in which the Cl—O bond distances were unequal. A careful examination of the structure of the ν_1 vibronic band in the $^2A_2 \leftarrow ^2B_1$ electronic band system by Brand *et al.*[77,78,80] showed that this interpretation was incorrect. In this excited state, ClO_2 has equal Cl—O internuclear distances of 1.619±0.016 Å and a bond angle of 107±0.5 degrees. An analysis of the ν_3 band of ClO_2 by Richardson[81] led to values for the rotational constants of the 0, 0, 1 level of $^{35}ClO_2$.

The paramagnetic properties of this odd electron molecule have been studied by the e.s.r. technique under a variety of conditions. Thus McClung and Kivelson[82] examined e.s.r. line widths as a function of temperature in a number of solvents, while Eachus *et al.*[59] were able to show that ClO_2 was one of the paramagnetic species produced by γ-irradiation of $KClO_3$. Gitsan and Panfilar[83] examined the kinetics of ClO_2 formation during the gas phase decomposition of perchloric acid using e.s.r.

Mass spectrometry gives a value of 103.2±9.7 kJ mol^{-1} for the first ionisation potential of ClO_2. In view of this relatively low value, the possibility of forming the ClO_2^+ ion has been discussed[84] and investigated. Carter *et al.*[85] showed that ClO_2 was oxidised by $S_2O_6F_2$ to chloryl fluorosulphate, which was shown by conductometric and spectroscopic measurements to yield the solvated ClO_2^+ ion in HSO_3F.

$$2ClO_2 + S_2O_6F_2 \xrightarrow{-40\,^{\circ}C} 2(ClO_2)OSO_2F$$

$$(ClO_2)OSO_2F \xrightarrow{HSO_3F} ClO_2^+ \text{ (solvated)} + SO_3F^- \text{ (solvated)}$$

Subsequently two groups of workers[86,87] characterised solid chloronium salts. As the ClO_2^+ ion is isoelectronic with SO_2 it would be expected to have a similar C_{2v} structure. This has been confirmed by infrared measurements which show three infrared active fundamentals.

The results of a general valence force field for ClO_2^+, calculated from isotopic frequency data, and assuming a bond angle in the range 115–125° are given in Table 8.9.

Table 8.9 Infrared fundamentals and GVFF for ClO_2^+ and SO_2

ClO_2^+ (AsF_6^- salt) ref[86]	ref[87]	*Assignment*	*Force constants* mdyn/Å ClO_2^+ (ref[86])	SO_2 (ref[88])
1296.4	1293	ν_3	k_1 8.96±0.06	10.01
1282.6	1280		k_2 −0.45±0.13	0.02
1043.7	1045	ν_1	k_3 0.24±0.13	0.19
1038.3			k_4 0.82±0.03	0.79
521.0	515	ν_2		
517 (sh)				

The mechanism of the ClO_2–NO reaction has been studied in a fast flow system at 294 °C at pressures of about 1 mmHg[89]. When atomic chlorine is quickly removed from the system, the mechanism is described by

$$NO + OClO \rightarrow NO_2 + ClO$$
$$NO + ClO \rightarrow NO_2 + Cl$$
$$2Cl + M \rightarrow Cl_2$$

However, if the rate of combination of chlorine atoms is slow, chlorine atoms are removed by a further fast reaction

$$Cl + OCl \rightarrow ClO + ClO$$

Kinetic studies have also been carried out on the ClO_2–O, Cl and Br systems[90,91].

8.3.6 Chlorine perchlorate Cl_2O_4

A second new oxide of chlorine with the empirical formula ClO_2 has recently been prepared and partially characterised by Schack and Pilipovich[92]. The reaction of either caesium or nitronium perchlorate with chlorine fluorosulphate at −45 °C led to the formation of Cl_2O_4 in high yields.

$$MClO_4 + ClSO_3F \rightarrow MSO_3F + ClOClO_3$$

The new product was separated by fractional condensation and isolated as a white solid. Table 8.10 contains results of physical measurements on the compound.

Cl_2O_4 is stable for only limited periods at room temperature and decomposes to yield chlorine, oxygen and dichlorine hexoxide. At higher temperatures complete degradation to the elements occurs, while irradiation in quartz produces chlorine, oxygen and dichlorine heptoxide.

Table 8.10 Physical data on Cl_2O_4 [92]

Vapour pressure/temperature relationship over the temperature range −47 to 21 °C	$7.8156 - \frac{1568.0}{TK}$
B.P. (calculated)	44.5 °C
M.P.	−117±2 °C
Heat of vaporisation	165.3 kJ mol^{-1}
Density (−78.8 to 21.2 °C)	$\rho = 1.806 - 2.30 \times 10^{-3}$ t °C
Trouton constant	22.6
Infrared bands (3000–400 cm^{-1})	1282(vs), 1041(s), 752(w), 661(sh), 652(s), 585(sh), 574(sh), 561(m), 511(w).

Reaction of Cl_2O_4 with HCl or AgCl produces chlorine and a perchlorate

$$MCl + Cl_2O_4 \rightarrow MClO_4 + Cl_2$$

consistent with the formulation of Cl_2O_4 as chlorine perchlorate ($ClOClO_3$).

Alternative structural formulations are ruled out by the infrared spectrum which exhibits strong bands at 1282 and 1041 cm^{-1} attributable to the ClO_3 antisymmetric and symmetric stretching vibrations, as well as other bands characteristic of the stretching modes of the Cl—O—Cl linkage.

Based on the known heats of formation of other oxides of fluorine and chlorine, the heat of formation of $ClOClO_3$ was estimated as approximately 180 kJ mol^{-1}.

8.3.7 Chlorine trioxide and dichlorine hexoxide, ClO_3 and Cl_2O_6

Chlorine trioxide readily dimerises in the solid and liquid state to Cl_2O_6. The process is almost thermoneutral.

$$2ClO_3 \rightleftharpoons Cl_2O_6$$

The position of the equilibrium depends upon physical conditions and can be followed by changes in the magnetic behaviour of the system. However, Cl_2O_6 is also thermally unstable with respect to ClO_2 and oxygen; consequently bulk susceptibility measurements will not differentiate between the two paramagnetic species, ClO_3 and ClO_2. Belevskii *et al.*[93] investigated the dissociation of liquid and solid Cl_2O_6 using e.s.r. measurements to determine equilibria. For both of these systems the e.s.r. signal was produced by ClO_2 radicals and no signal due to ClO_3 radicals was observed. Consequently, in the condensed phase Cl_2O_6 is virtually non-dissociated to ClO_3. U.V. irradiation of frozen solutions of Cl_2O_6 in oleum or anhydrous H_3PO_4 resulted in the formation of ClO radicals, but low temperature γ-irradiation did generate ClO_3 radicals. Low temperature γ-irradiation of the chlorate and perchlorate ions has also been shown to produce ClO_3 as one of the paramagnetic centres in the matrix[94–96].

In view of the very low dissociation energy of Cl_2O_6, the mass spectra of the hexoxide has not yet been obtained. Cordes and Smith[97] used an ion source of open structure which was effective at room temperature to record the mass spectra of this oxide, but at the low pressures used, no Cl_2O_6 parent ions were seen. All of the oxide was in the ClO_3 form which produced mainly ClO_2^+ and ClO^+ ions with smaller amounts of ClO_3^+.

8.3.8 Dichlorine heptoxide Cl_2O_7

Although the structure of gaseous chlorine heptoxide has been known for several years[98], there has been a considerable difference of opinion concerning the infrared assignments for the bridge part of this molecule.

Beagley *et al.*[99] treated Cl_2O_7 as a bent XOX system ($X = ClO_3$) and assigned two of the three vibrational fundamentals of the Cl—O—Cl skeleton to bands at 695 cm^{-1} (ν_a antisymmetric stretch) and 595 cm^{-1} (ν_s symmetric stretch) and estimated the third to be at ~195 cm^{-1} (ν_b symmetric bend). More recent work by Witt and Hammaker[100] on the Raman spectrum of Cl_2O_7 gave rise to values of ν_a (775) and ν_s (699) considerably higher than those above, and they observed the symmetric bending mode for the first time at 161 cm^{-1}. The correctness of either set of assignments must await the results of a full normal coordinate analysis.

A re-investigation of the pyrolysis of Cl_2O_7 by Fisher[55] resulted in the observation of no ClO_3 or ClO_4 radicals. At ≈400 °C a large yield of ClO_2 and a smaller yield of ClO radicals were observed. At higher temperatures, equal yields of ClO and ClO_2 radicals were produced. Assuming that the initial step in the Cl_2O_7 decomposition is

$$Cl_2O_7 \rightarrow ClO_3 + ClO_4$$

then the ClO_4 and ClO_3 radicals do not survive long enough to be ionised, but immediately form ClO_2 and ClO.

8.4 BROMINE OXIDES

8.4.1 Bromine monoxide Br_2O

Little is known about the structural or spectral properties of Br_2O in comparison with the other known halogen monoxides. One recent infrared examination of solid $Br_2{}^{16}O$–$Br_2{}^{18}O$–Br_2 mixtures at −196 °C[101] has led to the assignment of the three infrared fundamentals expected for a bent triatomic molecule. Assuming a bond angle of 113 degrees, approximate values for four force constants were calculated. A summary of these results is given in Table 8.11.

Table 8.11 Infrared fundamentals and force constants of Br_2O

Vibrational frequencies (cm^{-1})	*Assignments*	*Force constants*
583	ν_3 asym. str.	k_1 2.4 ± 0.2
504	ν_1 sym. str.	k_2 0.4 ± 0.2
197	ν_2 bending mode	k_3 0.2 ± 0.1
		k_4 0.4 ± 0.1

8.4.2 The bromine oxide radicals BrO, BrO_2 and BrO_3

The physical and chemical properties of the BrO radical have been investigated in the gas phase by various spectroscopic methods. Powell and Johnson[102] were able to derive the molecular constants of BrO by microwave spectroscopy, while Carrington *et al.*[103] analysed the e.s.r. spectrum of gaseous BrO in order to deduce these parameters. Agreement between both sets of results is good as can be seen in Table 8.12.

By using time-resolved electronic absorption spectrophotometry in a discharge flow system, Clyne and Cruse[51, 105] were able to study the kinetics

Table 8.12 Molecular constants of ^{79}BrO

Constant	*Microwave method* [102]	E.S.R. [103]
Beff	12,824.80 ± 0.11 (MHz) (0.42779 cm^{-1})	0.4282 ± 0.0005 cm^{-1} (B_0)
r_0 (Å)		1.720
h (MHz)	505.27 ± 1.53	504.5 ± 1.0
eq$Q$ (MHz)	658.13 ± 2.69	649.8 ± 1.9
μ (D)		1.55 [104]
$D_{0\ (Br—O)}$ (kJ mol^{-1})	231 ± 3 [56]	

of formation of BrO (from ozone and bromine atoms) as well as several of its reactions. BrO radicals were found to decay by only one of the two mechanisms found for the ClO radicals, namely

$$BrO + BrO \rightarrow Br + BrOO$$
$$BrOO + M \rightarrow Br + O_2 + M$$

However, there was evidence to suggest that the peroxy species has a much weaker halogen–oxygen linkage than in ClOO.

BrO was found to react rapidly with other free radicals such as NO and O, but was unreactive towards the singlet ground-state molecules H_2, CH_4, C_2H_6 and C_2H_4.

The generation and reactions of the BrO radical has also been studied in aqueous solution. Buxton *et al.*[106] examined the radiolysis of Br^- free hypobromite solutions and identified a short lived species with an absorption maximum at 360 nm in its u.v. spectrum as BrO:

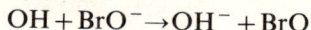

$$OH + BrO^- \rightarrow OH^- + BrO$$

The probable decay mechanism in alkaline solution was later shown to be

$$2BrO + 2OH^- \rightarrow BrO^- + BrO_2^- + H_2O$$

and possibly involved a Br_2O_2 intermediate[107].

Pulsing neutral or alkaline solutions of bromate or bromate and bromite ions gave a different species with an absorption maximum at 475 nm, which was identified as BrO_2:

$$e^-aq + BrO_3^- \rightarrow (BrO_3^{2-}) \rightarrow BrO_2 + 2OH^-$$
$$OH + BrO_2^- \rightarrow BrO_2 + OH^-$$

The decay kinetics were consistent with a mechanism in which the dimer of BrO_2 reverts to a monomer or is attacked by OH^- to give permanent products.

$$2BrO_2 \rightleftharpoons Br_2O_4$$
$$Br_2O_4 + OH^- \rightarrow \text{bromite and bromate ions}$$

Support for the above mechanism is given by studies on neutral solutions containing BrO_2 which decay in two detectable stages.

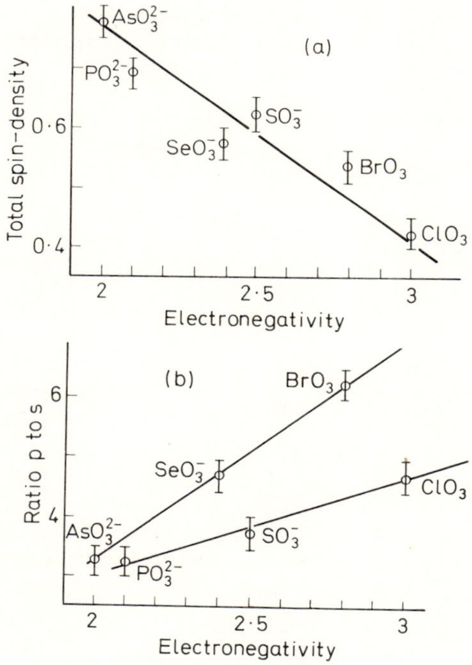

Figure 8.1 Plots of (a) total unpaired spin density and (b) ratio of p to s spin density on central atom, against the electronegativity of the central atom. (Reproduced by kind permission of the Chemical Society and author (Reference 112)).

Flash photolysis of aqueous solutions of halate ions has been shown to be a general method of generating all three XO_2 radicals[108, 109] and a detailed study by Amichai and Treinin[110] indicated that pulse radiolysis of BrO_3^- in water and its photolysis in boric acid glass led to the formation of BrO_3 (λ max 315 nm) as well as BrO_2 and BrO.

A preliminary communication by Collins *et al.*[111] gives an account of the electron spin resonance spectrum of the BrO_2 centre in irradiated zinc bromate, while a report by Begum *et al.*[112] deals with the magnetic parameters of BrO_3, produced by γ-radiolysis of $KBrO_3$ at 77 K. Spin population analysis of the bromine 4p and 4s orbitals showed that the unpaired electron has 46% Br 3p and 7% Br 3s character. The remaining 47% unpaired spin density is expected to be distributed equally on the three oxygen atoms. Determination of this data enables a comparison to be drawn with other isoelectronic species. Both the spin density on the central atom and the unpaired spin p to s ratio were found to vary linearly with electronegativity[112].

8.5 IODINE OXIDES

8.5.1 The iodine oxide radicals IO, IO_2 and IO_3

Only a few reports have been published recently on these radicals. One of the most important of these by Carrington *et al.*[103] deals with the e.s.r. spectrum of gaseous IO, and reports accurate molecular constants for this radical (Table 8.13).

Table 8.13 Molecular parameters for ^{127}IO

		Reference
B_o (cm^{-1})	0.3389 ± 0.0007	103
B_e (cm^{-1})	0.340 26	113
ω_e(cm^{-1})	681.4 (7)	113
$\omega_e x_e$ (cm^{-1})	4.2 (9)	113
r_o Å	1.871	103
r_e Å	1.867	113
h (MHz)	582.1 ± 2.3	103
eq$Q$ (MHz)	-1907.0 ± 13	103
D_o(I—O) kJ mol^{-1}	184 ± 19	114
	176 ± 21	104

Amichai and Treninin[115] showed that flash photolysis and pulse radiolysis techniques yield interesting information about these radicals. All three species are formed by flash photolysis of neutral, oxygen free, iodate solutions. IO_2 is generated by the primary reaction and IO_3 and IO by secondary reactions.

$$IO_3^- \xrightarrow{h\nu} IO_2 + O^-$$

$$IO_3^- + O^- \xrightarrow{H_2O} IO_3 + 2OH^-$$

$$IO_2 + IO_3^- \longrightarrow IO + IO_4^-$$

These radicals are also formed in pulse radiolysis of IO_3^- solutions, whereas only IO is formed in aqueous IO^-. The transition energies ($h\nu_{max}$) of the oxyiodine radicals were found to be in the order IO_3 (380 nm) > IO (490 nm) > IO_2 (715 nm), the same order as for the corresponding chlorine and bromine species. It was also noted that the transition energies of analogous oxyhalogen compounds increases almost linearly with the electronegativity of the halogen atom[115].

Finally, Clyne and Cruse[105] have shown that the decay of IO radicals is similar to that of BrO in mechanism and rate.

$$IO + IO \rightarrow I + IOO \rightarrow 2I + O_2$$

8.5.2 Iodine dioxide I_2O_4

Mössbauer studies of this oxide by Grushko *et al.*[116, 117] have shown the presence of two non-equivalent iodine atoms in the lattice and confirm previous predictions based on infrared evidence[118, 119], that this oxide has a polymeric structure composed of IO_3 and IO groups covalently bonded by oxygen bridges.

8.5.3 Iodine pentoxide I_2O_5

Despite the fact that I_2O_5 is the most commonplace of the halogen oxides*, its crystal structure was unknown until this year. Spectroscopic results have been interpreted in terms of a monomeric $O_2I—O—IO_2$ unit of C_{2v} symmetry by Duval and Lecompte[121], whereas a more comprehensive study by Sherwood and Turner[122] was compatible with a polymeric structure, although a molecular structure could not be completely ruled out. An x-ray diffraction study by Selte and Kjekshus[123] revealed that both points of view have some validity in that molecular I_2O_5 units are distinguishable in the structure,

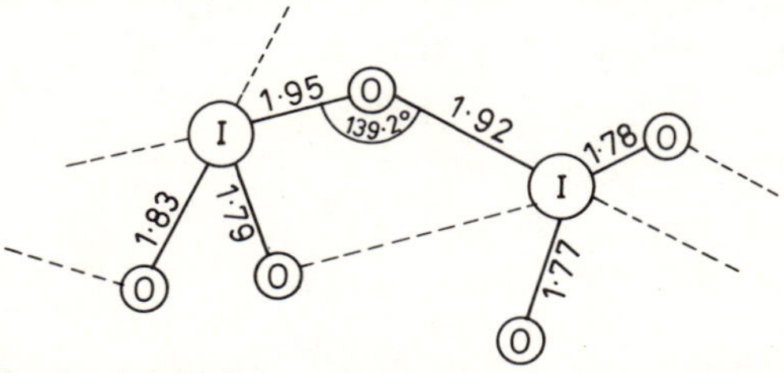

Figure 8.2 A molecular I_2O_5 unit in solid iodine pentoxide. The dashed lines represent weak inter and intramolecular iodine–oxygen bonds.

although fairly strong intermolecular forces are present in the lattice as shown by intermolecular distances as short as 2.23 Å. Thus the I_2O_5 structure may be regarded as being intermediate between the macromolecular and isolated molecular types.

Figure 8.2 shows the I_2O_5 unit which may be regarded as composed of

*It has been pointed out that commercial 'I_2O_5' is almost entirely HI_3O_8 [120].

two IO_3 pyramids sharing one common oxygen atom. The 'terminal' I—O distances are all in the range 1.77–1.83 Å whereas the bridging I—O distances are 1.92 and 1.95 Å respectively. All the intramolecular O—I—O bond angles are approximately 95 degrees. Each molecular unit is linked to adjacent molecular units via fairly short intermolecular I—O bonds which gives the structure the physical characteristics of a three-dimensional network. Individually, each iodine atom is linked at a distance of 2.72 Å or less to five oxygens occupying five of the apices of a distorted octahedron. Both bridge, and to an even greater extent, terminal I—O distances imply considerable π bonding character.

Interpretation of the mass spectrum of I_2O_5 has been shown to be more complex than was originally suspected[124]. As in a previous study[125], peaks due to I_2^+, I_2O^+ and $I_2O_3^+$ were observed, but peaks of higher mass number which were originally thought to be due to $I_2O_4^+$ and $I_2O_5^+$ were found to be due to copper–oxygen, copper–iodine and copper–iodine–oxygen species formed from the reaction of I_2O_5 vapour with copper in the source system of the mass spectrometer. These results have obvious implications in any system in which volatile metallic compounds may be generated by reaction of the source with the compound under investigation.

8.6 HYPOHALOUS ACIDS

8.6.1 Hypofluorous acid HOF

Although HOF is not credited with any permanence under normal conditions, it has been observed at low temperature by infrared studies[126].

Photolysis of mixtures of fluorine and water suspended in a nitrogen matrix at 20 K or lower, gave rise to new infrared absorptions, only some of which were assignable to F_2O, N_2O, O_2F, O_3 and HF. Growth and diffusion studies, isotopic labelling and normal coordinate analysis helped in the assignment of bands at 3483, 1393 and 884 cm^{-1} to HOF. Force constants for HOF and the other hypohalous acids are given in Table 8.14. The most likely mechanism of formation is

$$F_2 \xrightarrow{h\nu} 2F \tag{8.5}$$

$$H_2O + F \rightarrow HF + OH \tag{8.6}$$

$$OH + F \rightarrow HOF \tag{8.7}$$

Both reactions (8.6) and (8.7) are exothermic, and the product HOF appears stable when trapped in an unreactive medium. On diffusion a reaction such as

$$2HOF \rightarrow 2HF + O_2$$

will be highly exothermic.

8.6.2 Hypochlorous, hypobromous and hypoiodous acids

Of these three acids, only HOCl has been studied to any extent and very little information is available on the unstable iodine compound.

Infrared studies have recently been reported[127] on molecular HOCl and HOBr generated by the photolysis of mixtures of Ar—HX and O_3 at 4 K. The use of isotopically enriched materials gave sufficient data to enable the calculation of potential functions for these molecules. Table 8.14 contains details of these results.

Table 8.14

Frequencies and force constants	HOF* *Reference* 126	HOCl† *Reference* 127	HOCl (*gas phase*) *Reference* 128	HOBr *Reference* 127
ν_1	3483	3578	3626	3589
ν_2	1393	1239	1242	1164
ν_3	884	728	739	626
k_{OH} mdyn $Å^{-1}$	6.81	7.10	7.35	7.14
k_{OX} mdyn $Å^{-1}$	4.37	3.98	3.86	3.59
k_{HOX} mdyn Å rad^{-2}	0.96	0.78	0.77	0.70
$k_{HOX—OX}$ mdyn rad^{-1}	0.58	0.68	0.45	0.64

*based on the structural parameters of H_2O and OF_2 with a HOF bond angle of 104 degrees.
†based on r_{OH}, 0.96; r_{OCl}, 1.70; r_{OBr}, 1.85 Å. α_{HOCl} 113 degrees and α_{HOBr} 110 degrees.

Since these studies were carried out, the structure of HOCl has been elucidated by both high resolution infrared and microwave spectroscopy[129, 130]. The oxygen–chlorine and oxygen–hydrogen internuclear distances are 1.69 and 0.97 Å respectively and the H—O—Cl bond angle is 103–105 degrees.

In aqueous media, all three acids are unstable and only slightly dissociated as indicated by dissociation constants in the region 10^{-8}–10^{-11}. Accurate values for the pK_a of HOCl over the temperature range 5–35 °C have been determined recently using a spectrophotometric technique[131]. The overall best fit to the data was given by the equation,

$$pK_a = \frac{3000.00}{T} - 10.0686 + 0.0253\ T$$

which yields $pK_a = 7.537 \pm 0.005$ at 298 K.

8.7 HALOUS ACIDS

Although salts of the acids $HClO_2$ and $HBrO_2$ are isolable and have been well characterised, the parent acids are unstable and are known only in dilute solution. The existence of bromous acid has been disputed and Reference 132 summarises the history of its attempted preparation.

Kieffer and Gordon[133, 134] have extended the known chemistry of chlorous acid during their investigations on the redox reactions of acidified Cl^{III} solutions with various metal ions.

In the absence of added Cl^- ions the overall stoichiometry of the $HClO_2$ disproportionation approximates to

$$4HClO_2 = 2ClO_2 + ClO_3 + Cl^- + 2H^+ + H_2O$$

In the presence of appreciable concentrations of Cl^- ions the decomposition becomes

$$5HClO_2 = 4ClO_2 + Cl^- + H^+ + 2H_2O$$

Experimental observations were consistent with the following mechanisms.

Uncatalysed:

$$2HClO_2 \rightarrow HOCl + H^+ + ClO_3^- \text{ (rate determining)}$$
$$HOCl + HClO_2 \rightarrow [Cl_2O_2] + H_2O$$
$$HOCl + H^+ + Cl^- \rightleftharpoons Cl_2 + H_2O$$
$$Cl_2 + HClO_2 \rightarrow [Cl_2O_2] + H^+ + Cl^-$$
$$[Cl_2O_2] + H_2O \rightarrow Cl^- + ClO_3^- + 2H^+$$
$$2Cl_2O_2 \rightarrow Cl_2 + 2ClO_2$$

Catalysed by Cl^-:

$$HClO_2 + Cl^- \rightleftharpoons [HCl_2O_2^-]$$
$$[HCl_2O_2^-] + Cl^- \rightarrow \text{products (rate determining)}$$

Thompson and Gordon[135] have also shown that Cl_2 oxidises $HClO_2$ more rapidly than does HOCl, and in dilute solution the primary product is ClO_3^-, but in more concentrated solution the main product is ClO_2. Dichlorine dioxide is again suggested as the intermediate.

A voltammetric study of the system ClO_2^-–ClO_2–Cl^- [136] has led to the conclusion that the electroreduction of $HClO_2$ to Cl^- on platinised platinum in acid solution occurs via the heterogeneous decomposition of $HClO_2$ into ClO_2 and Cl^- and the simultaneous reduction of the liberated ClO_2 to $HClO_2$:

$$5HClO_2 \rightarrow 4ClO_2 + Cl^- + H^+ + 2H_2O \quad (8.8)$$
$$ClO_2 + H^+ + e^- \rightleftharpoons HClO_2 \quad (8.9)$$
$$HClO_2 + 3H^+ + 4e^- \rightarrow Cl^- + 2H_2O \quad (8.10)$$

Reaction (8.8) is first order in $HClO_2$ and its rate increases with H^+ concentration and reaches a maximum for pH values less than one.

In neutral or alkaline solutions the reduction of chlorite ion to chloride proceeds through the initial slow step,

$$ClO_2^- + H^+ + e^- \rightarrow ClO + OH^-$$

or

$$ClO_2^- + H^+ + e^- \rightarrow ClO^- + OH$$

8.8 HALIC ACIDS

Of the three halic acids only iodic acid is known in the free state and is the most stable oxygen containing acid of iodine. All three are powerful oxidising agents and strong acids in aqueous solution.

Pethybridge and Prue[137] re-examined the aqueous behaviour of iodic acid using conductimetric, potentiometric and kinetic methods to determine a reliable value for its acidity constant (0.157 mol l^{-1} at 25 °C). The stability constant of the complex ion $H(IO_3)_2^-$ formed in dilute solution by association of an iodic acid molecule with an iodate ion was found to be 4.1 mol^{-1} at 25 °C. In more concentrated solutions, cryoscopic and other measurements have suggested polymerised species, and the degree of dissociation of HIO_3 calculated from Raman, n.m.r. and conductance measurements do not agree. Dawber[138] has calculated acidity functions for HIO_3 and compared them with measured values. Agreement between calculated and measured values was closest when values of α from conductance data were employed.

Iodic acid has two modifications in the solid state. The crystal structure and configuration of the α-form has been determined by x-ray and neutron diffraction studies[139, 140], but β-HIO_3 has proved difficult to isolate and characterise[120].

Differential thermogravimetric and thermal analyses showed that α-HIO_3 decomposes in two stages[120], initially forming HI_3O_8 at 100–130 °C (although even at ~80 °C for prolonged periods some decomposition occurs) and finally I_2O_5:

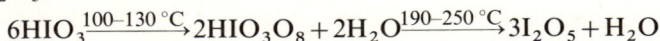

$$6HIO_3 \xrightarrow{100\text{–}130\,°C} 2HIO_3O_8 + 2H_2O \xrightarrow{190\text{–}250\,°C} 3I_2O_5 + H_2O$$

Recrystallisation of α-HIO_3 or I_2O_5 from acidic solutions also produces HI_3O_8 which Feikema and Vos[141] showed to be an addition compound of α-HIO_3 and I_2O_5 (Figure 8.3) in which the configurations of HIO_3 groups in both acids show only small differences.

Detailed vibrational spectra have been reported for both α-HIO_3 and HI_3O_8 [122]. The strong hydrogen bonding in the former produces chain

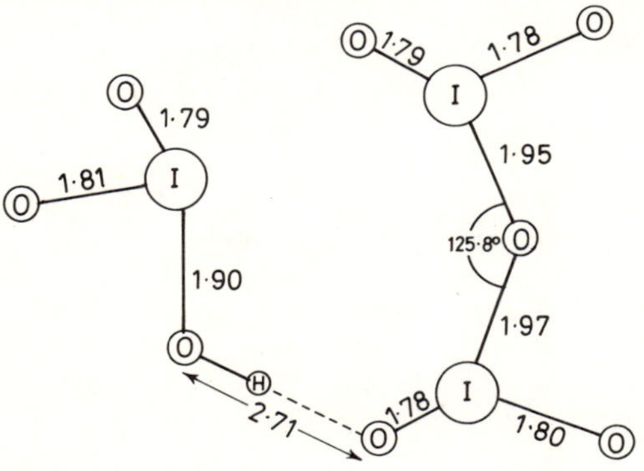

Figure 8.3 A molecular HI_3O_8 unit

polymers and leads to a doubling of the number of internal vibrations. Consequently a given vibrational mode may be split by static fields in the crystal as well as 'polymeric coupling'. In solution the spectrum of α-HIO_3 can be interpreted more easily; eight of the nine expected fundamentals have been assigned[142].

8.9 PERHALIC ACIDS

8.9.1 Perchloric acid $HClO_4$

Several informative reviews on perchloric acid are to be found in references[143–145]; the last of these emphasises the industrial importance of this acid. The thermal decomposition of ammonium perchlorate has also been reviewed, recently[146], and consequently this section will concern itself with recent advances in the chemistry of the free acid and its solutions.

8.9.1.1 *Anhydrous perchloric acid*

It now seems evident that anhydrous perchloric acid exists as such and the 'self dehydration' reaction indicated below occurs to a very limited extent, if at all. A summary of the evidence

$$3HClO_4 \rightleftharpoons Cl_2O_7 + H_3O^+ + ClO_4^-$$

for such a conclusion is given in reference 143.

The structure of $HClO_4$ in the vapour phase has been established by both electron diffraction and infrared spectroscopy[147–149]. The results of these studies are summarised in Table 8.15.

Although the internuclear distances are now well defined, the O—Cl—O bond angle cannot be regarded as unequivocal since the two electron diffraction studies given values differing by 4 degrees (well outside of experimental error). If structure (a), Table 8.15, is correct, the O—Cl—O angle is significantly smaller in perchloric acid than in either Cl_2O_7 (115.2 ± 0.2 [98]) or $FClO_3$ (116.5 ± 0.5 [151]), and the Cl—O distances in the —ClO_3 group are in the order $HClO_4$ (Cl—O, 1.408 ± 0.002) > Cl_2O_7 (Cl—O. 1.405 ± 0.002) ⩾ $FClO_3$ (Cl—O, 1.404 ± 0.002). These Cl—O distances are all ~0.3 Å shorter than the expected Cl—O single bond distance, implying considerable π bonding.

8.9.1.2 *The perchloric acid–water system*

The crystalline mono and dihydrates are known to be fully ionised in the solid state from x-ray diffraction and other studies[152–154], and are convenient sources for the study of hydronium ions. Vibrational studies on solid $HClO_4 \cdot 2H_2O$ [155] reinforce the x-ray diffraction study by Olovsson[152] and confirm the presence of centrosymmetric $[H_2O—H—OH_2]^+$ ions of *trans* configuration. Interference from the strong ClO_4^- bands prevented identification of all the fundamentals of $H_5O_2^+$. In the liquid phase the spectra of the dihydrate was observed to change which reflects the presence of other species, as confirmed by n.m.r. studies to be discussed subsequently.

In dilute solution (up to 6 M), perchloric acid is known to be completely ionised, whereas in nearly anhydrous solutions the water is completely protonated. Problems arise in the middle range of concentrations where the results of the two methods used to measure the dissociation of strong electrolytes (Raman intensities and n.m.r. shifts) show discrepancies for $HClO_4$ far greater than the inherent experimental errors. An extension of the n.m.r. method to the system $HClO_4$–$DClO_4$–H_2O–D_2O [156, 157] provided accurate chemical shift data over a large concentration range. These results were interpreted in terms of four species, H_3O^+, ClO_4^-, $HClO_4$ and $H_5O_2^+$ in the upper concentration range (⩾equimolar), and the degree of dissociation of the acid was then computed on this basis. On dilution, higher hydrates are

Table 8.15 Force constants and structural parameters of $HClO_4$ vapour

*Force constants**		$r_{Cl—O}$ (Å)	$r_{Cl—OH}$ (Å)	OClO angle	OCl(OH) angle	*Reference*
k_1 (Cl—O) str) [mdyn Å^{-1}]	9.20	(a)† 1.408 ± 0.002	1.635 ± 0.007	112.8 ± 0.5°	105.8 ± 0.5°	147
k_2 (Cl—OH str) [mdyn Å^{-1}]	3.85	or				
k_3 (OClO bend) [mdyn Å rad^{-2}]	1.90	(b) 1.410 ± 0.002	1.646 ± 0.007	117.3 ± 0.5°	99.6 ± 0.7°	
k_4 (OClOH bend) [mdyn Å rad^{-2}]	1.35	1.42 ± 0.01	1.64 ± 0.02	117.1 ± 1.2°		148
k_5 ClO—ClOH str–str interaction [mdyn Å^{-1}]	0.30					
k_6 ClOH—OClOH str–bend interaction [mdyn rad^{-1}]	0.46					

*based on the fundamental frequencies reported in Reference 149.

†of the two possible sets of structural parameters, set (a) was preferred[147] on the basis of Hamilton's significance test and the trends shown by the geometries of XO_4 tetrahedra[150].

formed and the degree of dissociation of the acid cannot be computed directly from n.m.r. data, although interpolated results may be obtained. Akitt *et al.*[158] have repeated ^{1}H, ^{35}Cl n.m.r. and Raman measurements in aqueous perchloric acid solutions over a limited range of concentrations and have confirmed and clarified previous observations[159, 160]. It is suggested that some of the discrepancies are due to refractive index effects[158].

The mass spectrum of perchloric acid has been reported several times. Heath and Majer[161] studied the heterogeneous decomposition of perchloric acid vapour (72%) on a hot wire outside the ionisation chamber of the mass spectrometer, and showed that the main mode of decomposition was

$$HClO_4 \rightarrow HCl + 2O_2$$

Fisher's results[162] on the same strength acid but under somewhat different conditions were compatible with a decomposition scheme.

$$HOClO_3 \rightarrow ClO_3 + OH$$
$$ClO_3 \rightarrow ClO_2 + O$$
$$ClO_3 \rightarrow ClO + O_2$$
$$ClO_2 \rightarrow ClO + O$$
$$ClO \rightarrow Cl + O$$

Cl and O dimerise

$$Cl, ClO, OH \xrightarrow{\text{wall}} HCl, ClOH, H_2O$$

Only a relatively small amount of HCl was detected.

The most recent investigation[97] was carried out on perchloric acid vapour generated from magnesium perchlorate and 95% sulphuric acid and employed an ion source working at room temperature. Under these conditions very little HCl^+ or O_2^+ was detected but large yields of the parent ion, ClO_3^+, ClO_2^+ and ClO^+ were formed.

8.9.2 Perbromic acid $HBrO_4$

Numerous unsuccessful attempts have been made to prepare perbromic acid and its salts, and reasons for its apparent non-existence have been suggested. In 1968 Appelman[163] was able to show that perbromates and aqueous solutions of perbromic acid were quite stable.

The initial synthesis involved a hot atom process whereby the β decay of ^{83}Se incorporated into a selenate produced perbromate. Co-precipitation of the perbromate with rubidium perchlorate indicated that the perbromate ion was reasonably stable. Thereafter successful attempts were made to oxidise bromate solutions either electrolytically or with xenon difluoride.

A new and more convenient synthesis for large amounts of perbromic acid was soon developed[164] in which fluorine was used to oxidise alkaline bromate solutions. After a lengthy working up procedure stable, perbromic acid solutions up to 6 M (55% $HBrO_4$) were readily obtained. At higher concentrations decomposition occurred, although very rapid removal of water resulted in crystallisation at a composition of approximately $HBrO_4 \cdot 2H_2O$ just before decomposition.

Titrimetric results showed that aqueous $HBrO_4$ is a strong monobasic acid which oxidises inorganic species such as I^- and Br^- only slowly at room temperature. Mass spectral studies[165] indicated the formation of BrO_2^+, $HBrO_2^+$, BrO_3^+ and $HBrO_3^+$ ions as well as the parent $HBrO_4^+$ ion. Under comparable conditions perchloric acid yielded ClO_2^+, ClO_3^+ and $HClO_4^+$ ions as the most prominent species.

Neutralisation of perbromic acid solutions resulted in the formation of stable alkali metal perbromates, solutions of which had the same u.v. and Raman spectra as the aqueous acid. The vibrational spectrum of BrO_4^- salts indicates that this ion is tetrahedral in the solid state and remains so in neutral and acidic solutions. Thus in its aqueous behaviour BrO_4^- resembles ClO_4^- rather than IO_4^-. This is confirmed by the lack of rapid oxygen exchange between BrO_4^- and water.

γ-Irradiation of $CsBrO_3$ [166] also resulted in the formation of perbromate ions as shown by vibrational spectra measurements. Chromatographic separations on the aqueous solutions of the radiolysed solid enabled the isolation of $RbBrO_4$ and Ph_4AsBrO_4 to be achieved. Table 8.16 summarises the vibrational frequencies and force constants for the ClO_4^-, BrO_4^- and IO_4^- ions.

Table 8.16 Vibrational frequencies and force constants for the XO_4^- ions (X = Cl, Br and I)

(Data from Reference 164 *except where otherwise indicated)*

	ClO_4^-	BrO_4^- [166]	BrO_4^-	IO_4^-
ν_1	935	801	798	791
ν_2	460	331	331	256
ν_3	1110	878	883	853
ν_4	630	410	410	325
k_1	8.24	6.05		5.90
k_2	0.87	0.48		0.30
k_3	−0.21	−0.12		−0.09
k_4	0.78	0.38		0.07

An x-ray diffraction study on $KBrO_4$ [167] showed the equivalence of all four Br—O bond lengths within experimental error (1.610 Å), whereas there are two slightly differing O—Br—O bond angles of approximately 110 and 109 degrees.

Thermodynamic data have also been presented for $KBrO_4$ which establish the stability of the perbromate ion. Initially Brand and Bunck[168] determined the enthalpies of formation and solution of $KBrO_4$ in order to derive standard thermodynamic properties for the BrO_4^- ion. Subsequent work[169] has given more accurate values for these quantities as well as heat capacity data[170] which enables a complete and accurate comparison to be made with other perhalate ions (Tables 8.17 and 8.18).

Potassium perbromate has a considerably higher Gibbs free energy of formation than either of the other two perhalates. This is attributable to the anomolous enthalpy of formation which in turn reflects the relatively weak Br—O bond in the perbromate ion. Nevertheless, ΔG_f° is still substantially negative and oxidising agents such as $K_2S_2O_8$ are thermodynamically cap-

Table 8.17 Thermodynamic properties of crystalline $KBrO_4$ and the aqueous BrO_4^- at 298–5[1] . Values are in kJ mol^{-1} or J deg^{-1} mol^{-1}.

Reference	168	169	170
Heat of solution of $KBrO_4$(c) (kJ mol^{-1})	48.15 ± 0.45	48.56 ± 0.08	
ΔH_f° ($KBrO_4$, c)	-291.8 ± 13.6	-287.61 ± 0.59	
S° ($KBrO_4$, c)		157.3 ± 8.4*	170.09 ± 0.2
ΔS° ($KBrO_4$, c)		-393.3 ± 8.4*	-380.7 ± 0.8
ΔG_f° ($KBrO_4$, c)		-170.33 ± 2.59	-174.10 ± 0.64
ΔH_f° (BrO_4^-, aq)	7.32 ± 14.60	13.34 ± 0.63	
S° (BrO_4^-, aq)	188.3 ± 20.9	187.0 ± 8.4*	199.8 ± 1.6
ΔG_f° (BrO_4^-, aq)	115.9 ± 20.9	122.09 ± 2.64	118.3 ± 0.8
Standard potential for $BrO_3^- + H_2O \rightarrow BrO_4^- + 2H^+ + 2e^-$	1.82 ± 0.1V	1.763 ± 0.014V	1.743 ± 0.006V

*These values were estimated, consequently quantities dependent upon entropy are approximate.

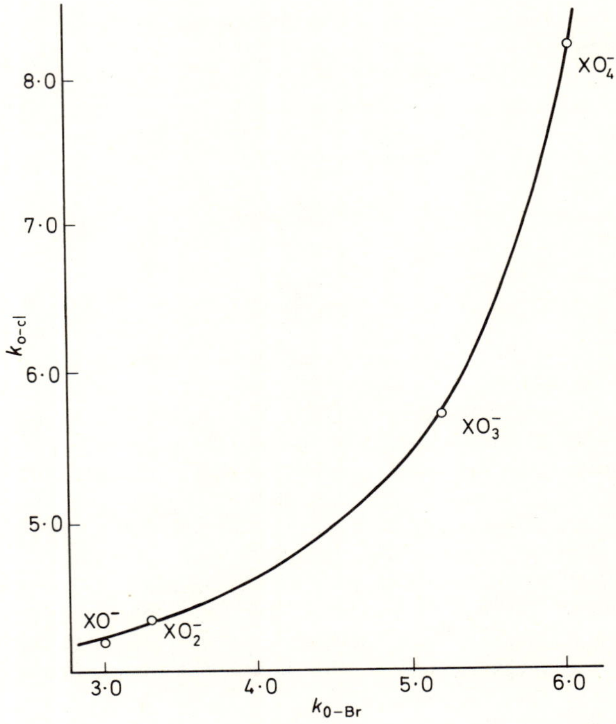

Figure 8.4 Plot of the chlorine–oxygen bond stretching force constants against the corresponding bromine–oxygen force constants, for the XO^-, XO_2^-, XO_3^- and XO_4^- ions

Force constant data from, Evans, J. C. and Lo, G. Y-S. (1967) *Inorg. Chem.*, **6**, 1483; Hovi, V. and Rasanen, V. (1967) *Ann. Acad. Sci. Fenn. Ser.* A. VI, **228**, 13; reference 41 able of oxidising Br^{V} to Br^{VII}. Consequently, the difficulties met in the synthesis of perbromates must reflect a high activation barrier between Br^{V} and Br^{VII}.

The relative weakening of the Br—O bond compared with the Cl—O bond as the formal oxidation state of the halogen atom increases from +1 to +7 is consistent with the curve obtained by plotting k_{O-Cl} versus

Table 8.18 Thermodynamic properties of potassium perhalates (KXO_4, X = Cl, Br and I)

	ΔH_f°	ΔS_f°	ΔG_f°	E°
$KClO_4$	−431.9	−435.1	−302.1	1.230
$KBrO_4$	−287.6	−380.7	−174.1	1.743
KIO_4^-	−460.6	−373.2 (est.)	−349.3	1.644

k_{O-Br} stretching force constants for the series XO^-, XO_2^-, XO_3^- and XO_4^-, where X = Cl and Br (Figure 8.4).

Some of the reactions of potassium perbromate are indicated below (data from References 163 and 171).

$$KBrO_3 + \tfrac{1}{2}O_2 \xleftarrow[275\text{–}280\ ^\circ C]{\text{thermal decomposition}} KBrO_4 \xrightarrow[HF]{SbF_5} BrO_3F$$

$$KBrO_3 + \tfrac{1}{2}O_2 \xrightarrow{390\text{–}395\ ^\circ C} KBr + \tfrac{3}{2}O_2$$

$$KBrO_4 \xrightarrow{HSO_3F} Br_2 \text{ and } O_2$$

$$BrO_3F \xrightarrow{OH^- \text{ at R.T.}} BrO_4^- + H_2O + F^-$$

The synthesis of perbromyl fluoride completes the XO_3F series and indicates the beginning of a fairly extensive Br^{VII} chemistry.

8.9.3 The periodic acids

Although there is a very extensive chemistry of oxysalts and oxyacids of iodine(VII), insufficient structural and physical characterisation of the various products has led to considerable uncertainty and ambiguity, particularly with regards to the condensed free acids.

Two recent reviews on these compounds were published in 1967 and 1968[172, 173] while a comprehensive survey of the organic chemistry of periodates is to be found in Reference 174.

8.9.3.1 Orthoperiodic acid H_5IO_6

This acid is the starting point for the other iodine(VII) oxyacids. Its solid state structure has been established by x-ray and neutron diffraction studies[175, 176].

Each molecular unit consists of a slightly distorted oxygen octahedron with an iodine atom at its centre. Five of the six I—O distances are 1.88 ± 0.03 Å whereas the 'terminal' I—O bond length is 0.1 Å shorter. All of the O—I—O angles are within the range 87–95 degrees. Each IO_6 unit is connected to neighbouring octahedra by ten hydrogen bonds, two of which are relatively short (2.60 Å) and the remainder are all approximately 2.78 Å.

H_5IO_6 is a weak acid in aqueous solution. The following equilibria have been suggested[177,178], with IO_4^- as the predominant form of the monoanion at room temperature:

$$H_5IO_6 \overset{k_1}{\rightleftharpoons} H_4IO_6^- + H^+ \qquad pK_1 = 3.3 \text{ at } 25\,°C$$
$$H_4IO_6^- \overset{k_2}{\rightleftharpoons} H_3IO_6^{2-} + H^+ \qquad pK_2 = 6.7$$
$$H_3IO_6^{2-} \overset{k_3}{\rightleftharpoons} H_2IO_6^{3-} + H^+ \qquad pK_3 = 12.2$$
$$H_4IO_6^- \overset{k_d}{\rightleftharpoons} IO_4^- + 2H_2O \qquad K_d = 40$$

Reinvestigation of this system have indicated the formation of a dimeric periodate ion in alkaline solution[178,179], which vibrational measurements show to be $H_4I_2O_{10}^{4-}$ rather than the dehydrated ion $I_2O_9^{4-}$ [178,180]. The heat of dimerisation for this ion was found to be 60.6 ± 4.2 kJ mol^{-1}, and values for the second and third dissociation constants of H_5IO_6 were redetermined[175]. A comparison of the Raman and infrared spectra of the dianion and trianion in solution with those of solid periodates of known structure gave rather inconclusive results for the dianion but better correlation for the trianion indicating that $H_2IO_6^{3-}$ is present as such in alkaline solution.

An n.m.r. study[181] of the equilibrium

$$H_4IO_6^- \text{ (aq)} \rightleftharpoons IO_4^- \text{ (aq)} + 2H_2O \text{ (1)}$$

over a temperature range leads to a new value of the equilibrium constant, $K_d = 29$ at 25 °C, in good agreement with the spectrophotometric result obtained by Salomaa and Vesala[182]. Values for the thermodynamic quantities ΔG_d° and ΔH_d° were also obtained from this study.

A calorimetric investigation of heats of reactions involving crystalline $NaIO_4$ have led to new values for the standard heats of formation of aqueous H_5IO_6 and $H_4IO_6^-$ of 789.6 ± 4.2 and 778.7 ± 4.2 kJ mol^{-1} respectively[183].

One of the most interesting chemical reactions of H_5IO_6 involves the interaction of the barium salt with excess fluorosulphonic acid to yield tetrafluoro-orthoperiodic acid[184]:

$$Ba_3H_4(IO_6)_2 + 14HSO_3F \rightarrow 2HOIOF_4 + 8H_2SO_4 + 3Ba(SO_3F)_2$$

^{19}F n.m.r. shows that both *cis* and *trans* forms are present in the mixture at room temperature. Treatment of the acid with 60% oleum liberates IO_2F_3 as yellow needles, melting point 41 °C. N.M.R. again indicates the presence of two isomers, assuming the fundamental structure of IO_2F_3 is a trigonal bipyramid. IO_2F_3 is a powerful oxidising agent which decomposes on exposure to sunlight forming IOF_3.

$$2IO_2F_3 \xrightarrow{h\nu} 2IOF_3 + O_2$$

8.9.3.2 Other periodic acids

The remaining I^{VII} oxyacids are obtained by the thermal decomposition of orthoperiodic acid. These reactions are summarised below (Reference 172 contains references to the original papers).

$$I_2O_5$$

$$\uparrow > 150\,°C$$

$$HIO_4 \xleftarrow[\text{12 mm Hg}]{100\,°C} H_5IO_6 \xrightarrow[\substack{\text{Atmospheric} \\ \text{pressure}}]{105–115\,°C} H_7I_3O_{14}$$

HIO$_4$ (sublimes)

metaperiodic acid

$$\downarrow \substack{105–117\,°C \\ \textit{in vacuo}}$$

H$_7$I$_3$O$_{14}$ triperiodic acid

[originally reported as H$_4$I$_2$O$_9$]

'$I_2O_7{\cdot}I_2O_5$'

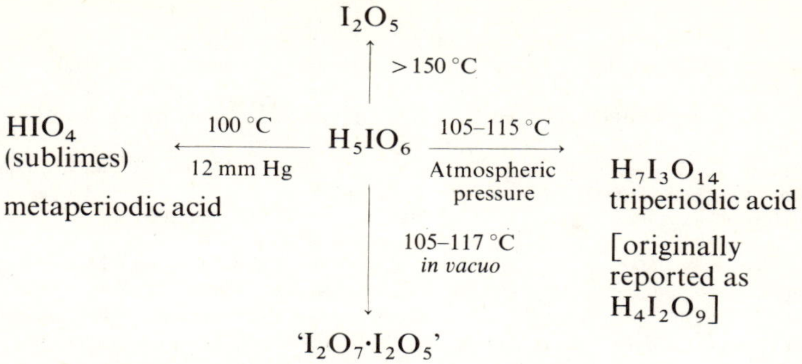

Both HIO_4 and $H_7I_3O_{14}$ exist in the free state although little is known of their structural or chemical properties.

Obviously the thermal decomposition of H_5IO_6 is complex and capable of yielding a variety of products. A careful reinvestigation of this reaction and its products would be timely, particularly as a great deal of structural work is now being done on the salts of these and other (hypothetical) periodic acids.

Acknowledgement: I should like to thank Dr. A. A. Woolf for helpful discussions and constructive criticism during the preparation of this article.

Note Added on Proof   (refers to Section 8.6.1, page 236)

A recent publication on hypofluorous acid (Studier, M. H. and Appelman, E. H. (1971). *J. Amer. Chem. Soc.*, **93**, 2349) indicates that this compound is significantly more stable than had been expected.

The free acid was formed by the rapid interaction of a stream of fluorine gas with water in a Kel–F vessel cooled in ice, and was trapped out at $-183\,°C$ in a fairly pure state after prior removal of water vapour. The preparation was monitored by mass spectrometry and the product estimated by its reaction with sodium iodide solution.

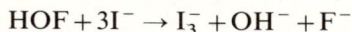

$$HOF + 3I^- \rightarrow I_3^- + OH^- + F^-$$

Hyperfluorous acid was reported as a white solid melting at approximately $-117\,°C$ to a colourless liquid, and with a half-life at room temperature which varied from 5 min to over 2 h depending upon conditions.

References

1. Schmeisser, M. and Brandle, K. (1963). *Advances in Inorganic Chemistry and Radiochemistry*, **5**, 41, (New York: Academic Press)
2. Turner, J. J. (1968). *Endeavour*, **27**, 42
3. Streng, A. G. (1963). *Chem. Rev.*, **63**, 607
4. Arkell, A. (1965). *J. Amer. Chem. Soc.*, **87**, 4057

5. Spratley, R. D., Turner, J. J. and Pimentel, G. C. (1966). *J. Chem. Phys.*, **64,** 2063
6. Noble, P. N. and Pimentel, G. C. (1966). *J. Chem. Phys.*, **44,** 3641
7. Fessenden, R. W. and Schuler, R. H. (1966). *J. Chem. Phys.*, **44,** 434
8. Adrian, F. J. (1967). *J. Chem. Phys.*, **46,** 1543
9. Lawrence, N. J., Ogden, J. S. and Turner, J. J. (1968). *J. Chem. Soc. (A)*, 3100
10. Malone, T. J. and McGee, H. A., Jr. (1966). *J. Phys. Chem.*, **70,** 316
11. Arkell, A., Reinhard, R. R. and Larson, L. P. (1965). *J. Amer. Chem. Soc.*, **87,** 1016
12. Arkell, A. (1969). *J. Phys. Chem.*, **73,** 3877
13. Dibeler, V. H., Reese, R. M. and Franklin, J. L. (1957). *J. Chem. Phys.*, **27,** 1296
14. Mortimer, F. S. (1966). *Advan. Chem. Ser.*, **54,** 39
15. O'Hare, P. A. G. and Wahl, A. C. (1970). *J. Chem. Phys.*, **53,** 2469
16. Borning, A. H. and Pullen, K. E. (1969). *Inorg. Chem.*, **8,** 1791
17. Nebgen, J. W., Metz, F. I. and Rose, W. B. (1966). *J. Mol. Spectrosc.*, **21,** 99
18. Ogden, J. S. and Turner, J. J. (1967). *J. Chem. Soc. (A)*, 1483
19. Pierce, L., DiCianni, N. and Jackson, R. H. (1963). *J. Chem. Phys.*, **38,** 730
20. Dauerman, L., Salser, G. E. and Tajima, Y. A. (1967). *J. Phys. Chem.*, **71,** 3999
21. Troe, J., Wagner, H. G. and Weden, G. (1968). *Z. Phys. Chem.*, **56,** 238
22. Blauer, J. A. and Solomon, W. C. (1968). *J. Phys. Chem.*, **72,** 2307
23. Solomon, W. C., Blauer, J. A. and Jaye, F. C. (1968). *J. Phys. Chem.*, **72,** 2311
24. Lin, M. C. and Bauer, S. H. (1969). *J. Amer. Chem. Soc.*, **91,** 7737
25. King, R. C. and Armstrong, G. T. (1968). *J. Res. Nat. Bur. Std.*, **72A,** 113
26. JANAF Thermochemical Tables. (1964). Dow Chemical, Midland, Mich.
27. Dibeler, V. H., Walker, J. A. and McCulloch, K. E. (1969). *J. Chem. Phys.*, **50,** 4592 and **51,** 4230
28. Solomon, I. J., Kacmarek, A. J. and Raney, J. (1968). *J. Phys. Chem.*, **72,** 2263
29. Gatti, R., Staricco, E. H., Sicre, J. E. and Schumacher, H. J. (1963). *Z. Phys. Chem.*, **36,** 211
30. Guisberg, V. A. and Tumanov, A. A. (1968). *Zh. Obshch. Khim.*, **38,** 1410
31. Anderson, L. R. and Fox, W. B. (1967). *J. Amer. Chem. Soc.*, **89,** 4313
32. Beal Jr., J. B., Pupp, C. and White, W. E. (1969). *Inorg. Chem.*, **8,** 829
33. Henrici, H., Lin, M. C. and Bauer, S. H. (1970). *J. Chem. Phys.*, **52,** 5834
34. Soria, D., DeStaricco, E. A. R. and Staricco, E. H. (1969). *Inorg. Nucl. Chem. Letters*, **5,** 35
35. Goetschel, C. T., Campanile, V. A. and Wagner, C. D. (1969). *J. Amer. Chem. Soc.*, **91,** 4702
36. Loos, K. R., Goetschel, C. T. and Campanile, V. A. (1970). *J. Chem. Phys.*, **52,** 4418
37. Malone, T. J. and McGee, H. A., Jr. (1965). *J. Phys. Chem.*, **69,** 4338
38. Nebgen, J. W., Metz, F. I. and Rose, W. B. (1967). *J. Amer. Chem. Soc.*, **89,** 3118
39. Solomon, I. J., Raney, J. K., Kacmarek, A. J., Maguire, R. G. and Noble, G. N. (1967). *J. Amer. Chem. Soc.*, **89,** 2015
40. Solomon, I. J., Brabets, R. I., Uenishi, R. K., Keith, J. N. and McDonough, J. M. (1964). *Inorg. Chem.*, **3,** 456
41. Keith, J. N., Solomon, I. J., Sheft, I. and Hyman, H. H. (1968). *Inorg. Chem.*, **7,** 230
42. Bantov, D. V., Sukoverhov, V. F. and Mikhailov, Yu. N. (1968). *Izr. Sib. Otd. Akad. SSSR. Ser. Khim. Nauk.*, **1,** 84
43. Solomon, I. J., Kacmarek, A. J. and Raney, J. (1968). *Inorg. Chem.*, **7,** 1221
44. Solomon, I. J. Kacmarek, A. J. and McDonough, J. M. (1968). *J. Chem. Eng. Data.*, **13,** 529
45. Solomon, I. J. (1967). U.S. Clearinghouse Fed. Sci. Tech. Information. AD-670531
46. Solomon, I. J., Keith, J. N., Kacmarek, A. J. and Raney, J. K. (1968). *J. Amer. Chem. Soc.*, **90,** 5408
47. Nicholas, J. E. and Norrish, R. G. W. (1968). *Proc. Roy. Soc.*, **307A,** 391
48. Clyne, M. A. A. and Coxon, J. A. (1968). *Proc. Roy. Soc.*, **303A,** 207
49. Morris, E. D., Jr. and Johnston, H. S. (1968). *J. Amer. Chem. Soc.*, **90,** 1918
50. Johnston, H. S., Morris, E. D., Jr. and van den Bogaerde, J. (1969). *J. Amer. Chem. Soc.*, **91,** 7712
51. Clyne, M. A. A. and Cruse, H. W. (1970). *Trans. Faraday Soc.*, **66,** 2214
52. Carrington, A., Dyer, P. N. and Levy, D. H. (1967). *J. Chem. Phys.*, **47,** 1756
53. Amono, T., Saito, S., Hirota, E., Morino, Y., Johnson, D. R. and Powell, P. X. (1969). *J. Mol. Spectrosc.*, **30,** 275
54. Rochkind, M. M. and Pimentel, G. C. (1967). *J. Chem. Phys.*, **46,** 4481

55. Fisher, I. P. (1968). *Trans. Faraday Soc.*, **64,** 1852
56. Durie, R. A. and Ramsay, D. A. (1958). *Canad. J. Phys.*, **36,** 35
57. Alcock, W. G. and Pimentel, G. C. (1968). *J. Chem. Phys.*, **48,** 2373
58. Arkell, A. and Schwager, I. (1967). *J. Amer. Chem. Soc.*, **89,** 5999
59. Eachus, R. S., Edwards, P. R., Subramanian, S. and Symons, M. C. R. (1968). *J. Chem. Soc. (A)*, 1704
60. Herberich, G. E., Jackson, R. H. and Millen, D. J. (1966). *J. Chem. Soc., (A)*, 337
61. Rochkind, M. M. and Pimentel, G. C. (1965). *J. Chem. Phys.*, **42,** 1361
62. Beagley, B., Clark, A. H. and Hewitt, T. G. (1968). *J. Chem. Soc. (A)*, 658
63. Christe, K. O. and Sawodny, W. (1969). *Inorg. Chem.*, **8,** 213
64. Freeman, C. G. and Phillips, L. F. (1968). *J. Phys. Chem.*, **72,** 3025
65. Freeman, C. G. and Phillips, L. F. (1968). *J. Phys. Chem.*, **72,** 3028
66. Freeman, C. G. and Phillips, L. F. (1968). *J. Phys. Chem.*, **72,** 3031
67. Perona, M. J., Setser, D. W. and Johnson, R. L. (1969). *J. Phys. Chem.*, **73,** 2091
68. Macheteau, Y. and Gillardeau, J. (1967). *Bull. Soc. Chem. France*, **11,** 4075
69. Schack, C. J. and Pilipovich, D. (1970). *Inorg. Chem.*, **9,** 387
70. McHale, E. T. and Von Elbe, G. (1967). *J. Amer. Chem. Soc.*, **89,** 2795
71. McHale, E. T. and Von Elbe, G. (1968). *J. Phys. Chem.*, **72,** 1849
72. Clark, A. H. and Beagley, B. (1970). *J. Chem. Soc. (A)*, 46
73. Walsh, A. D. (1953). *J. Chem. Soc.*, 2266
74. Curl, R. L., Jr., Kinsey, J. L., Baker, J. G., Baird, J. C., Heidelberg, R. F., Sugden, T. M., Jenkins, D. R. and Kenney, C. N. (1961). *Phys. Rev.*, **121,** 1119
75. Ward, J. K. (1954). *Phys. Rev.*, **96,** 845
76. Pillai, M. G. K. and Curl, R. F., Jr. (1962). *Spectrochim. Acta.*, **18,** 1382
77. Brand, J. C. D., Redding, R. W. and Richardson, A. W. (1970). *J. Mol. Spectrosc.*, **34,** 399
78. Richardson, A. W., Redding, R. W. and Brand, J. C. D. (1969). *J. Mol. Spectrosc.*, **29,** 93
79. Coon, J. B., Cesani, F. A. and Loyd, C. M. (1963). *Discussions. Faraday Soc.*, **35,** 118
80. Brand, J. C. D., Redding, R. W. and Richardson, A. W. (1969). *Chem. Commun.*, 618
81. Richardson, A. W. (1970). *J. Mol. Spectrosc.*, **35,** 43
82. McClung, R. E. D. and Kivelson, D. (1968). *J. Chem. Phys.*, **49,** 3380
83. Gitsan, V. I. and Panfilov, V. N. (1970). *Kinet. Katal*, **11,** 235
84. Woolf, A. A. (1966). *Advances in Inorganic Chemistry and Radiochemistry*. Volume 9, 217, (New York: Academic Press)
85. Carter, H. H. Qureshi, A. M. and Aubke, F. (1968). *Chem. Commun.*, 1461
86. Christe, K. O., Schack, C. J. S., Pilipovich, D. and Sawodny, W. (1969). *Inorg. Chem.*, **8,** 2489
87. Carter, H. A., Johnson, W. M. and Aubke, F. (1969). *Canad. J. Chem.*, **47,** 4619
88. Kivelson, D. (1954). *J. Chem. Phys.*, **22,** 904
89. Coxon, J. A. (1968). *Trans. Faraday Soc.*, **64,** 2118
90. Clyne, M. A. A. and Coxon, J. A. (1966). *Trans. Faraday Soc.*, **62,** 1175
91. Clyne, M. A. A. and Coxon, J. A. (1967). *Proc. Roy. Soc.*, **298A,** 424
92. Schack, C. J. and Pilipovich, D. (1970). *Inorg. Chem.*, **9,** 1387
93. Belevskii, V. N. and Bugaenko, L. T. (1967). *Zh. Neorg. Khim.*, **12,** 2277
94. Eachus, R. S. and Symons, M. C. R. (1968). *J. Chem. Soc. (A)*, 2433
95. Bloom, M. B. D., Eachus, R. S. and Symons, M. C. R. (1970). *J. Chem. Soc. (A)*, 1235
96. Patrick, P. F. and Sargent, F. P. (1968). *Canad. J. Chem.*, **46,** 1818
97. Cordes, H. F. and Smith, S. R. (1970). *J. Chem. Eng.*, **15,** 158
98. Beagley, B. (1965). *Trans. Faraday Soc.*, **61,** 1821
99. Witt, J. D. and Hammaker, R. M. (1970). *Chem. Commun.*, 667
100. Beagley, B., Clark, A. H. and Cruickshank, D. W. J. (1966). *Chem. Commun.*, 458
101. Campbell, C., Jones, J. P. M. and Turner, J. J. (1968). *Chem. Commun.*, 888
102. Powell, E. X. and Johnson, D. R. (1969). *J. Chem. Phys.*, **50,** 4596
103. Carrington, A., Dyer, P. N. and Levy, D. H. (1970). *J. Chem. Phys.*, **52,** 309
104. Carrington, A., Levy, D. H. and Miller, T. A. (1967). *J. Chem. Phys.*, **47,** 3801
105. Clyne, M. A. A. and Cruse, H. W. (1970). *Trans. Faraday Soc.*, **66,** 2227
106. Buxton, G. V., Dainton, F. S. and Wilkinson, F. (1966). *Chem. Commun.*, 320
107. Buxton, G. V. and Dainton, F. S. (1968). *Proc. Roy. Soc.*, **304A,** 427
108. Barat, F., Gilles, L., Hickel, B. and Sutton, J. (1969). *Chem. Commun.*, 1485

109. Amichai, O. and Treinin, A. (1969). *Chem. Phys. Letters,* **3,** 611
110. Amichai, O. and Treinin, A. (1970). *J. Phys. Chem.,* **74,** 3670
111. Collins, M. A., Cosgrave, M. M. and Betteridge, G. P. (1970). *J. Phys. B.,* **3,** L48-52
112. Begum, A., Subramanian, S. and Symons, M. C. R. (1970). *J. Chem. Soc. (A),* 918
113. Durie, R. A., Legay, F. and Ramsay, D. A. (1960). *Canad. J. Phys.,* **38,** 444
114. Coleman, E. H., Gaydon, A. G. and Vaidya, W. M. (1948). *Nature,* **162,** 108
115. Amichai, O. and Treinin, A. (1970). *J. Phys. Chem.,* **74,** 830
116. Grushko, Yu. S., Murin, A. N., Lur'e, B. G. and Motornyi, A. V. (1968). *Fiz. Tverd. Tela.,* **10,** 3704
117. Grushko, Yu. S., Lur'e, B. G. and Murin, A. N. (1969). *Fiz. Tverd. Tela.,* **11,** 2144
118. Dascent, W. E. and Waddington, T. C. (1960). *J. Chem. Soc.,* 3350
119. Dascent, W. E. and Waddington, T. C. (1963). *J. Inorg. Nucl. Chem.,* **25,** 132
120. Selte, K. and Kjikshus, A. (1968). *Acta. Chem. Scand.,* **22,** 3309
121. Duval, C. and Lecompte, J. (1960). *Rec. Trav. Chim.,* **79,** 523
122. Sherwood, P. M. A. and Turner, J. J. (1970). *Spectrochim. Acta.,* **26(A),** 1976
123. Selte, K. and Kjikshus, A. (1970). *Acta. Chem. Scand.,* **24,** 1912
124. Sherwood, P. M. A. and Turner, J. J. (1970). *J. Chem. Soc. (A),* 2349
125. Studier, M. H. and Huston, J. H. (1967). *J. Phys. Chem.,* **71,** 457
126. Noble, P. N. and Pimentel, G. C. (1968). *Spectrochim. Acta.,* **24A**, 797
127. Schwage, I. and Arkell, A. (1967). *J. Amer. Chem. Soc.,* **89,** 6006
128. Hedberg, K. and Badger, R. M. (1951). *J. Chem. Phys.,* **19,** 508
129. Lindsey, D. C., Lister, D. G. and Millen, D. J. (1969). *Chem. Commun.,* 950
130. Ashby, R. A. (1967). *J. Mol. Spectrosc.,* **23,** 439
131. Morris, J. C. (1966). *J. Phys. Chem.,* **70,** 3798
132. Jolles, Z. E. (1966). *Bromine and its Compounds,* (New York: Academic Press)
133. Kieffer, R. G. and Gordon, G. (1968). *Inorg. Chem.,* **7,** 235
134. Kieffer, R. G. and Gordon, G. (1968). *Inorg. Chem.,* **7,** 239
135. Thompson, R. and Gordon, G. (1967). *Inorg. Chem.,* **6**, 633
136. Pergola, F., Guidelli, R. and Raspi, G. (1970). *J. Amer. Chem. Soc.,* **92,** 2645
137. Pethybridge, A. D. and Prue, J. E. (1967). *Trans. Faraday Soc.,* **63,** 2019
138. Dawber, J. G. (1968). *J. Chem. Soc. (A),* 1532
139. Rogers, M. T. and Helmholtz, L. (1941). *J. Amer. Chem. Soc.,* **63,** 278
140. Garett, B. S. (1954). ORNL-1745, Oak Ridge National Lab., Tennessee; Diss. Abs. (1954). 14, 1152
141. Feikema, Y. D. and Vos, A. (1966). *Acta. Crystallogr.,* **20,** 769
142. Durig, J. R., Bonner, O. D. and Braeseak, W. H. (1965). *J. Phys. Chem.,* **69,** 3886
143. Pearson, G. S. (1966). *Advances in Inorganic Chemistry and Radiochemistry,* **8,** 117, (New York: Academic Press); see also, *Oxidation Combust. Rev.* (1969). **4,** 1
144. Fujimoto, S. (1970). *Kagaken Kogyo,* **21,** 641
145. Fujimoto, S. (1970). *Kagaken Kogyo,* **21,** 958
146. Jacobs, P. W. M. and Whitehead, H. M. (1969). *Chem. Rev.,* **69,** 551. See also Keenan, A. G. and Siegmund, R. F. (1969). *Quart. Rev.,* **23,** 430
147. Clark, A. M., Beagley, B., Cruickshank, D. W. J. and Hewitt, T. G. (1970). *J. Chem. Soc. (A),* 1613
148. Akishin, P. A., Vikov, L. V. and Rosolovskii, V. (1959). *Kristallografiya,* **4,** 353
149. Giguere, P. A. and Savoie, R. (1961). *Canad. J. Chem.,* **40,** 495
150. McDonald, W. S. and Cruickshank, D. W. J. (1967). *Acta. Crystallogr.,* **22,** 37
151. Clark, A. H., Beagley, B., Cruickshank, D. W. J. and Hewitt, T. G. (1970). *J. Chem. Soc. (A),* 872
152. Olovsson, I. (1968). *J. Chem. Phys.,* **49,** 1063
153. Lee, F. S. and Carpenter, G. B. (1959). *J. Phys. Chem.,* **63,** 279
154. Nordman, C. E. (1962). *Acta. Crystallogr.,* **15,** 18
155. Pavia, A. C. and Giguère, P. A. (1970). *J. Chem. Phys.,*
156. Duerst, R. W. (1968). *J. Chem. Phys.,* **48,** 2275
157. Redlich, O., Duerst, R. W. and Merbach, A. (1968). *J. Chem. Phys.,* **49,** 2986
158. Akitt, J. W. Carington, A. K., Freeman, J. G. and Lilley, T. H. (1969). *Trans. Faraday Soc.,* **65,** 2701
159. Heinzinger, K. and Weston, R. E. (1965). *J. Chem. Phys.,* **42,** 272
160. Covington, A. K., Tait, M. J. and Wynne-Jones, W. F. K. (1965). *Proc. Roy. Soc.,* **286A,** 235

161. Heath, G. A. and Majer, J. R. (1964). *Trans. Faraday Soc.*, **60,** 1783
162. Fisher, I. P. (1967). *Trans. Faraday Soc.*, **63,** 684
163. Appelman, E. H. (1968). *J. Amer. Chem. Soc.*, **90,** 1900
164. Appelman, E. H. (1969). *Inorg. Chem.*, **8,** 223
165. Studier, M. H. (1968). *J. Amer. Chem. Soc.*, **90,** 1901
166. Brown, L. C., Begun, G. M. and Boyd, G. E. (1969). *J. Amer. Chem. Soc.*, **91,** 2250
167. Siegel, S., Tani, B. and Appelman, E. (1969). *Inorg. Chem.*, **8,** 1190
168. Brand, J. R. and Bunck, S. A. (1969). *J. Amer. Chem. Soc.*, **91,** 6500
169. Johnson, G. K., Smith, P. N., Appelman, E. H. and Hubbard, W. N. (1970). *Inorg. Chem.*, **9,** 119
170. Schreiner, I., Osborne, D. W., Pocius, A. V. and Appelman, E. H. (1970). *Inorg. Chem.*, **9,** 2320
171. Appelman, E. H. and Studier, M. H. (1969). *J. Amer. Chem. Soc.*, **91,** 4561
172. Siebert, H. (1967). *Fortschr. Chem. Forsch.*, **8,** 470
173. Drátovský, M. and Pačesová, L. (1968). *Uspc. Khim.*, **37,** 537
174. Sklarz, B. (1967). *Quart. Rev. Chem. Soc.*, **1,** 21
175. Feikema, Y. D. (1961). *Acta. Crystallogr.*, **14,** 315
176. Feikema, Y. D. (1966). *Acta. Crystallogr.*, **20,** 765
177. Cronthamel, C. E., Hayes, A. M. and Martin, D. S. (1951). *J. Amer. Chem. Soc.*, **73,** 82
178. Buist, G. J. and Hipperson, W. C. P. (1969). *J. Chem. Soc. (A)*, 307
179. Buist, G. J. and Lewis, J. D. (1965). *Chem. Commun.*, 66
180. Aveston, J. (1969). *J. Chem. Soc. (A)*, 273
181. Kren, R. M., Dodgen, H. W. and Nyman, C. J. (1968). *Inorg. Chem.*, **7,** 446
182. Salomaa, P. and Vesala, A. (1966). *Acta. Chem. Scand.*, **20,** 1414
183. Mercier, E. E. and Farrer, D. T. (1968). *Canad. J. Chem.*, **46,** 2679
184. Engelbrecht, A. and Peterfy, P. (1969). *Angew. Chem. Intern. Edit. Engl.*, **8,** 768

9
Physical and Spectroscopic Properties of the Halogens

J. J. TURNER
University Chemical Laboratory, Cambridge, England

9.1 INTRODUCTION

In writing a review of the physical and spectroscopic properties of the halogens within a fairly short space there are two extreme courses. One is to attempt to list the most up-to-date values of all physical constants with some brief comment about each; the other is to seize on two or three items of current interest and discuss each at great length. The present article is a compromise and a personal one. I have chosen to expand on those aspects of halogen properties, mostly spectroscopic in various guises, which have led to the most exciting developments in the past few years. I hope, however, that I have not totally neglected any really important aspect.

It is also clear that it is not possible to elaborate on physical inorganic chemistry of the halogens. However, we are fortunate that Sharpe[1] has recently written a lucid article on this topic. Other valuable sources of data are to be found in Reference 2.

9.2 SPECTROSCOPIC PROPERTIES OF THE HALOGENS

9.2.1 Introductory comments on absorption and emission spectra of the halogens

Figure 9.1 summarises the main absorptions of the halogens in the ultraviolet and visible region of the spectrum[3]. There are many other important but weaker features in the spectra which will be discussed shortly.

Figure 9.2 is a schematic representation of just three of the energy levels for the halogens.

For a light molecule such as F_2, Hund's case (a)[5,6] is likely to apply and therefore the states are properly described as shown in the figure. The selection rules for absorption of light in this case are

$$\Delta\Lambda = 0, \pm 1 \quad \Delta\Omega = 0, \pm 1 \quad \Delta\Sigma = 0 \quad g \rightarrow u \quad + \rightarrow +$$

Thus of the possible transitions involving the ground state, ${}^1\Pi_u \leftarrow {}^1\Sigma_g^+$ and ${}^3\Pi_u \leftarrow {}^1\Sigma_g^+$, the former will be much more intense. Since the ${}^1\Pi_u$ state is repulsive the whole spectrum is a continuum as shown in Figure 9.1.

At the other extreme, for I_2, Hund's case (c)[5,6] is much more likely to apply. Thus the states are not correctly labelled; the correlations for (a) → (c) for the three states are

$${}^1\Sigma_g^+ \text{ to } 0_g^+,\ {}^1\Pi_u \text{ to } 1_u, \text{ to } {}^3\Pi_u \text{ to } 0_u^+ 0_u^- 1_u 2_u$$

It is important to note that whereas 0_u^-, 1_u and 2_u, derived from ${}^3\Pi_u$, can correlate with either two ground state atoms (${}^2P_{\frac{3}{2}}$) or one ground state atom

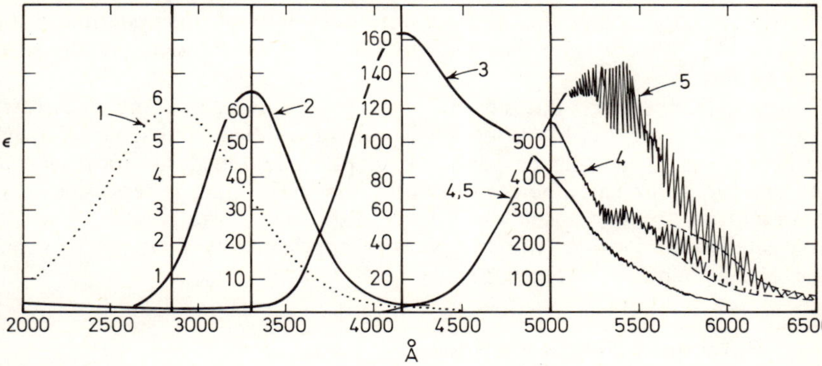

Figure 9.1 Absorption spectra of the halogens: (1) F_2(g), 25 °C; (2) Cl_2(g), 18 °C; (3) Br_2(g), 25 °C; (4) I_2(g), 70–80 °C; (5) I_2(g) plus 1 atm air, 70–80 °C; the weak banded region in Cl_2 which converges to the continuum at 4788 Å is not visible with the scale chosen. (From Calvert and Pitts[3], by courtesy of Wiley)

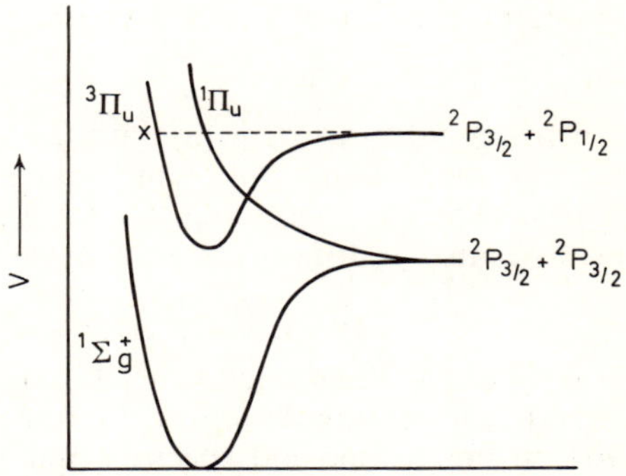

Figure 9.2 Schematic of three halogen energy levels – see reference 4

and one excited atom (${}^2P_{\frac{3}{2}}$ plus ${}^2P_{\frac{1}{2}}$) the 0_u^+ cannot correlate with two ground state atoms. Since in case (c) Λ and spin do not exist as rigid quantum numbers, the selection rules are

$$\Delta\Omega = 0, \pm 1 \quad g \rightarrow u \quad + \rightarrow +$$

Therefore, the allowed transitions in our example are

$$0_u^+ \leftarrow 0_g^+ \quad 1_u \leftarrow 0_g^+$$

More detailed consideration[4,5] shows that $0_u^+ \leftarrow 0_g^+$ is expected to be the more intense, as is the case for I_2, where the banded structure $\lambda \geqslant 4990$ Å is due to the transition from the ground state ($^1\Sigma(0_g^+)$) to $^3\Pi(0_u^+)$ below X in Figure 9.2. Absorption of shorter wavelength, i.e. to points on the upper curve above X, lead to dissociation and hence the spectrum in this region is continuous. The transition to $^1\Pi(1_u)$, dominant in F_2, is too weak to be observed.*

For the other two halogens the situation is intermediate. For Cl_2 the $^1\Pi_u \leftarrow {}^1\Sigma_g^+$ continuum dominates although the $^3\Pi(0_u^+) \leftarrow {}^1\Sigma_g^+$ bands and continuum are clearly seen at higher concentration, the dissociation limit being at ~4788 Å. For Br_2 the banded part of the $^3\Pi(0_u^+) \leftarrow {}^1\Sigma_g^+$ spectrum is clearly seen, the continuum being composed partly of this system and the $^1\Pi(1_u) \leftarrow {}^1\Sigma_g^+$ system. The dissociation limit for the banded system is at about 5148 Å.

Emission from excited states to either the ground state or some other low-lying state can be brought about in several ways – electric discharge, atom recombination, ultraviolet or visible absorption followed by fluorescence. The emission will clearly be continuous if the lower state is repulsive or if the transitions are, say, above the X level of the $^3\Pi(0_u^+)$ state.

We shall now consider those aspects of spectroscopy germane to other areas of the chemistry of the halogens.

9.2.2 Bond dissociation energies

For a recent discussion of the whole field of dissociation energies and spectra of diatomics the reader is referred to the third edition of Gaydon's book[6].

It should perhaps be emphasised that different physical techniques for obtaining the dissociation energy may measure different quantities:

for $$X_2 \rightarrow 2X$$

spectroscopic measurements give ΔE_0^0 ($\equiv \Delta H_0^0$) whereas, say, equilibrium thermal methods at room temperature give ΔH_{298}^0. Assuming that X_2 and X behave as ideal gases and that for X_2 the population of vibrational states higher than $v'' = 0$ is zero, the difference is calculated to be about 1 kcal mol^{-1}, i.e.

$$\Delta H_{298}^0 = \Delta H_0^0 + 1$$

(a) *Fluorine* – The value for the bond dissociation energy of F_2 is of particular interest – extrapolation of the other halogen data suggest a value near 90 kcal mol^{-1} and old spectroscopic studies on F_2 gave 63.3 and 70 ± 1. For several years, however, a value of 37–39 has been accepted based originally on deductions from the absorption spectrum of ClF and relevant thermodynamic quantities[6].

In view of the importance of this value in relation to the chemistry of F_2 there have been several more recent discussions some of which are considered below.

Since the main absorption spectrum of F_2 ($^1\Pi_u \leftarrow {}^1\Sigma_g^+$) is a continuum it is not possible to employ the most obvious method involving extrapolation

*See Note 1, p. 286

of banded structure to a dissociation limit. However, Rees[7], using the spectral data of Steunenberg and Vogel[8] and the vibrational Raman data of Andrychuk[9] estimated the dissociation limit from a consideration of the intensity distribution in the continuum – 37.1 ± 0.85 kcal.

Alternatively, Iczkowski and Margrave[10] observed several *band* systems for F_2 in the vacuum ultraviolet in the region 807–1035 Å. One particular progression of three bands allowed an estimate of the dissociation energy:

Å	ν (cm^{-1})	$\Delta\nu$
874.2	114 388	
		> 930*
867.2	115 318	
		> 800*
861.2	116 118	

*note the large anharmonicity

From a linear Birge-Sponer extrapolation the dissociation energy of the upper state is 3800 cm^{-1}. The dissociation energy of the ground state can be calculated from this and the energy of the dissociation products. The only acceptable dissociation products are ^2P_u and ^2P_g and since they are doublets they can be combined in four different ways to give possible values of the dissociation energy of 36.4, 37.4, 37.7, 38.6 – say 37.5 ± 2 kcal. However, Stricker and Krauss[11] from vacuum ultraviolet measurements in the same region extrapolated to give a value of 33.2 ± 1.1 kcal. Since this is lower than the currently accepted value, Margrave and Gole[12] have reinvestigated the spectrum under higher resolution and obtain 38.0 ± 0.5 meanwhile claiming that Stricker and Krauss's value results from a misinterpretation of the data.

There have been several other recent determinations using fairly novel techniques: De Corpo *et al.*[13] have very recently discussed some of these experiments.

By studying the flow of fluorine gas through an orifice Yates *et al.*[14] obtained a value of 41.8 ± 0.2 at 298 K. The quoted error is from an analysis of the variation in results; the authors acknowledge that the method has several inherent large errors.

At the other extreme Dibeler[15] from photoionisation and mass spectral measurements, obtained $D_0^0(F_2) = 30.9 \pm 0.7$ kcal. There is, however, a recent suggestion[16] that at higher sensitivity another process exists which may confuse the interpretation of data which ignores this process. This more recent work suggests 36.8 ± 0.2 kcal.

The most recent work[13] presents what is likely to prove the definitive mass spectrometric work on fluorine. Appearance potentials and translational energies of the electron impact products, both positive and negative ions, give four *independent* values for $D_0^0(F_2)$ – see Table 9.1.

NBS circular 270–3 [17] quotes for fluorine,
$\Delta H_{f0}^0(F) = 18.38$, $\Delta H_{f298}^0(F) = 18.88$ kcal mol^{-1} i.e.
$D_0^0(F_2) = 36.76$, $D_{298}^0(F_2) = 37.76$ kcal mol^{-1}.

(b) *Chlorine* – For chlorine, bromine and iodine we shall only consider spectroscopic determinations of the dissociation energy.

In 1963 Douglas *et al.*[18] examined the absorption band system of $^{35}Cl_2$ (to simplify the spectrum) in the region 4780–6000 Å, i.e. the $^3\Pi(0_u^+) \leftarrow {}^1\Sigma_g^+$

system. By working at -80 °C it was clear which absorptions originated in the vibrational ground state ($v''=0$) but they had to use the emission data of Khanna and Venkataswarlu and Khanna[19] to number the v' states. The convergence limit of the bands was at $20\,880\pm2$ cm^{-1} (4788 Å) above the $v''=0$ level. In addition, the observation that the rotational fine structure of the last four vibrational levels breaks off at approximately the same place is good evidence for there being no maximum in the $^3\Pi(0_u^+)$ state, which

Table 9.1 Mass spectral results on F_2
(From De Corpo *et al.*[13] by courtesy of American Institute of Physics)

Process		*Ion mode*	AP^a(eV)	E^{*b} (kcal mol^{-1})	$D^0(F_2)$ (kcal mol^{-1})
Dissociative electron attachment	$F_2+e\rightarrow F^-+F$	negative	0.0 ± 0.1	41.2 ± 2.0	38.3 ± 3.0
Ion-pair formation	$F_2+e\rightarrow F^++F^-+e$	negative	15.8 ± 0.1	6.0 ± 0.5	36.1 ± 2.4
	$F_2+e\rightarrow F^++F^-+e$	positive	15.8 ± 0.1	5.0 ± 0.5	37.1 ± 2.4
Dissociative ionisation	$F_2+e\rightarrow F^++F+2e$	positive	19.2 ± 0.2	0.0 ± 1.0	38.6 ± 4.8

^{a} AP = appearance potential; ^{b} E^* = excess energy

would give too high a value of the dissociation energy. Since $^3\Pi(0_u^+)$ dissociates to $^2P_{\frac{3}{2}}$ plus $^2P_{\frac{1}{2}}$, and $^2P_{\frac{1}{2}}$ is 881 cm^{-1} above $^2P_{\frac{3}{2}}$ this gives

$D_0^0(Cl_2) = 19\,999\pm2$ cm^{-1} (57.05 ± 0.01 kcal mol^{-1})*

A little earlier Rao and Venkataswarlu[20] had discovered in the region 1830–1400 Å a remarkable series of resonance doublets obtained by electrical excitation of Cl_2 vapour. (For a more detailed description of this sort of system see the iodine section.) They suggested that Cl_2 in $v''=0$, $J''=23$ of $^1\Sigma_g^+$ absorbs the 73 980.2 cm^{-1} line of atomic Cl and is excited to $J_r=22$ of v' of some unspecified excited state. They observed emission from this level to all v'' levels up to the dissociation limit of the ground state; this gives $v''=0$ to dissociation $=20\,062\pm10$ cm^{-1}—note that this measures directly dissociation energy to two ground state ($^2P_{\frac{3}{2}}$) atoms. Very recently Clyne and Coxon[21] suggest that in the RV[20] work the transition is to $J_r=21$ which leads to a slightly different value of the dissociation energy—$D_0^0(Cl_2) = 20\,040\pm20$. However, there is still a discrepancy between absorption and resonance conclusions perhaps because the ground state has a slight maximum.†

(c) *Bromine*—Horsley and Barrow[22] have examined the absorption spectrum of $^{79}Br^{79}Br$ and $^{81}Br^{81}Br$ in the 6200–5100 Å region—$^3\Pi(0_u^+)\leftarrow{}^1\Sigma_g^+$. Following the usual methods they obtained values for both isotopic species for dissociation from $v''=0$ to $^2P_{\frac{3}{2}}$ plus $^2P_{\frac{1}{2}}$:

$^{79}Br^{79}Br$ 19 577.2±0.5 cm^{-1} (5148 Å)
$^{81}Br^{81}Br$ 19 578.9±0.5 cm^{-1} (5148 Å)

Since $^2P_{\frac{1}{2}}$ is 3685 cm^{-1} above $^2P_{\frac{3}{2}}$

$$\left.\begin{aligned}D_0^0(^{79}Br^{79}Br) &= 15\,892.2\text{ cm}^{-1}\\ D_0^0(^{81}Br^{81}Br) &= 15\,893\text{ cm}^{-1}\end{aligned}\right\}\text{ mean}\quad\begin{aligned}&= 15\,893.1\pm1\text{ cm}^{-1}\\ &= 45.440+0.003\text{ kcal mol}^{-1}\end{aligned}$$

*See Note 2, p. 286
†See Note 3, p. 286

Since they were able to analyse in detail the last four vibrational bands (v', v'' 49,0–52,0) and show there was no maximum in the upper state the D_0^0 values are very accurate.*

(d) *Iodine*—There has been very considerable careful examination of the spectra of iodine because of this molecule's importance in photochemistry, lasers, quenching and energy transfer. Some of these topics will be considered

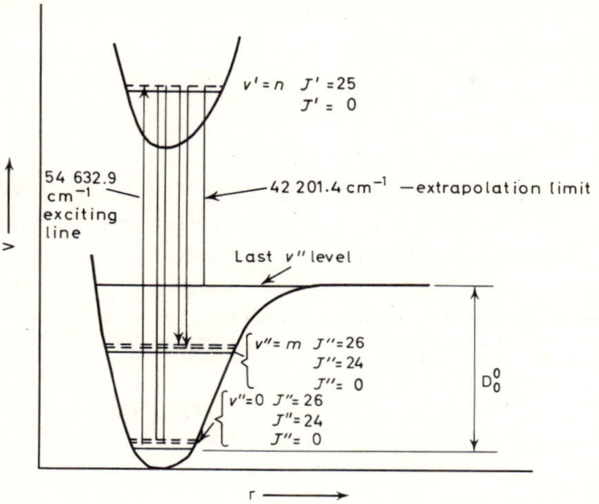

Figure 9.3 Schematic of some of levels observed in Verma's resonance study

later but in the present context some of these interesting spectral studies have provided values of the dissociation energy.

Because of the low vibrational frequency of I_2, at room temperature other levels than $v'' = 0$ are populated. Thus exact estimation of the convergence limit for bands of the $^3\Pi(0_u^+) \leftarrow {}^1\Sigma_g^+$ system is very difficult.

However, Verma[23] examined the u.v. resonance spectrum of I_2 in the 1830–2370 Å region. The transitions excited by the 1830.4 Å line of atomic iodine are

			Emission system
$v'' = 0, J'' = 23 \rightarrow$	$v' = n,$	$J' = 22$	IVa
24		$= 25$	IVb
47	$v' = n+1,$	$J' = 46$	II
48		49	III
86	$v' = n+4,$	$J' = 87$	I

In the resulting emission to the ground state the IVb series converges at 42 201.41 cm^{-1}. Figure 9.3 illustrates the relationship between the various energy terms.

$$D_0^0 = \text{exciting line (cm}^{-1}) + F''24(v'' = 0 \text{ level}) - \text{extrapolation limit (cm}^{-1})$$
$$= 54\,632.9 + 22.4 - 42\,201.4$$
$$= \underline{124\,53.9 \text{ cm}^{-1}}$$

This extrapolated value may however be too high by an amount equal to the

* See Note 2, p. 286

height of the $J'' = 24$ above $J'' = 0$ in the last v'' level. Estimating $B_{v''}$ in this level, from values for preceding levels, to be 2.9 cm^{-1}, gives a minimum value of $D_0^0 = 12451.0$ cm^{-1}. Taking the mean $D_0^0 = 12452.5 \pm 1.5$ cm^{-1} (35.603 ± 0.005 kcal mol^{-1}).

In the most recent examination[24] of the band/continuum transition in the $^3\Pi(0_u^+) \leftarrow {}^1\Sigma_g^+$ region the last discrete band observed was at 4990.59 Å (i.e. lower than the previously accepted value of Kuhn[25] – 4995 Å). Combining this with the $^2P_{\frac{1}{2}} - {}^2P_{\frac{3}{2}}$ separation for I gives:

$$D_0^0 = 20\,040 - 7603.15 \text{ cm}^{-1}$$
$$\sim 12\,436.85 \text{ cm}^{-1}$$
$$= 35.55_8 \text{ kcal mol}^{-1}*$$

Steinfeld *et al.*[24] suggested that the discrepancy† in the two values (16 cm^{-1} ~0.05 kcal mol^{-1}) was due to vibrational overlap in the absorption studies but they also suggested, contrary to all previous views, that the continuum

Table 9.2 Values for the dissociation energies of the halogens

F_2	Cl_2	Br_2	I_2
37.1 ± 0.85 [7]	57.05 ± 0.01 [18]	45.440 ± 0.003 [22]	35.603 ± 0.005 [23]
37.5 ± 2 [9]	57.00 ± 0.005 [157]		35.569 ± 0.003 [157]
38.0 ± 0.5 [12]	57.30 ± 0.06 [20]		35.55_8 [24]
36.8 ± 0.2 [16]			
37.5 ± 2.3 [13]			
NBS Note 270-3 [17]			
D_0^0 36.76	57.36	45.46	35.60
D_{298}^0 37.76	58.16	46.09	36.15

may not be due to transitions to the $^3\Pi(0_u^+)$ state. Fortunately, a very detailed study[26] of the spectroscopic data disproves this latter point.‡ Table 9.2 collects the data discussed. For a discussion of their significance see Section 9.2.3(b).

9.2.3 Spectroscopic constants and potential curves

(a) *Ground State* ($X^1\Sigma_g^+$) *and* $B^3\Pi(0_u^+)$ – In Table 9.3 are collected some spectroscopic constants determined, in the past few years for the ground states of the halogens, and also the ionisation potentials. Table 9.4 lists corresponding data for the $^3\Pi(0_u^+)$ state.

(b) *Other states of* X_2 (*and* X_2^+) – The banded emission spectrum of F_2, first studied by Gale and Monk[32] and Aars[33], has been re-examined by Porter[34] by passage of a low-frequency, high-voltage discharge in fluorine flowing through a sapphire tube. The variation in intensity in the fine structure shows that the transition must be either $C^1\Sigma_u^+ \rightarrow B^1\Pi_g$ or $C^1\Sigma_g^- \rightarrow B^1\Pi_u$. Eight of the nine bands analysed give the rotational constants for

* See Note 2, p. 286
† See Note 4, p. 287
‡ See Note 1, p. 286

Table 9.3 Spectroscopic constants for ground states $X(^1\Sigma_g^+)$ of the halogens; ionisation potentials

	F_2	$^{35}Cl_2$					Br_2 (HB)[22]			I_2	
	R[27]	RV[20]	RB[28]	DMS[18]	CC(70)[21]	CC(67)[29]	$^{79}Br^{79}Br$	$^{81}Br^{81}Br$	$^{79}Br^{81}Br$ (calc.)	V[23]	RR[30]
$\omega_e{}^a$ (cm^{-1})	923	560.50	559.78	559.71	559.72	559.8	325.366	321.29	323.33	214.52	214.51886
$x_e\omega_e$	15.6	2.904	2.707	2.70	2.675	2.675	1.0985	1.064	1.081		0.60738
$y_e\omega_e$		0.2093			−0.0067	−0.006					−0.001307
$z_e\omega_e$		−0.0013798									0.00000504
$t_e\omega_e$		0.00002852									−0.00000016
$s_e\omega_e$		−0.0000002553									
$B_e{}^b$		0.2438	0.2436_9	0.24407	0.2439_9		0.082114	0.080088	0.081101	0.03734	0.037389
α_e		0.00168	0.0015_6	0.00153	0.0014_9		0.000322	0.000319	0.000321	0.0001208	0.000120
γ_e		−0.0000198			-0.000001_7					4.44×10^{-7}	1.90×10^{-8}
D_e		1.84×10^{-7}			$\sim 1.85 \times 10^{-7}$		2.09×10^{-8}	1.99×10^{-8}	2.05×10^{-8}	1.99×10^{-9}	4.54×10^{-9}
β_e										1.236×10^{-10}	1.20×10^{-11}
r_e(Å)	1.418	1.988	1.989_0	1.9878	1.9881		2.2809	2.2809	2.2809	2.668	2.660
Ionisation[c] $X_2 \rightarrow X_2^+$	D_0^0 365.1 D_{298}^0 366.6	D_0^0 264.8	D_{298}^0 266.3				D_0^0 244.7	D_{298}^0 246.1		D_0^0 214.8	D_{298}^0 216.3
potentials $X_2 \rightarrow X_2^{2+}$ (kcal mol^{-1})		752	755								

[a] $G_v = \omega_e(v+\frac{1}{2}) - x_e\omega_e(v+\frac{1}{2})^2 + y_e\omega_e(v+\frac{1}{2})^3$ etc.

[b] $\left.\begin{array}{l} B_v = B_e - \alpha_e(v+\frac{1}{2}) + \gamma_e(v+\frac{1}{2})^2 \\ D_v = D_e + \beta_e(v+\frac{1}{2}) + \delta_e(v+\frac{1}{2})^2 \end{array}\right\} F_v(J) = B_vJ(J+1) - D_vJ^2(J+1)^2$

[c] Ref. 17

Coxon[159] has very recently re-examined the $^{79}Br^{79}Br$ spectrum: ω_e, 325.29; $x_e\omega_e$, 1.072; B_e, 0.082121; α_e, 0.000323; D_e, 2.095×10^{-8}; r_e(Å), 2.2808

Table 9.4 Spectroscopic constants for $B(^3\Pi(0_u^+))$ states of the halogens

	Cl_2				Br_2		I_2
	RB[28]	DMS[18]	CC(70)[21]	$CC(67_1)$[29]	$^{79}Br^{81}Br$(HB)[22]	$CC(67_2)$[31]	SCW[24]
$\omega_e\ (cm^{-1})$	259.57	261.9	259.5	259.5	167.85	167.9	125.531
$x_e\omega_e$	4.753	5.45	5.3	5.3	1.73	1.84	0.73389
$y_e\omega_e$	-0.067_7				−0.0059		0.004133
$z_e\omega_e$	−0.00212						0.000001195
$t_e\omega_e$							2.208×10^{-7}
B_e	0.1688	0.168_0	0.162_6		0.0585		0.028873
α_e	0.0031	0.0037_2	0.0021_2		0.00041		0.0001345
γ_e	0.0000527		-0.00009_1		−0.00001		-1.148×10^{-6}
D_e			2.365×10^{-7}		2.8×10^{-8}		3.5×10^{-9}
β_e			0.225				3.9×10^{-10}
r_e(Å)	2.39_0	2.39_6	2.43_5		2.68		3.034

*Coxon's very recent work on $^{79}Br^{79}Br$ gives: ω_e, 167.55; $x_e\omega_e$, 1.625; B_e, 0.059579, α_e, 0.0004868; D_e, 3.09×10^{-8}; r_e(Å), 2.6777

the $B^1\Pi$ and $C^1\Sigma$ states; the other band (5852.63 Å) is ascribed to a new transition $C^1\Sigma \rightarrow B'^1\Pi$. The analysis gives:

State	$B^1\Pi$	$B'^1\Pi$	$C^1\Sigma$
Number of vibrational levels observed	8	1	2
B_e (cm^{-1})	1.047*	1.005*	0.804
α_e (cm^{-1})	0.012*		0.008
r_e(Å)	1.302*	1.329*	1.485
ω_0 (cm^{-1})	1100*		1110

* Approximate values because of uncertainty in vibrational numbering

Of the two possible assignments involving B and C, a $^1\Pi_g$ state for B, because of excitation from antibonding orbital, would be expected to have a smaller r than ground state F_2 ($X^1\Sigma_g^+$) and a $^1\Pi_u$ state a larger r. Since $r_0 \sim 1.435$ Å and $r_e(B^1\Pi) \sim 1.302$ Å, this shows the transition to be

$$C^1\Sigma_u^+ \rightarrow B^1\Pi_g$$

Porter also showed that many of the bands observed were due to the $A^2\Pi \rightarrow X^2\Pi$ systems of F_2^+ – the first observation of the emission spectrum of this molecule; although Stricker[35] claimed to have observed F_2^+ in emission, the more detailed treatment by Porter indicates persuasively that Stricker actually observed bands due to F_2.

Analysis of the F_2^+ bands gives:

State	$X^2\Pi$	$A^2\Pi$
Number of vibrational levels observed	5	3
B (cm^{-1}) for lowest v observed	1.010	
r (Å)	1.326	
α (cm^{-1})	~0.01	
Corrected ΔG for lowest v observed	1054.5	498.6
$\omega_e x_e$ estimated	9.1	7.3

Table 9.5 collects some data on ground electronic and vibrational states of X_2 and X_2^+ for comparison. This table draws attention to some interesting features. The bond dissociation energies and force constants of I_2, Br_2 and Cl_2 increase steadily with decreasing atomic weight. Removal of an antibonding electron to give X_2^+ is expected to lead to increase in bond energy and force constant and a decrease in bond length; again Cl_2^+, Br_2^+ and I_2^+ behave entirely as expected. However, although the force constants for F_2 and F_2^+ follow the same trend the bond energies do not. Thus F_2^+ is more tightly bound than F_2, as expected, but both F_2 and F_2^+ have lower D_0^0 values than Cl_2 and Cl_2^+ respectively. For comparison it would be valuable to have data for the corresponding negative ions X_2^-. Unfortunately, although these species have been examined by pulse radiolysis[38], irradiation or photolysis of substrates at low temperature[39] and reaction of sodium atoms with halogen at low temperature[40], the resulting ultraviolet or e.s.r. spectra do not give information directly about force constants and bond energies. However, Haas and Griscom[40a] have obtained the Raman spectrum of Cl_2^- following

γ irradiation of borate glass with small amounts of alkali chloride. The frequency, 265 cm^{-1}, is about half the free neutral molecule (554 cm^{-1}) as expected. It is slightly higher than the V_k centre value (235 cm^{-1}) in KCl and Griscom[40b] argues, from optical data, that the Cl_2^- in the glass is closer

Table 9.5 Some data on X_2 and X_2^+ species

	F_2	F_2^+	Cl_2	Cl_2^+	Br_2	Br_2^+	I_2	I_2^+
ω_e (cm^{-1})	923	~1100	559.7	645.61, 644.77*	323.3	376.0†	214.5	~238‡
$x_e\omega_e$ (cm^{-1})	15.6	~91	2.70	3.015, 2.998	1.08	1.25	0.61	
B_e (cm^{-1})		1.010		0.26950, 0.2697				
r_e(Å)	1.418	1.326	1.988	1.8917, 1.8910	2.281		2.666	
k§ (mdyne $Å^{-1}$)	~4.45	6.77	3.23	4.29	2.46	3.33	1.72	2.12
D_0^0 (kcal mol^{-1})‖	36.8	73.5	57.2	92.6, 99.3¶	45.4	73.8	35.6	61.9

* From two different systems; see Reference 36
§ Calculated from ω_e
‖ $D_0^0(X_2^+) = D_0^0(X_2) + IP(X)$ (Table 9.8) $- IP(X_2)$ (Table 9.3)
† Reference 36
‡ Reference 37
¶ Birge-Sponer extrapolation, Reference 36

to the value which the 'free' ion would have. The dissociation bond energy of X_2^- into X^- and X is related to that of X_2 by

$$D(X_2^-) = +\text{electron affinity } (X_2) + D(X_2) - \text{electron affinity } (X)$$

By semi-empirical methods, Person[41] estimated the adiabatic* electron affinities of molecular Cl_2, Br_2 and I_2: $55\pm7, 57\pm7, 52\pm7$ kcal mol^{-1} respectively. This gives approximate values for $D(X_2^-)$ of Cl_2^- (28 kcal mol^{-1}), Br_2^- (23), I_2^- (16), i.e. substantially weaker than the corresponding halogen as expected, and also in the expected order.

The iodine value has been confirmed in an elegant pulse radiolysis/flash photolysis study by Baxendale and Bevan[41a]. Radiolysis of air-saturated I^- solution gives OH which reacts with I^-, $OH + I^- \rightarrow OH^- + I$, the I then reacting with I^- to give the unstable I_2^-. The concentration of I_2^-, monitored spectrophotometrically, allows estimation of the equilibrium constant, and by temperature variation, the heat of reaction for

$$I_{aq} + I_{aq}^- \rightleftharpoons I_{2aq}^- \quad \Delta H^0 = -5.6 \text{ kcal mol}^{-1}$$

Making a reasonable estimate for ΔH(solvation) of I_2^-,

$$I_g + I_g^- \rightleftharpoons I_{2g}^- \quad \Delta H^0 = -28\pm10 \text{ kcal mol}^{-1}$$

This, of course, leads to 64 ± 10 kcal for the electron affinity of I_2 (cf. Person's 52 ± 7).

For fluorine, however, Balint-Kurti and Karplus[42] have recently performed 'multistructure valence-bond and atoms-in-molecules' calculations. These suggest that, although the equilibrium bond length in F_2^- is greater than F_2, the bond strength is *greater* by some 17 kcal mol^{-1}.

The anomalous behaviour of fluorine has been much discussed in the past and this is not the place to take up the whole issue again. It is, however,

*i.e. from minimum of X_2^- ground state potential to minimum of X_2 ground state potential curve.

tempting to suggest that the above behaviour is exactly what one would expect if the low bond energy of F_2 is due to repulsion of non-bonding electrons: removal of an electron to give F_2^+ does not relieve this effect so that F_2^+ is also anomalous, but addition of an electron to give F_2^- increases the bond length, relieves the non-bonding repulsion and the net effect is that F_2^- fits in the trend of I_2^-, Br_2^- and Cl_2^-. However, the recent article by Politzer[43] argues that the fluorine atom or ion is anomalous with respect to the other halogens as measured by dissociation energy of alkali halides, hydrogen halides and methyl halides and even electron affinity of the halogen itself. This 'destabilisation energy' is about 26 kcal mol^{-1} for either F *or* F^-. Thus in F_2 it should be $2 \times 26 = 52$ kcal mol^{-1}; extrapolation of $D_0^0(X_2)$ for I_2, Br_2 and Cl_2 gives about 90 for F_2, i.e. some 52 kcal above the observed value! Politzer suggests we should be more cautious in suggesting that the low bond energy of F_2 is due to the non-bonding electron repulsion.

It should be added for completeness that in addition to the well known blue I_2^+ species, observed in oleum and fluorosulphuric acid by Gillespie[44], there is growing evidence for other X_2^+ species in condensed phases. For example, the e.s.r. spectrum[45] of ClF in SbF_5 suggests the presence of Cl_2^+, although Symons and co-workers[45a] have argued persuasively that the spectrum was actually due to Cl_2O^+; from Br_2 in SbF_5 and BrF_5, Edwards *et al.*[46] have isolated a scarlet solid, $Br_2^+(Sb_3F_{16})^-$ and an x-ray examination of this gives a Br—Br bond length of 2.13 Å.

From the selection rules given in Section 9.2.1, $0_g^+ \rightarrow 0_u^+$ and 1_u for case (c), we can expect to see for the heavier halogens in absorption

$$^3\Pi(0_u^+),\ ^1\Pi(1_u),\ ^3\Pi(1_u) \leftarrow 1\Sigma(0_g^+)$$

So far we have only considered transitions to the first two states; the third transition is not seen in F_2 or Cl_2* but was first observed for Br_2 and I_2 by Brown in 1931[47]. Recent analyses of the $A^3\Pi(1_u)$ state for Br_2 in absorption (6450–7600 Å) and emission and for I_2 give:

	Br_2	I_2
ω_e (cm^{-1})	153 [31]	44.0 [5]
$x_e\omega_e$ (cm^{-1})	2.7 [31]	1.01 [5]
$y_e\omega_e$ (cm^{-1})	−0.015 [31]	+0.008 [5]
B_e (cm^{-1})	0.0657 [48]	
α_e (cm^{-1})	0.00155 [48]	
r_e (Å)	2.55 [48]	

We have not considered so far the determination of potential curves by any of the well known methods.† For details the reader is referred to Herzberg[5], Mulliken and for some recent considerations of the halogens, Cl_2 [49], Br_2 [49], I_2 [50].

9.2.4 Resonance fluorescence and resonance Raman effect

The resonance fluorescence (RF) spectra of the halogens in the ultraviolet and visible have been studied for many years and we have so far considered

* Although Clyne and Coxon[29] see evidence for the state in emission
† See Note 2, p. 286

examples relevant to the determination of bond dissociation energies although making no comment about lifetimes of the states involved.

Gas-phase Raman spectroscopy until very recently was restricted to studies of molecules with no electronic absorption overlapping the exciting line(s). In the past few years, in the liquid and solid phase, largely with the advent of laser sources, the Raman spectra of compounds with overlapping absorptions have been investigated. This phenomenon, known as the Resonance Raman Effect (RRE) has been discussed by Behringer[51]. A nice example in halogen chemistry is Gillespie's and Morton's[37] work on I_2^+. Figure 9.4 shows the Raman spectrum for 10^{-2} molar I_2^+ in fluorosulphuric acid. The spectrum is clearly distinguished from RF since the same Raman shifts are observed in the other He/Ne line at 6118 Å. Note the increasing width of the overtones and their relatively high intensities. Gillespie and Morton suggest that the RRE spectrum is so intense that it might form a sensitive method for the estimation of I_2 and I^- by oxidation to I_2^+ using 65% oleum.

More recently, Holzer *et al.*[52] have examined the RF and RRE in halogen *gases* using a variety of laser sources. The different characteristics of RF and RRE as quoted by the authors are:

	RF	RRE
Band envelope	Very sharp doublet lines at low pressure	Broad regular Q branch and rotational wings
Overtone pattern	Irregular overtone sequence of doublets, some lines might be completely missing	Continuous broadening of the Q branch and continuous decrease in peak intensity with higher order.
Depolarisation ratio	Depolarised	Bands are polarised if expected to be polarised in Raman.
Behaviour with increasing gas pressure	Intensity decreases, (quenching) doublet of sharp lines changes into multiple structure	Intensity increases, shape of the band does not change considerably.
Behaviour with foreign gases added	As above	Intensity and band shape do not change considerably.

Holzer *et al.* observed the effects, shown in Table 9.6, with different laser frequencies. Thus at energies below the dissociation limit, i.e. in the banded region, RF occurs, we ignore for the moment any predissociation. At energies above the dissociation limit, i.e. in the continuum, RRE occurs. A theoretical explanation of this phenomenon has been given by Behringer[53]. It should perhaps be added that in the Raman scattering process it is not correct to think of the molecule as 'absorbing' the radiation – if this were the case, then in the continuum the molecule would simply dissociate to $^2P_{\frac{1}{2}}$ and $^2P_{\frac{3}{2}}$ atoms. Full data on overtone frequencies, polarisation data, relative Raman scattering cross-sections are given in the original paper[52].

The RF is usually several orders of magnitude more intense than RRE. However, recently Kiefer and Schrötter[54] have observed simultaneously the RF and

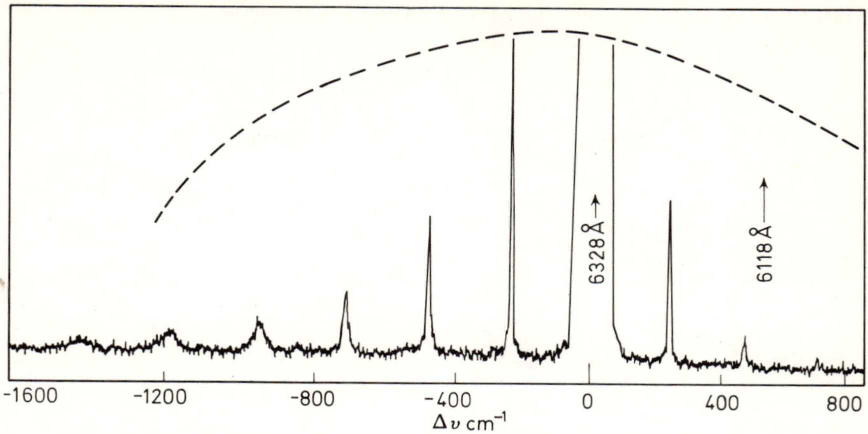

Figure 9.4 Resonance Raman spectrum of a 10^{-2} m solution of I_2^+ in fluorosulphuric acid: (---), contour of visible absorption band. (Fron Gillespie and Morton[37], by courtesy of Academic Press)

Table 9.6 RF and RRE effects in halogen gases

(From Holzer *et al.*[52], by courtesy of American Institute of Physics)

Molecule	*Wavelength of excitation* (Å) 5145	5017	4965	4880	4765	*Approx. dissociation limit for* $^3\Pi(0_u^+) \leftarrow {}^1\Sigma_g^+$ (Å)
Cl_2	R*	R	R	R, RF	R	4788
Br_2	RF	RRE	RRE	RRE	RRE	5148
I_2	RF	RF	RRE	RRE	RRE	4990
BrCl	RRE	RRE	RRE	RRE	RRE	5344
ICl	RRE	RRE	RRE	RRE	RRE	5510
IBr	RRE	RRE	RRE	RRE	RRE	5477

* Raman effect only, probably because absorption very weak

RRE in gaseous bromine excited with a quasi-continuous ruby laser at 6943 Å. Note that the energy corresponding to this wavelength is too small to excite Br_2 from $v'' = 0$ to the $^3\Pi(0_u^+)$ level; however, it does fall in the $^3\Pi(1_u) \leftarrow {}^1\Sigma_g^+$ region (6450–7000 Å). KS[54] suggest that the RRE occurs via 'absorption' into this state but that the RF occurs via transitions from excited vibrational states ($v'' \geqslant 5,6,7..$) to $^3\Pi(0_u^+)$. This leads them to suggest that Stammreich's spectrum[55] was essentially fluorescence and Delhaye[56] could not distinguish between RF and RRE. Table 9.7 collects the relevant data. The discrepancy in the Br_2 values is because of vibrational overlap.

The properties so far considered have been essentially molecular—the only atomic properties of importance have been the energies of the $^2P_{\frac{1}{2}}$ states relative to the $^2P_{\frac{3}{2}}$ states. Before considering some phenomena dependent on both molecular and atomic properties we consider some properties of the isolated atom.

Table 9.7 Vibrational data for halogens (and interhalogens)

	Raman gas[52] (cm^{-1})	*Gas (literature)*	*Raman in* CCl_4 solution*	*I.R. gas**	$\Delta G\frac{1}{2}$ (i.e. $v'' = 0 \rightarrow v'' = 1$) *from visible absorption*
$^{35}Cl_2$	554 ± 0.5	557.5 ± 1 [52]	548		554.3†
Br_2	318.5 ± 0.5	316.8 [55]			321.17†
	319	316 [56]			
I_2	213 ± 0.5				213.3†
$Br^{35}Cl$	440 ± 1		431	439.5	439.9*
$I^{35}Cl$	381 ± 1		375	381.5	381.3*
IBr	265 ± 1		261.5		266.0*

* *See* Table 5 in Reference 52
† *See* Table 9.3

9.2.5 Spectral properties of halogen atoms and related data

(a) *Term values and ionisation potentials*—Moore[57] has tabulated an extensive set of term values for the halogen atoms observed in emission. Table 9.8 gives data for some low-lying states and also successive ionisation potentials. The table contains no surprises but there are some observations of interest and significance. Firstly, note that the spin-orbit splitting in the ground state increases markedly from F (404 cm^{-1}) to I (7603 cm^{-1})—the importance of this will be apparent in Section 9.2.6(c). Secondly, there are no transitions to the ground state with energies corresponding to the usual ultraviolet/visible region of the spectrum—the longest wavelength transition will be I $(5s^25p^4(^3P)6s(^4P_{\frac{5}{2}})) \leftarrow 5s^25p^5(^2P_{\frac{3}{2}})$ at 54633 cm^{-1} (= 1830 Å). In the past few years Husain and Donovan and collaborators[59] have employed kinetic flash spectroscopy to measure transitions in absorption in the vacuum ultraviolet from *both* spin-orbit components of the ground state. This technique, a development of the early work of Norrish and Porter[60], employs a high energy photolytic flash in halogen gas to generate halogen atoms, followed by a spectroscopic flash within the lifetime of the atoms. Table 9.8 contains some of their absorption data from the $^2P_{\frac{3}{2}}$ states for comparison with Moore's data. Husain and Donovan have used this technique to monitor the reactions of $^2P_{\frac{3}{2}}$ and particularly $^2P_{\frac{1}{2}}$ atoms (see Section 9.2.6(c).

(b) *Electron affinities*—This is the appropriate section to consider this topic since the most accurate methods of determining the electron affinities have been spectroscopic. In one method, developed by Branscombe and co-workers[61] a mass-selected beam of negative halogen atoms is crossed with a perpendicular beam of light and the onset of photodetachment (i.e. $F^- \xrightarrow{hv} F + e^-$) measured. In another technique[62] the ultraviolet spectrum of gaseous halide ion is measured directly following shock tube dissociation of alkali halide vapours. Using higher temperatures in the shock tube it is possible to observe emission corresponding to $X + e^- \rightarrow X^- + hv$. It is interesting that for fluorine, two absorption thresholds[63] were observed at 3595 ± 2 Å and 3542 ± 2 Å, whereas in emission[64] two thresholds were

Table 9.8 Term values and successive ionisation potentials of halogen atoms

(From Moore[57], by courtesy of the National Bureau of Standards, Washington)

		F		Cl		Br		I	
		eV	cm^{-1}	eV	cm^{-1}	eV	cm^{-1}	eV	cm^{-1}
A*		185.139	1493656	114.27	921902	103.0			
ns		157.117	1267581	96.7	780000	88.6			
ns^2		114.214	921450	67.80	547000	59.7			
ns^2np^1		87.14	703020	53.5	431226	47.3			
ns^2np^2		62.646	505410	39.90	321936	35.9	289529	~43	
ns^2np^3		34.98	282190.2	23.80	192000	21.6	174119	19.09	154050
ns^2np^4		17.418	140524.5	13.01†	104995.46†	11.84	94440	10.454†	84340†
		Absorption	*Emission*	*Absorption*	*Emission*‡	*Absorption*	*Emission*	*Absorption*	*Emission*
$ns^2np^4(^3P)(n+1)s$	$^2P_{\frac{1}{2}}$		105057.10	74861	74865.667	69209	68963.52	—	63186.76
	$^2P_{\frac{3}{2}}$		104731.86	74222	74225.846	—	67176.87	56091	56092.88
	$^4P_{\frac{1}{2}}$		102841.20		72827.038	—	66877.16	60898	60896.27
	$^4P_{\frac{3}{2}}$		102681.24		72488.568	64901	64900.50	61820	61819.81
	$^4P_{\frac{5}{2}}$		102406.50		71958.363	63432	63429.82	54633	54633.46
ns^2np^5	$^2P_{\frac{1}{2}}$		404.0		822.36		3685		7603.15
	$^2P_{\frac{3}{2}}$		0		0		0		0

* A≡inert gas configuration under valence electrons; the number opposite, say, ns^2np^1 equals ionisation potential of ns^2np^2

† Reference 58; revised values for Cl 12.967 eV, 104 588 cm^{-1}, for I 10.451 eV, 84 294 cm^{-1}

‡ Reference 57a – this also gives 104 591 $\pm$0.3 cm^{-1} for limit

observed at 3646 ± 2 Å and 3595 ± 2 Å. The separation of the two absorption thresholds corresponds to ~410 cm^{-1}, i.e. the two processes are $F^- \rightarrow F(^2P_{\frac{1}{2}})+e$ and $F^- \rightarrow F(^2P_{\frac{3}{2}})+e$ (see Table 9.8). However, the separation of the two emission bands is also ~400 cm^{-1} suggesting the two processes $F(^2P_{\frac{1}{2}})+e \rightarrow F^-$ and $F(^2P_{\frac{3}{2}})+e \rightarrow F^-$. Thus two different values of the electron affinity for ground state fluorine atoms ($^2P_{\frac{3}{2}}$) are obtained; as far as the author is aware this discrepancy has not been resolved.*

The most accurate values of the electron affinities of the halogens, taken from the recent review article by Berry[62], are:

	F	Cl	Br	I
eV	3.448* ± 0.005	3.613 ± 0.003*	3.363 ± 0.003*	3.063 ± 0.003*
	3.400† ± 0.002	3.616 ± 0.003†		3.076 ± 0.005‡
kcal*	79.50 ± 0.12	83.32 ± 0.07	77.54 ± 0.07	70.62 ± 0.007

* absorption; Reference 63 † emission; Reference 64 ‡ crossed beam; Reference 62

Although the formation of X^{2-} from X^- is presumably highly endothermic there is a suggestion that the ions F^{2-}, Cl^{2-}, Br^{2-} have been detected in an omegatron mass spectrometer[65]. More evidence will be needed before this is generally accepted.

9.2.6 Some recent photochemical studies

(a) *Introduction – ground state atoms* – At the simplest level irradiation of all the halogens in certain regions of the spectrum leads to formation of atoms which may then react with some substrate. A great deal of inorganic preparative chemistry relies on this; perhaps the most striking demonstration of such a preparation[66] is that exposure to sunlight of a Pyrex bulb containing gaseous fluorine and xenon leads to slow formation of XeF_2 – presumably via

$$F_2 \xrightarrow{h\nu} 2F$$
$$2F + Xe \rightarrow XeF_2$$

The same reaction can, of course, be promoted by heating the gaseous mixture to produce the halogen atoms. In so far as the rate of synthesis depends on the concentration of halogen atoms this will depend on the absorption extinction coefficient, intensity and frequency of the photolysis source, quantum yield, recombination, wall reactions etc. Photochemical reactions may also be studied at very low temperatures by the matrix isolation technique, developed largely by Pimentel[67]. In one version of this method a parent compound, isolated in a large excess of inert solid rigid matrix (e.g. argon at 20 K), is photolysed; reactive intermediates formed on photolysis may be stabilised, partly because of the low temperature and partly because of isolation, and their spectroscopic properties examined at leisure. Some halogen containing species generated this way include O_2F (from photolysis of F_2 and O_2 in argon at 4 K or 20 K)[68a], ClCO (Cl_2 in CO)[68b], FCO(F_2 + CO in argon)[68c]. The technique has recently been reviewed[69].

The reactions of halogen atoms have been reviewed by Fettis and Knox[70],

*See Note 5, p.269

some photochemical reactions by Calvert and Pitts[3] and the reactions of iodine atoms by Golden and Benson[71].

Although not strictly photochemical, there is little doubt that the most interesting development in studies of reactions of ground state halogen atoms has been the molecular-beam experiments of Herschbach and co-workers[72]. They have obtained evidence for 'sticky collisions' in reactions of $Cl+IBr$, $Cl+I_2$, $Cl+Br_2$. Although long postulated, the only positive previous evidence for a three atom species was the matrix detection of Cl_3 by Nelson and Pimentel[73].

(b) *Predissociation; photodissociation translational spectroscopy*—It should be recalled that photodissociation, as outlined previously, does not only occur in continuous spectral regions where the transition is to a repulsive state (e.g. F_2, $^1\Pi_u \leftarrow {}^1\Sigma_g^+$, to give $2\times{}^2P_{\frac{3}{2}}$) or above the dissociating level of a bound state (e.g. I_2 $^3\Pi(0_u^+) \leftarrow {}^1\Sigma_g^+$, to give $^2P_{\frac{1}{2}}+{}^2P_{\frac{3}{2}}$). At relatively high pressures collision induced predissociation can transfer, say, an I_2 molecule in the banded region of the $^3\Pi(0_u^+)$ state to a dissociating state. However, until recently there was no evidence for *direct* predissociation of iodine; since these experiments illustrate interesting techniques they are considered in more detail, but the whole field of halogen predissociation will not be discussed.

Wasserman *et al.*[74] subjected gaseous I_2, in a quartz vessel in an e.p.r. spectrometer, to irradiation with $\lambda \geqslant 5000$ Å or $\lambda \leqslant 5000$ Å (dissociation limit of $B^3\Pi(0_u^+)$ state, 4990 Å). The concentration of $^2P_{\frac{3}{2}}$ atoms produced was monitored by their X-band spectrum; concentration of $^2P_{\frac{1}{2}}$ atoms was below the detection threshold. Both above and below the dissociation threshold high concentration of atoms was observed and the authors concluded that at least 30% of the B-state molecules undergo direct predissociation. From a comparison of the *radiative*[75] and *total*[76] decay rate of $3\Pi(0_u^+)$ molecules, Chutjian and James[75] concluded that approximately 66% of these B-state molecules predissociate to $2\times{}^2P_{\frac{3}{2}}$ (probably via the $^1\Pi(1_u)$ state) and approximately 33% radiate to the ground state. At higher pressures there is substantial ($\sim$70%) collisional predissociation.

In photodissociation translational spectroscopy, developed by Wilson and colleagues[77], a molecular beam is crossed at right angles with an intense pulse of laser light and photodissociation fragments are detected in the third perpendicular direction by a quadrupole mass spectrometer. By measuring the mass spectrometer signal as a function of time after the laser pulse, it is possible to obtain the distribution of total translational energy in the fragments. Wilson[77] points out that such measurements for diatomics are likely to provide information about dissociative states such as symmetry, shape of potential curve, curve crossings, lifetimes and correlation with atomic states. Work is still in the early stages, but in the present context Wilson has shown that by crossing a beam of I_2 with 5310 Å light, dissociation to $2\times{}^2P_{\frac{3}{2}}$ certainly occurs.

(c) *Excited atom chemistry and the chemical laser*—Although the striking differences in the chemical behaviour of first singlet and ground state oxygen and sulphur atoms have been known and studied for some time, it is only in the past few years that attention has focused on possible differences in the behaviour of ground state ($^2P_{\frac{3}{2}}$) and first excited state ($^2P_{\frac{1}{2}}$) halogen atoms—

in fact, in 1966, Calvert and Pitts[3] could say there was no evidence for any difference. By using kinetic flash spectroscopy (*see* 9.2.5(i)) Husain and Donovan[78] have been able to monitor the concentration of $^2P_{\frac{3}{2}}$ and, more importantly, $^2P_{\frac{1}{2}}$ halogen atoms in their reactions with a wide range of substances. This method depends on the fact that the $^2P_{\frac{1}{2}} \rightarrow {}^2P_{\frac{3}{2}}$ transition is forbidden and hence the $^2P_{\frac{1}{2}}$ radiative lifetimes are long: F 830 s, Cl 83 s, Br 1.1 s, I 0.13 s. Thus the $^2P_{\frac{1}{2}}$ state exists long enough to undergo collisions and hence reaction. The question of lifetimes of excited states is of immense importance for the discovery and development of chemical lasers. Before considering some of Husain's and Donovan's results we shall briefly outline the principles of the chemical laser.

For laser action from an excited state $^2P_{\frac{1}{2}}$ to the ground state $^2P_{\frac{3}{2}}$ to occur certain conditions must be satisfied[78]: (a) a population inversion must be achieved – this means for the halogen atoms $[^2P_{\frac{1}{2}}]/[^2P_{\frac{3}{2}}]$ must be at least $\frac{2}{4}$ because of the degeneracies; (b) a threshold concentration of excited atoms is necessary to overcome losses in the optical cavity – this in practice means that the $^2P_{\frac{1}{2}} \rightarrow {}^2P_{\frac{3}{2}}$ transition is forbidden in order to build up sufficient concentration of $^2P_{\frac{1}{2}}$ during the duration of the photolytic flash; (c) the population inversion must be produced in the primary photolytic step. In 1964 Kasper and Pimentel[79], who were actually investigating intermediates by rapid scan infrared techniques, observed stimulated emission during flash photolysis of a number of alkyl and perfluoroalkyl iodides.

$$RI \xrightarrow[\text{u.v.}]{h\nu} R + I(5^2P_{\frac{1}{2}})$$
$$I(5^2P_{\frac{1}{2}}) + h\nu\ (1.315\ \mu = 7603\ \text{cm}^{-1}) \rightarrow I(5^2P_{\frac{3}{2}}) + 2h\nu(1.315\ \mu)$$

The optical gain in these systems is so high that emission of the order of a kilowatt can be achieved. Pimentel and co-workers[80] and Pollack[81] have developed this work considerably. One interesting observation is that, contrary to what might be supposed, no laser action is observed on flash photolysis of I_2 – dissociation of I_2 from the $^3\Pi(0_u^+)$ state should give an equal mixture of $^2P_{\frac{3}{2}}$ and $^2P_{\frac{1}{2}}$ atoms, a considerable population inversion*. However, the I_2 molecule is very efficient at quenching the $^2P_{\frac{1}{2}}$ atom[59] so not enough concentration is built up for laser action. Recently, laser action in the bromine atom has also been observed[82]. The studies on such laser systems are of value in elucidating primary photochemical processes and energy transfer mechanisms.

The lasers described above are not strictly chemical and are more accurately described as photodissociation. There are no continuous-wave lasers in which the excited atom is populated by a chemical reaction. On the other hand, there are lasers in which a *molecular* excited lasing state is populated by a reaction involving a halogen[83]:

$$F_2 + NO \rightarrow ONF + F \qquad \Delta H = -18\ \text{kcal mol}^{-1}$$
$$F + H_2 \rightarrow HF^* + H \qquad \Delta H = -32\ \text{kcal mol}^{-1}$$

the HF*, which is in the ground electronic state $v'' = 2$, lases by emission to $v'' = 1$.

Husain and Donovan[84] have recently reviewed the reactions of excited atoms and have also described in detail the behaviour of excited halogen

* See Note 1, p. 286

atoms[78]. Since the spin-orbit separation (Table 9.8) is greatest for the iodine atom we might expect any differences in chemical behaviour of $^2P_{\frac{3}{2}}$ and $^2P_{\frac{1}{2}}$ atoms to be most marked for this halogen. This should be particularly striking when reaction with $^2P_{\frac{3}{2}}$ is endothermic but with $^2P_{\frac{1}{2}}$ is exothermic – this is the case with the reaction of I with Cl_2 Br_2 ICl and IBr – see Table 9.9. Hydrogen atom abstraction from paraffins is also much more rapid for

Table 9.9 Rates of reaction of $I(^2P_{\frac{1}{2}})$ and $I(^2P_{\frac{3}{2}})$

Species	$I(^2P_{\frac{1}{2}})$	k(cm³ mol⁻¹ s⁻¹) $I(^2P_{\frac{3}{2}})$
Cl_2	2.1×10^{-13}	6.1×10^{-21}
Br_2	1.5×10^{-12}	2.7×10^{-14}
ICl	3.4×10^{-12}	—
IBr	5.0×10^{-12}	$\sim 10^{-17}$

$I(^2P_{\frac{1}{2}})$ than $I(^2P_{\frac{3}{2}})$[85]. Similarly, it would be expected on energetic grounds that $I(^2P_{\frac{1}{2}})$ would rapidly abstract I from RI – it is known that abstraction by $I(^2P_{\frac{3}{2}})$ is endothermic and slow. However, $I(^2P_{\frac{1}{2}})$ rapidly decays to $^2P_{\frac{3}{2}}$ in such systems because of collisional quenching[86], so care is always needed in interpreting data from these experiments.

(d) *Molecular emission on atom recombination* – The emission spectra produced on halogen recombination have been examined by a number of workers, particularly Ogryzlo, Palmer and Clyne and their respective co-workers.

The present situation seems to be that for the Cl_2 afterglow[87], combination occurs between two $^2P_{\frac{3}{2}}$ atoms (any $^2P_{\frac{1}{2}}$ atoms formed are quenched too rapidly to react) to give an unstable intermediate followed by a radiationless transition into the emitting $^3\Pi(0_u^+)$ state. It will be recalled that two ground state atoms cannot correlate *directly* with a $^3\Pi(0_u^+)$ state. Interestingly, the $^3\Pi(0_u^+)$ state can be quenched by halogen atoms so that the overall mechanism is:

$$Cl(^2P_{\frac{3}{2}})^+Cl(^2P_{\frac{3}{2}}) + M \rightarrow Cl_2(^3\Pi(0_u^+)) + M$$
$$Cl_2(^3\Pi(0_u^+)) \rightarrow Cl_2(^1\Sigma_g^+) + h\nu\lambda > 5000\text{ Å})$$
$$Cl(^2P_{\frac{3}{2}}) + Cl_2(^3\Pi(0_u^+)) \rightarrow Cl_2(\text{non-radiative}) + Cl(^2P_{\frac{3}{2}})$$

For Br_2, although Ogryzlo[88] originally suggested that the emission was due to $^3\Pi(0_u^+) \rightarrow {}^1\Sigma_g^+$, it has been shown[31] that the main emission is due to $^3\Pi(1_u) \rightarrow {}^1\Sigma_g^+$ although there is evidence for a weak emission from the 0_u^+ state. Since the zero level of the $^3\Pi(0_u^+)$ state of Br_2 is close to the energy of two ground state atoms population of the $^3\Pi(0_u^+)$ state via the same mechanism as Cl_2 is very inefficient. However, the two $^2P_{\frac{3}{2}}$ atoms readily combine into the $^3\Pi(1_u)$ state*. For further details see Husain's and Donovan's review[78].

9.2.7 Complexes

(a) *Charge-transfer complexes* – No discussion of halogen spectroscopy would be complete without at least some mention of 'charge-transfer

* See Note 6, p. 287

complexes'. One of the most frequently studied is the complex of iodine with pyridine; Table 9.10 lists some data from the measurements of Bist and Person[89], confirming earlier work of Reid and Mulliken[90].

Interaction between I_2 and pyridine thus affects the I_2 absorptions, produces a new band and also lowers the vibrational frequency of the I—I stretch

Table 9.10 Some spectroscopic properties of the pyridine-iodine complex in n-heptane at 25°C

Band	$\nu_m(cm^{-1})$ *free* I_2	λ_m(Å) *free* I_2	$\nu_m(cm^{-1})$ *complex*	λ_m(Å) *complex*	ε_m *free*	ε_m *complex*
$^3\Pi_1 \leftarrow {}^1\Sigma_g^+$	14 300	6993	18 500	5405	40	45
$^3\Pi(0_u^+) \leftarrow {}^1\Sigma_g^+$	19 225	5215	23 700	4219	900	1540
Charge-transfer band			42 300	2364		51 730

(167 cm^{-1} *v.* 213 cm^{-1} in gas). This is an example of an n-$a\sigma$ (lone pair on donor N, σ acceptor I_2) complex and is particularly strong as evidenced for example by the intensity of the charge-transfer band. Other types of complex are conveniently labelled [92]$b\pi$–$a\pi$ (π donor, π acceptor) and $b\pi$–$a\sigma$(π donor, σ acceptor). These two types are in general much weaker than the n–$a\sigma$ complexes, e.g. the ε_m of the charge-transfer band of I_2/hexamethylbenzene is 6700 [93].

There has been considerable recent discussion of the extent to which charge-transfer forces contribute to these complexes, as opposed to classical electrostatic forces. There is little doubt that for the n-$a\sigma$ complexes the charge-transfer contribution is large, i.e. in

$$\Psi(D\text{–}A) = a\Psi_0(D,A) + b\Psi_1(D^+\text{–}A^-) + \ldots$$

$$(D = \text{donor},\ A = \text{acceptor})$$

the coefficient of b is relatively large. However, the value of b in the weaker complexes is uncertain and it is likely that Coulomb and polarisation interactions, including quadrupole effects[94], are the most important. In view of this *all* such complexes are probably more correctly described as Electron Donor–Acceptor (EDA)[92, 93].

There has been a growing interest in comparison of data obtained in solution with gas phase values[95]. There are some striking and unusual differences which do not yet appear to be understood.

(b) *Gas-phase dimers*—Kokovin[96], using diaphragm-gauge PVT data to $3\frac{1}{2}$ atmospheres attributed apparent gas imperfections in bromine to the presence of Br_4 and deduced for $2Br_2 \rightleftharpoons Br_4$,

$$\log K_{\text{Torr}} = 576T^{-1} - 6.16$$

over the range 368–458 K. Thus $\Delta H^0 = -2.640$ kcal and $\Delta S^0 = -15$ e.u. Passchier *et al.*[97] have examined with great care the absorbance in the spectrum of bromine vapour 2000–7500 Å, between 298 and 713 K. They conclude that deviation from Beer's law is due to association to Br_4. From the spectral data, $\Delta H^0 = -2.270 \pm 0.25$ kcal.

Similarly, Passchier and Gregory[98] have examined the spectrum of iodine vapour in the 2650 Å region. Again they ascribe Beer's law deviation to dimerisation and from the data deduce

$$\log K = 368.4T^{-1} - 1.007 \text{ at } 605 \text{ K},$$
$$\Delta H^0 = -2.900 \text{ kcal}; \Delta S^0 = -14.4 \pm 2 \text{ e.u.}$$

This agrees closely, on extrapolation, with De Maine's[99] solution results on I_2 in CCl_4: $\Delta H^0 = -1.960$ at 298 K. The nature of the bonding in the dimers is probably simply Van der Waals attraction and the X_4 absorptions are reasonably assigned to charge-transfer bands.

Measurements[99a] of the temperature variation of the infrared binary absorption coefficient give evidence for the existence of $(Cl_2)_2$ in the gas phase.

9.3 NUCLEAR PROPERTIES AND SPECTROSCOPY

9.3.1 Some nuclear properties

Table 9.11 lists some properties of the halogens relative to the topics to be discussed in this section.

9.3.2 E.S.R. spectra of halogen atoms

Figure 9.5 shows a schematic representation of the diagonal energy levels of the free $^2P_{\frac{3}{2}}$ Cl atom ($I = \frac{3}{2}$), assuming the uncoupled notation (JIM_JM_I)

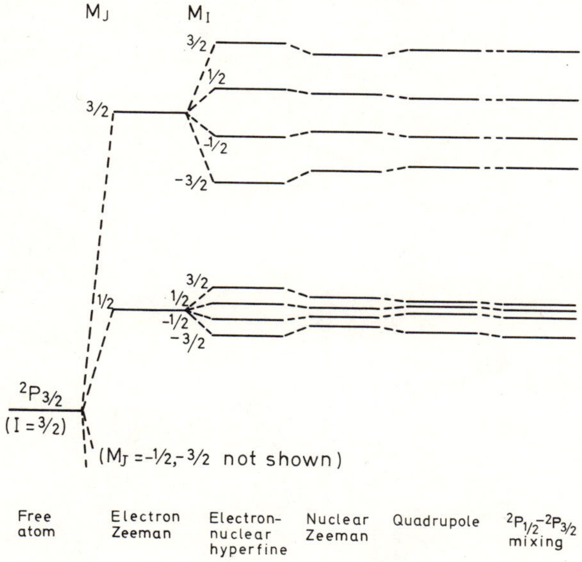

Figure 9.5 Schematic of $^2P_{\frac{3}{2}}$ ($I = \frac{3}{2}$) Cl atom energy levels in magnetic field

Table 9.11

	^{19}F	^{35}Cl	^{37}Cl	^{79}Br	^{81}Br	^{127}I	^{129}I
Nuclear spin I (units of $h/2\pi$)	$\frac{1}{2}$	$\frac{3}{2}$	$\frac{3}{2}$	$\frac{3}{2}$	$\frac{3}{2}$	$\frac{5}{2}$	$\frac{7}{2}$
Nuclear g_I factor (dimensionless)	5.255	0.547	0.456	1.399	1.508	1.118	0.748
g_II (units of nuclear magnetons)	2.627	0.821	0.683	2.099	2.263	2.794	2.617
Quadrupole moment Q (multiples of $e \times 10^{-24}$ cm^2)	—	−0.0797	−0.0621	0.33	0.28	−0.75	−0.43
e^2qQ/h atom		−109.746	−86.510	769.756	643.032	−2292.712	—
Mössbauer excited states:		energy eV				59.0	26.8
		Spin I				$\frac{7}{2}$	$\frac{5}{2}$
		$Q(e \times 10^{-24})$				0.71	0.68
		lifetime (s × 10^9)				2.68	26.8
		Parent				^{127}Te	^{129}Te

rather than the coupled ($FJIM_F$) notation; the former is more appropriate in the presence of a large magnetic field (see below).

The energy of the transitions shown thus depends on several factors; in addition there are the off-diagonal terms which perturb the levels so that determination of the hyperfine parameters from observed gas phase spectra is not easy.

Halogen atoms are readily generated in flow systems by microwave discharge either in the halogen itself or in, for example, CF_4 for F. The e.s.r. spectrum can thus be obtained by pumping the discharge products through the spectrometer cavity; alternatively for iodine, the I_2 molecule can be photolysed in a stationary system. Thus the e.s.r. spectra of all the ground state atoms (i.e. $^2P_{\frac{3}{2}}$ state of ^{19}F, ^{35}Cl, ^{37}Cl, ^{79}Br, ^{81}Br, ^{127}I, ^{129}I) have been obtained – for details see the original papers[100].

Since the e.s.r. spectrum can also provide an estimate of the concentration of atoms, as discussed earlier for iodine in section 9.2.6(b), the possibilities for kinetic studies are clear. Westenberg[101] has used e.s.r. to measure absolute concentrations of Cl, Br and I atoms with O_2 as reference. Further, he and De Haas[102] have monitored Cl and H concentrations by e.s.r. to measure separately the rate constants for the forward and backward reactions:

$$Cl + H_2 \rightleftharpoons HCl + H$$

In this context and of relevance to the work of Husain and Donovan[78], it would be of interest to detect the presence of excited states of the halogens. Carrington and co-workers[103] have been able to obtain the spectra of $^2P_{\frac{1}{2}}$ F and Cl atoms by discharge in CF_4 and Cl_2 plus CF_3Cl respectively. Figure 9.6 shows a schematic of the energy levels for F, ignoring Zeeman nuclear interactions, in both uncoupled and coupled notation. The spectrum of $^2P_{\frac{1}{2}}$ F consists of two lines at 2732 and 13 330 gauss at a frequency of 9131.3

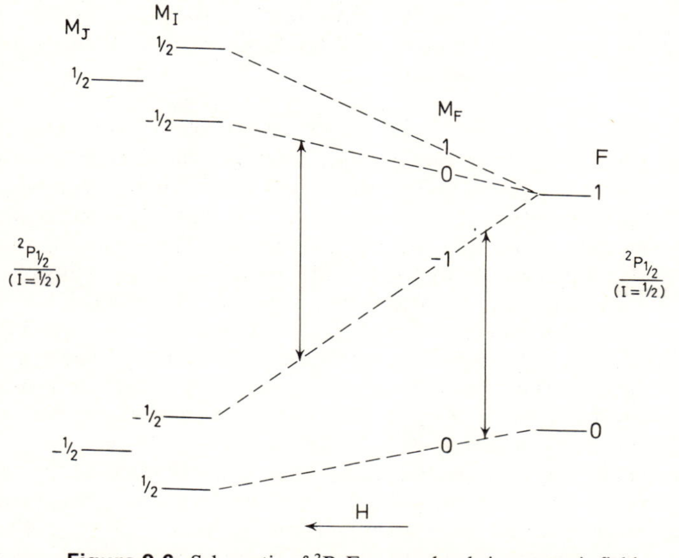

Figure 9.6 Schematic of $^2P_{\frac{1}{2}}F$ energy levels in magnetic field

MHz. Interestingly, in view of Donovan and Husain[78], there is a non-Boltzmann distribution between $^2P_{\frac{3}{2}}$ and $^2P_{\frac{1}{2}}$ states.

There have been several attempts[104] to observe e.s.r. spectra of halogen atoms in condensed phases, particularly using the matrix isolation technique mentioned previously. These have been unsuccessful presumably because strong coupling between the atoms and the environment broadens any spectral lines beyond detection. It should be added that halogen atoms have been trapped in irradiated halide crystals, but these are more correctly described as X_2^-, i.e. V centres.

The spectra of halogen containing radicals can, of course, be very simple, particularly in the liquid phase. For example, the radical O_2F, easily formed in any condensed oxygen fluoride, shows a doublet e.s.r. spectrum due to coupling with the fluorine nucleus which has spin $\frac{1}{2}$ [105]. On the other hand, e.s.r. spectra of radicals in the gas phase can be very complex because of coupling with molecular rotation. The analysis of such spectra is more akin to microwave spectroscopy than conventional magnetic resonance. However, Carrington and his colleagues[106] have performed comprehensive studies of a number of such species, e.g. ClO, BrO, IO, SF etc., to obtain very great detail about structure and properties.

9.3.3 N.M.R. spectroscopy

Since the fluorine nucleus has no quadrupole moment and the relaxation terms are just right, fluorine n.m.r. spectra usually consist of narrow lines and hence fine structure details due to chemical shift and coupling constant effects are readily seen. The use of ^{19}F n.m.r. in inorganic structure determination is well known and has been discussed elsewhere[107].

The Cl, Br, I nuclei have spin $I > \frac{1}{2}$ and hence have quadrupole moments which, through the fluctuations in orientation and magnitude of the electric field gradient q, provide an efficient relaxation mechanism. The contribution to the line-width from this cause is

$$\Delta\nu = K(e^2qQ)^2\,\tau_c$$

where K is a function of the nuclear spin and τ_c is the correlation time for molecular motion; $\Delta\nu$ is the line-width at half height in Hz. For X^- anions in dilute aqueous solution the ion solvation is essentially symmetrical and hence $q \sim 0$ so there is no quadrupole broadening. Thus the Cl^- signal for NaCl in dilute solution is about 10 Hz wide. However, in, for example, liquid CCl_4, the Cl environment is asymmetric, $e^2qQ = 81.3$ MHz [109], $\tau_c \sim 1.7 \times 10^{-12}$ s, and the line-width is $\sim 12\,000$ Hz [109]. Thus, generally speaking, fine structure details are not observable for Cl, Br and I and most spectra of these nuclei have been virtually useless in conventional structure determination. However, there have been many studies involving such nuclei and we consider just two aspects of this work.

From above it is to be expected that variation in chemical shifts and line-widths as a function of cation, solvent, concentration and temperature will provide information about solute-solvent interactions[110].

In a situation where Cl^- is exchanging rapidly between a symmetrical environment and an asymmetrical environment the line-width depends on the relative concentration at each site, the values of e^2Qq and τ_c at each site:

$$\Delta v = \Delta v_a P_a + \Delta v_b P_b$$

where P_a and P_b are the relative concentrations at sites *a* and *b*. 2 M NaCl has a Cl n.m.r. line-width of 16 Hz; on making the solution 10^{-3} M in $HgCl_4^{2-}$ the line-width increases to 34 Hz. A similar effect is observed with RSHgCl [111] where R is almost any organic group, which thus provides a method for determining τ_c. For example Bryant[112] has used this technique to study the helix-coil transition in poly-L-glutamate.

The most detailed and comprehensive exchange studies so far have been those of Richards and co-workers[113], particularly on the equilibrium

$$Cl^- + Cl_2 \rightleftharpoons Cl_3^-$$

They have been able to obtain a rate constant at 25 °C $-(8 \pm 4) \times 10^6$ s^{-1}.

9.3.4 Spectral effects of quadrupole moments and nuclear quadrupole resonance (N.Q.R.) [114]

We have so far noted the effects of quadrupole moments on e.s.r. and n.m.r. spectra and shall shortly see effects in Mössbauer spectroscopy. It is thus possible in principle if not always in practice to determine quadrupole coupling constants from these spectral studies. In addition quadrupole effects can be examined in atomic and molecular beams, rotational spectra of gases and fine structure in optical spectra.

For solids it is also possible to observe n.q.r. directly—the asymmetric electric environment of a particular quadrupolar nucleus splits the appropriate energy levels and transitions between these energy levels can be measured directly using, usually, a superregenerative oscillator. In the past few years, with increase in experimental sophistication, there have been many studies of halogen n.q.r. in a wide variety of compounds. One of the most direct applications of this technique is to simple structural studies. For example, two Cl nuclei in different environments will give two n.q.r. resonances. Thus, whereas Et_4NHCl_2 shows a single ^{35}Cl signal at 12 MHz, Me_4NHCl_2 shows two signals at 20 and <6 MHz [115]. By following the changes in the ^{35}Cl n.q.r. of $PCl_4^+PCl_6^-$ and PCl_5, phase transitions were identified at 102.3 K for the ionic solid and 183 K for the molecular solid[116].

Although for complex molecules quantitative interpretation of the magnitudes of coupling constants is beyond present quantum mechanics, qualitative interpretation can provide considerable insight into chemical bonding. Table 9.12 repeats some familiar numbers for diatomic halogen systems, mostly from Lucken[114]. Such data are usually discussed in terms of the ionic character of the bond. The simplest possible relationship for the halogen atom, provided it is the more electronegative partner is

$$\frac{e^2qQ}{e^2qQ\ (\text{atomic})} = (1-i)$$

where i is fractional ionic character. Thus in X_2, where $i = 0$, the value is nearly the same as in the atom (ignoring difference of phase, which does have an effect on the values); whereas in KCl where i is presumably 1, the Cl^- ion, as expected because of spherical symmetry, has $eqQ \sim 0$. However, this equation ignores, among others, the percentage s character (s orbitals are

Table 9.12 Quadruple coupling constants (e^2qQ) in MHz for some halogen systems

(From Lucken[114], by courtesy of Academic Press)

	^{35}Cl	^{79}Br	^{127}I
Free atom	−109.746	769.756	−2292.712
X_2*	±108.95	±765.86	±2156
X_2^-*	50±6	224±2	~930
FCl	−146		
FBr		1089.0	
BrCl	−103.6	876.8	
ICl	−82.5		−2944
IBr		722±15	−2731±60
KX*	~0.04	10.2	−60

* Solid data; rest are gases

spherically symmetrical so make no contribution), amount of multiple bonding, effect of contraction in orbitals. Townes and Dailey[117] and Gordy[118] have provided more detailed consideration of these effects. That the problem is not easy can be seen from the X_2^- data where most simple schemes would predict values about half that of the corresponding X_2 molecule. The seventh chapter in Lucken's book[114] is strongly recommended for more detailed discussion.

In compounds where the atom environment is not axially symmetrical (i.e. $q_{xx} \neq q_{yy}$) the interpretation of experimental data is more complicated. For example for a nucleus with $I = \frac{3}{2}$

$$\nu = \frac{e^2qQ}{2h}\left(1+\frac{\eta^2}{3}\right)^{\frac{1}{2}}$$

where η is the asymmetry parameter $(q_{xx}-q_{yy}/q_{zz})$ and ν is the frequency of the n.q.r. transition. To determine the coupling constant and η in *this* case only requires a magnetic field to lift the degeneracy of the M_I states.

Solid iodine shows a large asymmetry parameter[114]; this can only arise from additional intermolecular bond formation, which is consistent with the known crystal structure (see Section 9.4.1). Values of η, not simply due to intermolecular effects have been used to deduce the degree of p_π–d_π bonding in Group IV chlorides[119].

9.3.5 Mössbauer spectroscopy of ^{127}I and ^{129}I

A particularly readable introduction to Mössbauer spectroscopy is Greenwood's article[120]. Perlow[121] has discussed iodine Mössbauer, particularly using ^{127}I.

Iodine is the only halogen for which Mössbauer spectroscopy is possible. The effect for ^{129}I was demonstrated in 1962[122] and for ^{127}I in 1964[123], the classic papers on the chemical applications being those of Hafemeister *et al.*[124] for ^{129}I and Perlow and Perlow[125] for ^{127}I. Although ^{127}I is much more convenient to use, since it is non-radioactive and occurs with 100% abundance, the lifetime of the excited state of ^{129}I is longer than the corresponding ^{127}I level, so that ^{129}I Mössbauer spectra give narrower lines and hence finer detail can be resolved. One very interesting difference between the two nuclei is in the effect of relative sizes of ground and excited nuclear states on the chemical shift:

$$\delta = \text{const.}\frac{\delta r}{r}\,\delta|\psi_s(\mathrm{O})^2|$$

where δ = chemical shift, $\delta|\psi_s(\mathrm{O})^2|$ is the difference in s electron density at the nucleus between source and absorber and $\delta r/r$ is the relative size change between ground and excited nuclear states.

For ^{127}I the excited state is smaller ($\delta r/r$ negative) but the reverse is true for ^{129}I. Thus for the same compound, $Na_3H_2IO_6$, the ^{127}I chemical shift versus ZnTe is $+1.19$ mm s^{-1} but the ^{129}I shift is -3.35 mm s^{-1} [126].

From the above equation it is clear that the actual values of chemical shifts will provide information on chemical bonding.

Since in the ground and excited states of both ^{127}I and ^{129}I the nuclear spin $I > \frac{1}{2}$ we expect the degeneracy of the nuclear substates to be lifted due to quadrupole interaction, provided, of course, the nucleus is in an asymmetric electric environment. The consequent spectral splittings can be used to measure quadrupole coupling constants, one of the advantages over n.q.r. being that it is possible to determine the sign as well as the magnitude of the field gradient. For example, the quadrupole interaction in the ground and excited states of ^{129}I splits the Mössbauer spectrum of solid I_2 into eight lines. Analysis[127] of these gives $e^2qQ = -1426 \pm 15$ MHz. Since $Q(^{129}I)/Q(^{127}I) = 0.701$, this gives for ^{127}I, $e^2qQ = -2085 \pm 20$ MHz, which is close to the value from n.q.r. measurements (Table 9.12).

A neat example of non-observation of quadrupole splitting is in $IF_6^+AsF_6^-$ where the ^{129}I resonance from the cation is a single line; the environment must therefore be symmetrical, i.e. IF_6^+ is octahedral[128]. However, from the broad line n.m.r. of ^{127}I, it has been concluded that e^2qQ is between 1.25 and 2.9 MHz; this indicates a very slight distortion[129].

More sophisticated uses of both chemical shift and quadrupole data are becoming more common for iodine-containing compounds. For example, Pasternak and Sonnino[130], from ^{129}I data in ICl and IBr, conclude from the chemical shift values that there is pure p bonding, i.e. no s-p hybridisation and from both chemical shift and quadrupole values that in IBr there is a movement of 0.28 e from the I shell into the IBr bond.

Although not strictly relevant to this section, some of the most interesting Mössbauer studies have been those of xenon compounds using ^{129}I as the parent for the excited xenon state. For example with $K^{129}ICl_4 \cdot H_2O$ and $K^{129}ICl_2 \cdot H_2O$ as sources and Na_4XeO_6 as absorber, the spectra of $XeCl_4$ and $XeCl_2$ respectively were obtained[121]. Attempts to observe the Mössbauer of krypton compounds using radioactive Br have not been successful.

9.4 SOME ASPECTS OF THE HALOGEN SOLID STATE

In the past few years there has been a growing interest in the degree and nature of the interaction between molecular halogens in the condensed phase. This interaction is expected to be manifest in certain physical properties such as crystal structures, heats of sublimation, n.q.r. and spectral data.

9.4.1 Structures of solid halogens

The solid halogens: Cl_2 at about $-160\,°C$[131], Br_2 at $-150\,°C$[132] and I_2 at room temperature[133], crystallise in the orthorhombic space group *Cmca* with two molecules per unit cell. The Cl_2 single-crystal data of Collin[131] has been refined by Donohue and Goodman[134] who made allowance for anisotropic thermal motion. The crystal structure of I_2 at $-163\,°C$ has been investigated by Bolhuis *et al.*[135] with the significant result that the I—I molecular bond length is longer than previously believed[133]. Below is a selection of the data:

	Bond length (Å) *Solid*	*gas*	*Intralayer*		*Interlayer*			*Twice Van der Waals radius*[139]
Cl_2	1.98[134]	1.986[136]	3.32	3.82[134]	3.74	3.84	3.97[134]	3.60
Br_2	2.27[134]	2.285[137]	3.31	3.79[134]	3.99	4.02	4.14[134]	3.50
I_2	2.715[135]	2.676[138]	3.496	3.972[135]	4.269	0.337	4.412[135]	4.30

Essentially the halogens crystallise in layers with Van der Waals interaction between layers cf. shortest interlayer distance with the Van der Waals distances. Within the layers the difference between the shortest interlayer distance and the molecular bond length decreases from Cl_2 to I_2 (1.34, 1.04, 0.78 Å for Cl_2, Br_2, I_2 respectively). This suggests that intermolecular forces increase from Cl_2 to I_2. It is to be expected that such interactions will have an effect on the molecular bond length compared with the gas phase value. Only I_2 shows a really significant difference.

Figure 9.7 shows the 2-dimensional network of I_2 molecules in the (100) plane; one layer thus consists of approximately linear I.....I–I.....I chains, each I atom being involved in two nearly perpendicular chains. The angles of approximately 90 degrees suggest a simple model for the bonding via the p orbitals of iodine, i.e. each I.....I—I.....I unit is a four-centre six-electron system[140] which results in an order for the central bond of less than one.

Simple homonuclear diatomics or pseudo-diatomics (N_2, CO, C_2H_2, CO_2) crystallise in the cubic Pa3 structure. Thus it must be supposed that it is the particular nature of the interhalogen attraction that causes them to crystallise in the orthorhombic form.

For fluorine, Lipscomb and co-workers[141] have carried out an x-ray study of a single crystal of βF_2 (stable 45.55 K–53.54 K). and shown that it has a similar structure to γO_2. The structure is disordered but is certainly cubic and probably space group *Pm3n* with eight molecules per unit cell—$a = 6.67 \pm 0.07$ Å. These workers were unable to obtain single crystals of αF_2

(stable below 45.55 K) but Meyer *et al.*[142] have examined the x-ray powder pattern. This phase is monoclinic, $C2/m$ probably, but $C2/c$ cannot be ruled out; there are four molecules per unit cell and structural parameters are— $a = 5.50 \pm 0.01$ Å, $b = 3.28 \pm 0.01$ Å, $c = 10.01 \pm 0.01$, $\beta = 134.66 \pm 0.02$ degrees. This structure is similar to αO_2, the main difference being that the F_2 molecules are tilted by about 11 degrees.

9.4.2 Some thermochemical and N.Q.R. data

For the halogens some approximate thermochemical data[143] are:

	F_2	Cl_2	Br_2	I_2
Approximate heat of sublimation (kcal mol^{-1})	2	6	10	15
Bond dissociation energy (kcal mol^{-1})—see Table 9.2	37	57	45	36

Bersohn[144] has used the ratio of these figures, namely 2/37 = 0.05, 6/57 = 0.11, 10/45 = 0.22, 15/36 = 0.42, to indicate increasing intermolecular attraction.

Since the n.q.r. asymmetry parameter is zero in an axially symmetrical

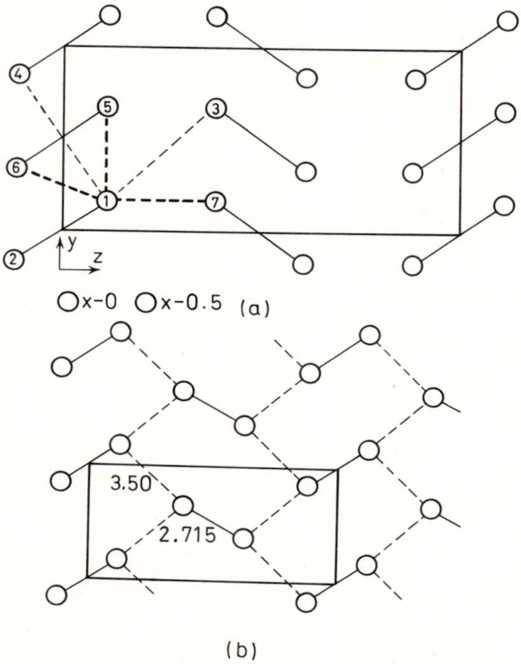

Figure 9.7 (a) The structure of solid iodine in (100) projection. (b) The two-dimensional network of iodine molecules in the (100) plane. (From Bolhuis *et al.*[135], by courtesy of Munksgaard)

field (Section 9.3.4), the difference from zero for solid halogen must reflect the anisotropy of the intermolecular interaction; for solid Cl_2, Br_2, I_2 the asymmetry parameters are $\eta \geqslant 0.08$ [145a], $= 0.20$, $= 0.175$ [145b] respectively. (The previously accepted value for Cl_2 ($\eta \leqslant 0.03$) [145c] is too small to explain the structure of the solid.)

9.4.3 Spectroscopic properties

The intermolecular forces in the solid will have an effect on both vibrational and electronic spectroscopic properties – there is no evidence for free rotation of halogen molecules in the solid.

In vibrational spectroscopy this will be shown in the frequencies of translational and vibrational lattice modes as well as the internal vibrations. The space group for Cl_2, Br_2, I_2 is *Cmca* (D_{2h}^{18}); therefore the point group isomorphous with the factor group is D_{2h} so that the $k = 0$ vibrational modes can be classified under the labelling of this point group:

Translation	$A_u + B_{1u} + B_{2u}$
Libration	$A_g + B_{1g} + B_{2g} + B_{3g}$
Molecular (internal)	$A_g + B_{3g}$

(excluding the three pure translations of the lattice). Thus there are two infrared-active modes (B_{1u} and B_{2u}) and six Raman-active modes (all the g vibrations). Walmsley and Anderson[146] examined the far infrared spectra of solid Cl_2, Br_2 and I_2 at 77 K – since the solids were prepared by deposition from the gas, they consisted of randomly oriented crystallites. The frequencies observed were (in cm^{-1}):

Cl_2	62	90	
Br_2	49	74	298
I_2	41	65	211

The high frequency molecular modes at 298 and 211 cm^{-1} probably appear because of breakdown of the simple $k = 0$ selection rules. The other two frequencies clearly arise from the allowed translational modes. From this data the authors calculate the self-force constants (F_{11}) for translation:

		Low frequency	High frequency
F_{11} in mdynes $Å^{-1}$	Cl_2	0.040	0.084
	Br_2	0.057	0.130
	I_2	0.063	0.159

The increasing values of F_{11} imply increasing interaction from Cl_2 to I_2.

More recently, Cahill and Leroi[147] and Suzuki *et al.*[148] have examined the Raman spectra of solid Cl_2 and Br_2 and, with some slight differences between the two laboratories, have obtained and assigned the remaining vibrational modes. The lattice and internal frequencies and force constants support the view that the intermolecular forces are greater in Br_2 than Cl_2. Also the two internal modes (A_g and B_{3g}) have been observed[147,148]; polarisation studies on single crystals of bromine[149] show that the two components, separated by 8 cm^{-1}, do belong to different symmetry representations, thus confirming the assignment of this splitting to the effect of the factor group.

The effect on electronic energy levels is less easy to interpret. By considering the p orbitals in the plane of one of the layers in I_2, Rosenberg[150] constructed a crystal band model. Schnepp *et al.*[151] have used this model to interpret the spectrum of oriented single crystals of I_2 in the 7000–3000 Å region.

There have been various other attempts to interpret the structures of halogen solids. One of the problems is that any spherically symmetrical interaction will not explain the orientation or 'tilt' of the molecules. Probably the most successful of these attempts has been Hillier's and Rice's[152] calculations on Cl_2; they find that approximately 25% of the intermolecular attraction is due to contribution from charge-transfer states.

No doubt there will continue to be great interest in the theoretical interpretation of solid halogens.

9.5 SOME PHYSICAL DATA COLLECTED*

This section is simply an attempt to provide a collection of those physical properties most often encountered in inorganic writings. Further details, particularly of thermophysical properties may be found in Gmelin[2], Mellor[2] and in the new series 'Thermophysical Properties of Matter'[153].

Table 9.13 lists some properties mostly gleaned from a variety of common references. The values quoted for the polarisability are from Fajans' recent paper[154] – he argues that Wilson and Curtis's[155] values, particularly for F^-, are not a consistent set.

The electronegativity values are derived from Pauling. The present author is not yet convinced that more sophisticated ways of apportioning numbers add greatly to Pauling's original idea of the 'power of an atom in a molecule to attract electrons to itself'.

Table 9.13 Some physical properties of the halogens

	F	Cl	Br	I
m.p.(K)	53.5	172.2	265.7	382.1
b.p.(K)	85.0	239.1	331	456
$r(X_2(g))$Å	1.418	1.988	2.284	2.666
$r(X^-)$NaCl structure	1.33	1.81	1.96	2.19
$\Delta H^0(X_2(g))$ (kcal mol^{-1})	0	0	7.3	14.9
$\Delta G^0(X_2(g))$ (kcal mol^{-1})	0	0	0.751	4.627
$\Delta G^0(X(g))$ (kcal mol^{-1})	14.8	25.4	19.7	16.8
$\Delta H^0(\frac{1}{2}X_2(s.s.) \rightarrow X^-(g)$ (kcal mol^{-1})	−62.1	−55.7	−52.3	−46.6
$\Delta H^0(X^-(g) \rightarrow X^-(aq.)$ (kcal mol^{-1})	−121	−88	−80	−70
$E^0(\frac{1}{2}X_2(s.s.) \rightarrow X^-(aq.))$V	~2.9	1.356	1.065	0.535
Polarisability of X^-	0.952	3.475	4.821	7.216
Electronegativity	4.0	3.2	3.0	2.7
Ionic mobility of X^- in water (18 °C)	46.6	65.3	67.3	66.5

* See Note 7, p. 287

Acknowledgements

I should like to acknowledge advice and help from the following, who have read the manuscript in whole or in part and made many suggestions: Drs. H. J. Bernstein, R. S. Berry, M. A. A. Clyne, E. A. C. Lucken, A. G. Maddock, T. L. Porter, A. G. Sharpe, P. M. A. Sherwood and Messrs. J. K. Burdett, D. J. Gardiner, M. A. Graham and M. Poliakoff.

Notes added in proof.

1. Wilson and co-workers[156] have very recently applied the technique of photodissociation translational spectroscopy (see Section 9.2.6(b)) to the continuum in the iodine absorption spectrum. They convincingly demonstrate that this continuum is *not* entirely due to $^3\Pi(0_u^+) \leftarrow {}^1\Sigma_g^+$ but that the $^1\Pi(1_u) \leftarrow {}^1\Sigma_g^+$ transition is at least as important! The consequencies for a considerable amount of iodine spectroscopy need hardly be emphasised. For example, photolysis in the continuum will not necessarily lead to population inversion since the $^1\Pi(1_u)$ state correlates with ground state atoms, hence the non-observation of laser action on I_2 photolysis (Section 9.2.6(c)).
2. Leroy and Bernstein[157] have recently introduced a new WKB-based method of extrapolating the vibrational data of an electronic state to the dissociation limit. This appears to be much more reliable than the more conventional Birge–Sponer plot[6] which generally shows curvature near the limit. Applying this method to Douglas *et al.*'s data[18] they obtain

$$D_0^0(^{35}Cl^{35}Cl) = 19997.25(\pm 0.15)\ cm^{-1}$$

Applying their extrapolations technique to Horsley and Barrow's[22] data for Br_2 they obtain D_0^0 values some 2 cm^{-1} higher. However, as pointed out in NSRDS News[158], their 'analysis is correct only if Horsley and Barrow missed many upper rotational levels of higher vibrational states in their interpretation of the absorption spectra'.

For I_2 they calculate 12440.9 ($\pm$1.1)cm^{-1}.
3. By comparison with very recent data on $^{79}Br^{79}Br$ by Coxon[159] it appears likely that there may be a systematic error in Rao and Venkataswarlu's chlorine work.
4. Another interpretation of the discrepancy has been convincingly suggested by Leroy[160]. From a detailed consideration of Verma's data he concludes that in fact the IVb series does *not* involve emission to the ground state but rather to the $^3\Pi(0_g^+)$ state. This latter state, although dissociating to two ground state atoms, does have a slight maximum which consequently makes Verma's value too high.
5. Berry (private communication to the author) has obtained some *preliminary* evidence that there may be a threshold in absorption at 3.400 eV (3646 Å); thus, if this work is confirmed, the electron affinity of fluorine can be more positively given a value 3.400 eV.
6. Coxon, Woon-Fat and Clyne (private communication) have recently examined the Br_2 recombination emission in more detail and conclude that (a) Most emission comes from the $^3\Pi(1_u)$ state (b) the $^3\Pi(0_u^+)$ state is populated from $^2P_{\frac{3}{2}} + {}^2P_{\frac{3}{2}}$ atoms via a potential maximum of about 600 cm^{-1} height.

7. Since this article was prepared the National Bureau of Standards, in the Technical News Bulletin for January 1971[158], has produced the latest set of recommended thermochemical data for the halogens. For completeness the following table lists the data; readers are strongly recommended to consult this publication for a discussion of the values.

Table 9.14 Table of proposed key values for thermodynamics[158]

Substance	*State**	$\Delta Hf^{\circ}_{298.15}$ kJ·mol^{-1}	$S^{\circ}_{298.15}$ J·mol^{-1}·K^{-1}	$H^{\circ}_{298.15}-H_0^{\circ}$ kJ·mol^{-1}
Cl	g	121.290 ±0.008	165.076 ±0.020	6.272 ±0.003
Cl^-	aq	−167.08 ±0.14	56.78 ±0.60	—
Cl_2	g	0	222.965 ±0.040	9.180 ±0.008
HCl	g	−92.31 ±0.13	186.786 ±0.033	8.640 ±0.004
Br	g	111.84 ±0.12	174.904 ±0.020	6.197 ±0.002
Br^-	aq	−121.50 ±0.17	83.3 ±1.3	—
Br_2	l	0	152.210 ±0.040	24.52 ±0.13
Br_2	g	30.91 ±0.11	245.350 ±0.054	9.724 ±0.012
HBr	g	−36.38 ±0.17	198.585 ±0.033	8.648 ±0.004
I	g	106.762 ±0.040	180.673 ±0.020	6.197 ±0.002
I^-	aq	−56.90 ±0.84	107.11 ±0.42	—
I_2	c	0	116.139 ±0.080	13.196 ±0.040
I_2	g	62.421 ±0.080	260.567 ±0.063	10.117 ±0.012
HI	g	26.36 ±0.80	206.480 ±0.040	8.657 ±0.006

1 cal = 4.1860 J
* All phases at pressure 101 325 Pa (formerly known as 1 standard atmosphere)

References

1. Sharpe, A. G. (1967). *Halogen Chemistry,* Vol. 1, Ed. by Gutmann, V. (London: Academic Press)
2. *Mellor's Comprehensive Treatise on Inorganic and Theoretical Chemistry,* (1956). Supplement II Part I. (London: Longmans); Pascal, P. (Editor), (1960). *Nouveau Traité de Chimie Minérale,* Tome XVI. (Paris: Masson); *Gmelins Handbuch der Anorganischen Chemie* Fluor. Chlor. Brom. Jod. (Verlag Chemie: Weinheim)
3. Calvert, J. G. and Pitts, J. N. (1966). *Photochemistry,* 184. (New York: John Wiley)
4. see e.g. Mulliken, R. S. (1940). *Phys. Rev.,* **57,** 500, which also has earlier references
5. Herzberg, G. (1950). *Spectra of Diatomic Molecules.* (Princeton: D. Van Nostrand)
6. Gaydon, A. G. (1968), *Dissociation Energies and Spectra of Diatomic Molecules,* 3rd edn. (London: Chapman and Hall)

7. Rees, A. L. G. (1957). *J. Chem. Phys.*, **26,** 1567
8. Steunenberg, R. K. and Vogel, R. C. (1956). *J. Amer. Chem. Soc.*, **78,** 901
9. Andrychuk, D. (1951). *Canad. J. Phys.*, **29,** 151
10. Iczkowski, R. P. and Margrave, J. L. (1959). *J. Chem. Phys.*, **30,** 403
11. Stricker, W. and Krauss, L. (1968). *Z. Naturforsch.*, **23a,** 486
12. Gole, J. L. and Margrave, J. L., unpublished observations quoted in Reference 13
13. De Corpo, J. J., Steiger, R. P., Franklin, J. L. and Margrave, J. L. (1970). *J. Chem. Phys.*, **53,** 936
14. Yates, R. E., Blauer, J. A., Greenbaum, M. A. and Farber, M. (1966). *J. Chem. Phys.*, **44,** 498
15. Dibeler, V. H., Walker, J. A. and McCulloch, K. E. (1969). *J. Chem. Phys.*, **50,** 4592
16. Berkowitz, J. and Chupka, W. A., unpublished observations quoted in Reference 13
17. National Bureau of Standards (1968). Technical Note 270–3, *Selected Values of Chemical Thermodynamic Properties*
18. Douglas, A. E., Moller, C. K. and Stoicheff, B. P. (1963). *Canad. J. Phys.*, **41,** 1174
19. Khanna, B. N. (1959). *Proc. Ind. Acad. Sci.*, **A49,** 293; Venkataswarlu, P. and Khanna, B. N. *ibid.* **A49,** 117
20. Rao, Y. V. and Venkataswarlu, P. (1962). *J. Molec. Spectrosc.*, **9,** 173
21. Clyne, M. A. A. and Coxon, J. A. (1970). *J. Molec. Spectrosc.*, **33,** 381
22. Horsley, J. A. and Barrow, R. F. (1967). *Trans. Faraday Soc.*, **63,** 32
23. Verma, R. D. (1960). *J. Chem. Phys.*, **32,** 738
24. Steinfeld, J. I., Campbell, J. D. and Weiss, N. A. (1969). *J. Molec. Spectrosc.*, **29,** 204
25. Kuhn, H. (1926). *Z. Phys.*, **39,** 77
26. Kroll, M. (1970). *J. Molec. Spectrosc.*, **36,** 44
27. *See* Reference 7.
28. Richards, W. G. and Barrow, R. F. (1962). *Proc. Chem. Soc.*, 297
29. Clyne, M. A. A. and Coxon, J. N. (1967). *Proc. Roy. Soc.*, **A298,** 404, 424
30. Rank, D. H. and Rao, B. S. (1964). *J. Molec. Spectrosc.*, **13,** 34
31. Clyne, M. A. A. and Coxon, J. N. (1967). *J. Molec. Spectrosc.*, **23,** 258
32. Gale, H. G. and Monk, G. S. (1929). *Astrophys. J.*, **69,** 77
33. Aars, J. (1932). *Z. Phys.*, **79,** 122
34. Porter, T. L. (1968). *J. Chem. Phys.*, **48,** 2071
35. Stricker, W. (1966). *Z. Naturforsch.*, **21a,** 1518
36. Huberman, F. P. (1966). *J. Molec. Spectrosc.*, **20,** 29
37. Gillespie, R. J. and Morton, J. B. (1969). *J. Molec. Spectrosc.*, **30,** 178
38. Anbar, M. and Thomas, J. K. (1964). *J. Phys. Chem.*, **68,** 3829; Cercek, B., Ebert, M., Keene, J. P. and Swallow, A. J. (1964). *Science*, **145,** 919
39. Brown, D. M. and Dainton, F. S. (1966). *Nature*, **209,** 195; Roncin, J. (1968). *J. Chem. Phys.*, **49,** 2876; Zvi, E. B., Beaudet, R. A. and Wilmarth, W. K. (1969). *J. Chem. Phys.*, **51,** 4166
40. Bennett, J. E., Mile, B. and Ward, B. (1968). *J. Chem. Phys.*, **49,** 5556
40a. Haas, M. and Griscom, D. L. (1969). *J. Chem. Phys.*, **51,** 5185
40b. Griscom, D. L. (1969). *ibid.*, 5186
41. Person, W. B. (1963). *J. Chem. Phys.*, **38,** 109
41a. Baxendale, J. H. and Bevan, P. L. T (1969). *J. Chem. Soc. A.*, 2240
42. Balint-Kurti, G. G. and Karplus, M. (1969). *J. Chem. Phys.*, **50,** 478
43. Politzer, P. (1969). *J. Amer. Chem. Soc.*, **91,** 6235
44. Gillespie, R. J. and Milne, J. B. (1966). *Chem. Commun.*, 158; *Inorg. Chem.*, **5,** 1577; *see also* Kemmitt, R. D. W., Murray, M., McRae, V. M., Peacock, R. D., Symons, M. C. R. and O'Donnell, T. A. (1968). *J. Chem. Soc. A.*, 862
45. Olah, G. A. and Comisarov, H. B. (1968). *J. Amer. Chem. Soc.*, **90,** 5033
45a. Eachus, R. S., Sleight, T. P. and Symons, M. C. R., (1969). *Nature*, **222,** 769
46. Edwards, A. J., Jones, G. R. and Sills, R. J. C. (1968). *Chem. Commun.*, 1527
47. Brown, W. G. (1931). *Phys. Rev.*, **38,** 709, 1179
48. Horsley, J. A. (1967). *J. Molec. Spectrosc.*, **22,** 479
49. Todd, J. A. C., Richards, W. G. and Byrne, M. A. (1967). *Trans. Faraday Soc.*, **63,** 2081
50. Mathieson, L. and Rees, A. L. G. (1955). *J. Chem. Phys.*, **25,** 753
51. Behringer, J. (1967), in *Raman Spectroscopy*, Ed. by Szymanski, H. A. (New York: Plenum Press)
52. Holzer, W., Murphy, W. F. and Bernstein, H. J. (1970). *J. Chem. Phys.*, **52,** 399

53. Behringer, J. (1969). *Z. Phys.*, **229**, 209
54. Kiefer, W. and Schrötter, H. W. (1970). *J. Chem. Phys.*, **53**, 1612
55. Stammreich, H. (1950). *Phys. Rev.*, **78**, 79
56. Delhaye, M. (1968). *Appl. Opt.*, **7**, 2195
57. Moore, C. E. (1949, 1952, 1958). *Atomic Energy Levels*, (National Bureau of Standards, Washington)
57a. Radziemski, L. J. and Kaufman, V. (1969). *J. Opt. Soc. Am.*, **59**, 424
58. Huffman, R. E., Larrabee, J. C. and Tanaka, Y. (1967). *J. Chem. Phys.*, **47**, 856
59. Donovan, R. J. and Husain, D. (1966). *Trans. Faraday. Soc.*, **62**, 11, 2643; *Nature*, **209**, 609; (1965). *Nature*, **206**, 171; and unpublished data.
60. *See* e.g. Porter, G. (1967). *Photochemistry and Reaction Kinetics* (Ed. by Ashmore, P. G., Dainton, F. S. and Sugden, T. M. (Cambridge University Press)
61. *See e.g.* Branscombe, L. M. (1966). *Phys. Rev.*, **148**, 11
62. *See* Berry, R. S. (1969). *Chem. Rev.*, **69**, 533
63. Berry, R. S. and Reinmann, C. W. (1963). *J. Chem. Phys.*, **38**, 1540
64. Popp, H. P. (1965). *Z. Naturforsch*, **20a**, 642; (1967). *Z. Naturforsch*, **22a**, 254
65. Stuckey, W. K. and Kiser, R. W. (1966). *Nature*, **211**, 963
66. Holloway, J. H. (1966). *Chem. Commun.*, 22
67. *See* e.g. Pimentel, G. C. (1960). *Formation and Trapping of Free Radicals*, Ed. by Bass, A. M. and Broida, H. P. (New York: Academic Press)
68a. Spratley, R. D., Turner, J. J. and Pimentel, G. C. (1966). *J. Chem. Phys.*, **44**, 2063: Arkell, A. (1965). *J. Amer. Chem. Soc.*, **87**, 4057
68b. Jacox, M. and Milligan, D. E. (1965). *J. Chem. Phys.*, **43**, 866
68c. Cochran, E. L., Adrian, F. J. and Bowers, V. A. (1970). *J. Phys. Chem.*, **74**, 2083
69. Barnes, A. J. and Hallam, H. E. (1969). *Quart. Rev. Chem. Soc.*, **23**, 392; Hastie, J. W. and Margrave, J. L. (1970). *Spectroscopy in Inorganic Chemistry*, Ed. by Rao, C. N. R. and Ferraro, J. R. (New York: Academic Press); Ogden, J. S. and Turner, J. J. (1971). *Chem. Brit.*, **7**, 186
70. Fettis, G. C. and Knox, J. H. (1964). *Progress in Reaction Kinetics*, Vol. 2, Ed. by Porter, G. (Oxford: Pergamon)
71. Golden, D. M. and Benson, S. W. (1969). *Chem. Rev.*, **69**, 125
72. Lee, Y. T., McDonald, J. D., Le Breton, P. R. and Herschbach, D. R. (1968). *J. Chem. Phys.*, **49**, 2447; (1969). *ibid.*, 455
73. Nelson, L. Y. and Pimentel, G. C. (1967). *J. Chem. Phys.*, **47**, 3671; (1968). *Inorg. Chem.*, **7**, 1695
74. Wasserman, E., Falconer, W. E. and Yager, W. A. (1968). *J. Chem. Phys.*, **49**, 1971
75. Chutjian, A. and James, T. C. (1969). *J. Chem. Phys.*, **51**, 1242
76. Chutjian, A., Link, J. K. and Brewer, L. (1967). *J. Chem. Phys.*, **46**, 2666
77. *See* e.g. Busch, G. E., Mahoney, R. T., Morse, R. I. and Wilson, K. R. (1969). *J. Chem. Phys.*, **51**, 449, 837
78. *See* Husain, D. and Donovan, R. J. (1971). *Advances in Photochemistry*, Vol. 8, Ed. by Noyes, W. A., Hammond, G. S. and Pitts, J. N. (New York: Interscience)
79. Kasper, J. V. V. and Pimentel, G. C. (1964). *Appl. Phys. Lett.*, **5**, 231
80. Kasper, J. V. V., Parker, J. H. and Pimentel, G. C. (1965). *J. Chem. Phys.*, **43**, 1827
81. Pollack, M. A. (1966). *Appl. Phys. Lett.*, **8**, 36
82. Guilano, C. R. and Hess, L. D. (1969). *J. Appl. Phys.*, **40**, 2428
83. Cool, T. A., Stephens, R. R. and Shirley, J. A. (1970). *J. Appl. Phys.*, **41**, 4038
84. Donovan, R. J. and Husain, D. (1970). *Chem. Rev.*, **70**, 489
85. Callear, A. B. and Wilson, J. H. (1966). *Nature*, **211**, 517; (1967). *Trans. Faraday Soc.*, **63**, 1358, 1983
86. Donovan, R. J. and Husain, D. (1965). *Nature*, **206**, 171
87. Clyne, M. A. A. and Stedman, D. H. (1968). *Trans. Faraday Soc.*, **64**, 1816
88. Gibbs, D. B. and Ogryzlo, E. A. (1965). *Canad. J. Chem.*, **43**, 1905
89. Bist, H. D. and Person, W. B. (1967). *J. Phys. Chem.*, **71**, 2750
90. Reid, C. and Mulliken, R. S. (1954). *J. Amer. Chem. Soc.*, **76**, 3869
91. Klaboe, P. (1967). *J. Amer. Chem. Soc.*, **89**, 3667
92. Mulliken, R. S. and Person, W. B. (1969). *Molecular Complexes, A Lecture and Reprint Volume*. (New York: Wiley); *idem.*, *J. Amer. Chem. Soc.*, **91**, 3409
93. Briegleb, G. (1961). *Electronen–Donator–Acceptor–Komplexe*. (Springer-Verlag: Berlin-Göttingen-Heidelberg)

94. Hanna, M. W. (1968). *J. Amer. Chem. Soc.*, **90,** 285
95. *See* e.g. Rao, C. N. R., Chaturvedi, G. C. and Bhat, S. N. (1970). *J. Molec. Spectrosc.*, **33,** 556
96. Kokovin, G. A. (1965). *Russ. J. Inorg. Chem.*, **10,** 750
97. Passchier, A. A., Christian, J. D. and Gregory, N. W. (1967). *J. Phys. Chem.*, **71,** 537
98. Passchier, A. A. and Gregory, N. W. (1968). *J. Phys. Chem.*, **72,** 2697
99. De Maine, F. A. D. (1957). *Canad. J. Chem.*, **35,** 573
99a. Winkel, R. G., Hunt, J. L. and Clouker, M. J., *J. Chem. Phys.* (1969). **50,** 1298
100. Radford, H. E., Hughes, V. W. and Beltram-Lopez, V. (1961). *Phys. Rev.*, **123,** 153; Beltram-Lopez, V. and Robinson, H. G. (1961). *Phys. Rev.*, **123,** 161; Vanderkooi, N. and Mackenzie, J. S. (1962). *Adv. Chem. Ser.*, **36,** 98; Aditya, S. and Willard, J. E. (1966). *J. Chem. Phys.*, **44,** 833
101. Westenberg, A. A. (1965). *J. Chem. Phys.*, **43,** 1544
102. Westenberg, A. A. and de Haas, N. (1968). *J. Chem. Phys.*, **48,** 4405
103. Carrington, A., Levy, D. H. and Miller, T. A. (1966). *J. Chem. Phys.*, **45,** 4093
104. Jen, C. K., Foner, S. N., Cochran, E. L. and Bowers, V. A. (1958). *Phys. Rev.*, **112,** 1169; Cochran, E. L., Bowers, V. A., Foner, S. N. and Jen, C. K. (1959). *Phys. Rev. Lett.*, **2,** 43
105. *See* e.g. Lawrence, N. J., Ogden, J. S. and Turner, J. J. (1968). *J. Chem. Soc. A.*, 3100
106. *See* Carrington, A. and Levy, D. H. (1967). *J. Phys. Chem.*, **71,** 2; Carrington, A. (1970). *Chem. Brit.*, **6,** 71
107. For a recent review *see* Mooney, E. F. and Winson, P. H. (1968) in *Annual Review of NMR Spectroscopy*. (London: Academic Press)
108. Hooper, H. O. and Bray, P. J. (1960). *J. Chem. Phys.*, **33,** 334
109. Johnson, K. J., Hunt, J. P. and Dodgen, H. W. (1969). *J. Chem. Phys.*, **51,** 4493
110. *See* e.g. Hall, C., Haller, G. L. and Richards, R. E. (1969). *Molec. Physics*, **16,** 377; Deverell, C. and Richards, R. E. (1969) ibid., 421
111. Stengle, T. R. and Baldeschwieler, J. D. (1966). *Proc. Nat. Acad. Sci.*, **55,** 1020
112. Bryant, R. G. (1967). *J. Amer. Chem. Soc.*, **89,** 2497
113. Hall, C., Kydon, D. W., Richards, R. E. and Sharp, R. R. (1970). *Proc. Roy. Soc.*, **A318,** 119
114. *See* e.g. Das, T. P. and Hahn, E. L. (1958). *Nuclear Quadrupole Resonance Spectroscopy*. (New York: Academic Press): Lucken, E. A. C. (1969). *Nuclear Quadrupole Coupling Constants*. (London: Academic Press)
115. Evans, J. C. and Lo, G. Y-S. (1966). *J. Phys. Chem.*, **70,** 2702; (1967). ibid., **71,** 3697
116. Chihara, H., Nakamura, N. and Seki, S. (1967). *Bull. Chem. Soc. Japan.*, **40,** 50
117. Townes, C. H. and Dailey, B. P. (1949). *J. Chem. Phys.*, **17,** 782; Dailey, B. P. and Townes, C. H. (1953). *J. Chem. Phys.*, **23,** 118
118. Gordy, W. (1953). *Microwave Spectroscopy*. (New York: John Wiley and Sons)
119. Graybeal, J. D. and Green, P. J. (1969). *J. Phys. Chem.*, **73,** 2948
120. Greenwood, N. N. (1967). *Chem. Brit.*, **3,** 56
121. Perlow, G. J. (1968). *Chemical Applications of Mössbauer Spectroscopy*, Ch. 7. (New York: Academic Press)
122. Jha, S., Seguan, R. and Lang, G. (1962). *Phys. Rev.*, **128,** 1160
123. Perlow, G. J. and Ruby, S. L. *Phys. Lett.*, **13,** 198
124. Hafemeister, D. W., de Pasquali, G. and de Waard, H. (1964). *Phys. Rev.*, **135,** B1089
125. Perlow, G. J. and Perlow, M. R. (1966). *J. Chem. Phys.*, **45,** 2193
126. Rama Reddy, K., Barros, F. de S. and Debenedetti, S. (1966). *Phys. Lett.*, **20,** 297
127. Pasternak, M., Simopoulos, A. and Hazony, Y. (1965). *Phys. Rev.*, **140,** A1892
128. Bukshpan, S., Soriano, J. and Shamir, J. (1969). *Chem. Phys. Lett.*, **4,** 241
129. Hon, J. F. and Christe, K. O. (1970). *J. Chem. Phys.*, **52,** 1960
130. Pasternak, M. and Sonnino, T. (1968). *J. Chem. Phys.*, **48,** 1997
131. Collin, R. L. (1952). *Acta Crystallogr.*, **5,** 431; (1956) ibid., **9,** 539
132. Vonnegut, B. and Warren, B. E. (1936). *J. Amer. Chem. Soc.*, **58,** 2459
133. Kitaigorodskii, I. I., Khotsyanova, V. and Struchkov, M. (1953). *Zh. Fiz. Khim.*, **27,** 780
134. Donohue, J. and Goodman, S. H. (1965). *Acta Crystallogr.*, **18,** 568
135. van Bolhuis, F., Koster, P. B. and Migchelson, T. (1967). *Acta Crystallogr.*, **23,** 90
136. Shibata, S. (1963). *J. Phys. Chem.*, **67,** 2256
137. Hanson, H. P. (1962). *J. Chem. Phys.*, **36,** 1043
138. Ukaji, T. and Kuchitsa, K. (1966). *Bull. Chem. Soc. Japan*, **39,** 2153

139. Pauling, L. (1960). *The Nature of the Chemical Bond,* 3rd ed. (Ithaca: Cornell University Press)
140. Pimentel, G. C. (1951). *J. Chem. Phys.,* **19,** 446; Hach, R. J. and Rundle, R. E. (1951). *J. Amer. Chem. Soc.,* **73,** 4321
141. Jordan, T. H., Streib, W. E. and Lipscomb, W. N. (1964). *J. Chem. Phys.,* **41,** 760
142. Meyer, L., Barrett, C. S. and Greer, S. C. (1968). *J. Chem. Phys.,* **49,** 1902
143. N.B.S. Circular 500 *Selected Values of Chemical Thermodynamic Properties,* Washington D.C. (1952)
144. Bersohn, R. (1962). *J. Chem. Phys.,* **36,** 3445
145a. Nakamura, N. and Chihara, H. (1967). *J. Phys. Soc. Japan.,* **22,** 201
145b. *See* Reference 114
145c. Adrian, F. J. (1963). *J. Chem. Phys.,* **38,** 1258
146. Walmsley, S. H. and Anderson, A. (1964). *Molec. Phys.,* **7,** 411
147. Cahill, J. E. and Leroi, G. E. (1969). *J. Chem. Phys.,* **51,** 4514
148. Suzuki, M., Yokoyama, T. and Ito, M. (1969). *J. Chem. Phys.,* **50,** 3392; **51,** 1929
149. Melvegger, A. J., Brasch, J. W. and Lippincott, E. R. (1970). *Applied Optics,* **9,** 11
150. Rosenberg, J. L. (1964). *J. Chem. Phys.,* **40,** 1707
151. Schnepp, O., Rosenberg, J. L. and Gouterman, M. (1965). *J. Chem. Phys.,* **43,** 2767
152. Hillier, I. H. and Rice, S. A. (1967). *J. Chem. Phys.,* **46,** 3881
153. Touloukian, Y. S. and Ho, C. Y. eds. *Thermophysical Properties of Matter.* (London: Heyden and Son)
154. Fajans, K. (1970). *J. Phys. Chem.,* **74,** 3407
155. Wilson, J. N. and Curtis, R. M. (1970). *J. Phys. Chem.,* **74,** 187
156. Oldman, R. J., Sander, R. K. and Wilson, K. R. (1971). *J. Chem. Phys.,* **54,** 4127
157. Leroy, R. J. and Bernstein, R. B. (1970). *J. Chem. Phys.,* **52,** 3869; *Chem. Phys. Letters,* **5,** 42
158. National Bureau of Standards, Technical News Bulletin, (January 1971) (Washington D.C.)
159. Coxon, J. A. (1971). *J. Molec. Spectrosc.,* **37,** 39
160. Leroy, R. J. (1970). *J. Chem. Phys.,* **52,** 2678